Electrochemistry of Cleaner Environments

Electrochemistry of Cleaner Environments

Edited by
John O'M. Bockris
Electrochemistry Laboratory
John Harrison Laboratory of Chemistry
University of Pennsylvania
Philadelphia, Pennsylvania

PLENUM PRESS • NEW YORK–LONDON • 1972

Library of Congress Catalog Card Number 72-179762
ISBN 0-306-30560-7

A Division of Plenum Publishing Corporation
227 West 17th Street, New York, N.Y. 10011

United Kingdom edition published by Plenum Press, London
A Division of Plenum Publishing Company, Ltd.
Davis House (4th Floor), 8 Scrubs Lane, Harlesden, NW10 6SE,
London, England

Printed in the United States of America

CONTRIBUTORS TO THIS VOLUME

J. O'M. BOCKRIS	The Electrochemistry Laboratory, University of Pennsylvania, Philadelphia, Pennsylvania
B. D. EPSTEIN	Gulf General Atomic Company, San Diego, California
D. P. GREGORY	Institute of Gas Technology, Chicago, Illinois
R. PHILIP HAMMOND	Oak Ridge National Laboratory, Oak Ridge, Tennessee
T. A. HENRIE	Bureau of Mines, U.S. Department of Interior, Washington, D.C.
E. H. HIETBRINK	Electrochemistry Department, Research Laboratories, General Motors Corporation, Warren, Michigan
ANSELM T. KUHN	Department of Chemistry, University of Salford, Salford, Lancashire, England
R. E. LINDSTROM	Bureau of Mines, U.S. Department of Interior, Washington, D.C.
G. M. LONG	Institute of Gas Technology, Chicago, Illinois
J. MCBREEN	Electrochemistry Department, Research Laboratories, General Motors Corporation, Warren, Michigan
D. Y. C. NG	Institute of Gas Technology, Chicago, Illinois
GILBERT N. PLASS	Department of Physics, Texas A. & M. University, College Station, Texas
EDMUND C. POTTER	Division of Mineral Chemistry, Commonwealth Scientific and Industrial Research Organization, Sydney, Australia

S. M. Selis	Electrochemistry Department, Research Laboratories, General Motors Corporation, Warren, Michigan
S. B. Tricklebank	Electrochemistry Department, Research Laboratories, General Motors Corporation, Warren, Michigan
R. R. Witherspoon	Electrochemistry Department, Research Laboratories, General Motors Corporation, Warren, Michigan

PREFACE

Of the societal ills which are recognized as present in the Western countries during the 1970's, pollution of air and water is among the first. Whether the breathing of noxious gases acts biochemically as a source of mental irritation is not yet known. But it is not in doubt that reduction of the grime, smog, fouled water, and acrid air would lead to an increase in a feeling of well-being. Nor is it speculative to state that a reversal of the present trend to poison the atmosphere and the water is essential if man is to survive in a technological society.

It was partly the lack of realization of the intrinsic nature of overpotential in electrochemical reactions,* and hence the failure of the early fuel cells to come up to expectations, which led engineers at the turn of the century to rely upon the combustion of oil and coal for the production of energy, with the associated era of increasing atmospheric pollution.

There is a clean electrochemical route for carrying out many of the tasks of chemical technology, in energy production and in cleanup and antipollution chemistry. Electrochemical reactions may occur spontaneously, electrogeneratively, or may need electricity to drive them. By using them on a large scale, one could continue to develop the high technology which supports man in affluence, without his drowning in the effluents which result from much of present rather than electrochemical technology, or being overcome by the inhalation of the air fouled by the products of the present means of transportation. Here, in the electrification of chemical processes, is a field of great relevance to the foundations of a clean future for man.

The object of this book is to illustrate a few specific ways in which a cleaner world might be approached by developing the technology seen as potentially present behind presently existing electrochemical science.

J. O'M. Bockris

Sam Lord's Castle, Barbados
April 1971

*The shift in the Fermi level from that for zero current to cause a current *i* to flow.

CONTENTS

Chapter 3

Electrochemical Power Sources for Vehicle Propulsion

E. H. Hietbrink, J. McBreen, S. M. Selis, S. B. Tricklebank, and R. R. Witherspoon

Chapter 4

The Electrochemical Treatment of Aqueous Effluent Streams

Anselm T. Kuhn

Chapter 5

The Electrofiltration of Particulates from Gases

Edmund C. Potter

Chapter 6

Electrochemical Methods of Pollution Analysis

B. D. Epstein

Chapter 1

THE ELECTROCHEMICAL FUTURE

J. O'M. Bockris

The Electrochemistry Laboratory
University of Pennsylvania
Philadelphia, Pennsylvania

I. INTRODUCTION

During the preindustrial era, pollutants (e.g., from fires burned for warmth) were of trivial total content, partly because of the relatively smaller population (600 million up to about 1600 A.D.). The first fogs arose after the Industrial Revolution, because of the increased burning of coal.

The beginning of the nineteenth century saw the birth of the electric battery (Volta, 1800). Sir Humphrey Davy showed in 1802 the possibility of a fuel cell when he built a carbon cell to operate at room temperature, using nitric acid as electrolyte. In 1839, Grove constructed a successful hydrogen–oxygen fuel cell. The first practical device for the storage of electrical energy was constructed by Planté in 1860, with two lead sheets and sulfuric acid (7.25 A·h/lb of lead).

Work on the polluting internal combustion engine was also making progress during the nineteenth century. This gathered momentum from 1867 onwards (with the invention of the Otto engine). In the 1880's, Drake showed how *oil* could be obtained by digging deep wells (in western Pennsylvania). The development of the spark-ignition engines around the turn of the century resulted in a demand for gasoline that has been on the increase ever since. Thus, in the late nineteenth century, both the electrochemical and internal combustion approaches were available to those who wanted to develop the power sources of the twentieth century. In 1894, Ostwald, the president of the Bunsengesellschaft, at that time the most prestigeous of scientific societies, gave a presidential address in which he pointed out two important things: there is an *intrinsic* increase in efficiency of energy conversion of electrochemical over chemical energy conversion; and if we did develop a thermal and not an electrochemical technology, there would be intolerable air pollution in cities. Ostwald's speech of 1894 could come largely out of a *Scientific American* article of 1971.

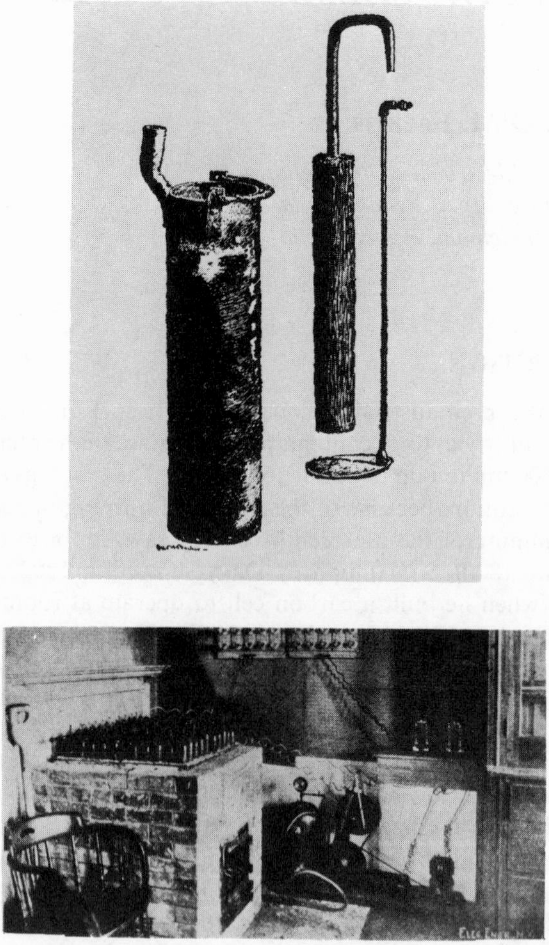

Figure 1. Jacques air fuel cell.

Between 1839, when Grove made his fuel cell, and 1910, when thermal combustion can be considered to have won, however temporarily, the battle, good progress was made in the development of fuel cells. Surprising indeed is the article by Jacques in *Harper's Magazine* of 1897. He had built a carbonate fuel cell, a 1.5-kW battery (Fig. 1). Jacques included a design of the powering of a ship by an electrochemical engine (the fuel cell–electric motor combination). He stressed the advantage of the lesser weight of the electrochemical situation by working out what it would take to drive a ship across the Atlantic using electrochemical engines compared with that using external combustion and a heat engine.

This bright introductory situation for electrochemical sources, leading away from the pollution of the twentieth century, was intellectually hindered by the work of the great Nernst. He stressed *thermodynamics* in his well-known treatment of electrochemical cells, in which it is implied that the electron transfer reaction is thermodynamically reversible, i.e., rapid and giving no overpotential. It was the slowness in realization of the physical interpretation of overpotential* in electrochemical cells which caused the suspension of animation of the electrochemical path to energy (for engineers did not understand the origin of the losses). Thus, the tendency to convert chemical to electrochemical technology was abandoned early in the century. As a consequence, we have experienced (as Ostwald had predicted) a century of atmospheric pollution.

The situation is illustrated in Fig. 2.

II. TRANSPORTATION

We must be careful not to condemn atmospheric pollution as evil without reviewing scientific knowledge concerning what it does to us. The overt results of pollution, the darkening of the sky, smog and its side effects, e.g., eye smarting, etc., are facts of experience for many. A number of effects exist which are less well known. Thus, automobile exhausts cause emphysema; cancer due to exhaust products has been established, although it is less than that caused by cigarettes; an increase in the number of people who die of heart disease has been correlated with effluents of the automobile. Figures 3 and 4 illustrate two of these statements.

The mechanism of the formation of smog is well understood. One must have unsaturated hydrocarbons, and these come from a lack of complete combustion to CO_2 in the internal combustion engine. There must also be NO_2 to allow the photochemical production of O, so that there may be the addition of compounds formed, as shown in Fig. 5. Correspondingly, solid

*Remarkably, some physical chemists of the present time are still not familiar with the concept of overpotential.

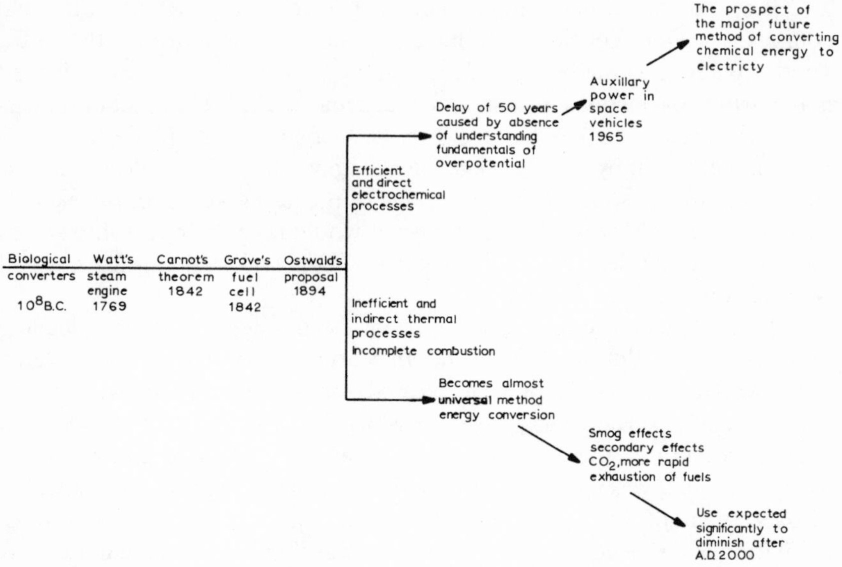

Figure 2. Energy conversion took the wrong path in 1894.

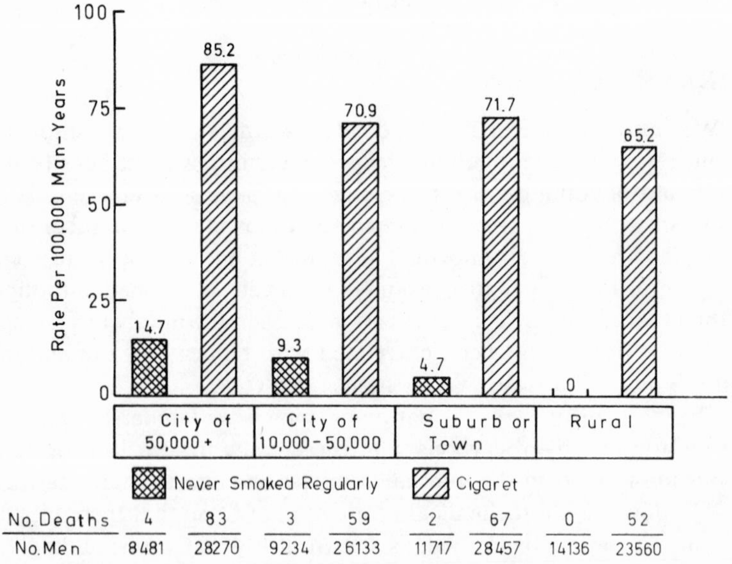

Figure 3. Relative difference of lung cancer in different population groups.

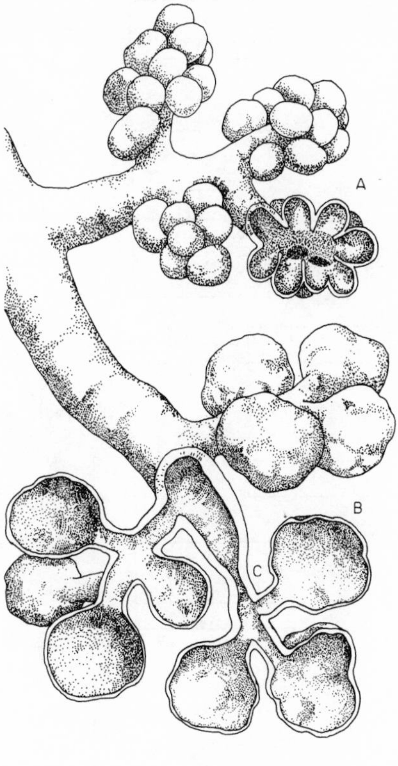

Figure 4. Effect of air pollution on lungs (aggravating bronchitis–emphysema).

particles must be present on which peroxyacylnitrates, formaldehyde, ozone, and water vapor condense. The sunlight is needed for the photochemical decomposition of NO_2 (cf. Fig. 5). It is easy to understand why the Smoggy State developed firstly in Los Angeles, for there one finds not only the screening of the city from east winds, due to a mountain range, but also the presence of intense sunlight, combined with a very spread-out city containing automobiles emmitting pollutants.

Is a cleanup of the internal combustion engine possible? A degree of cleaner exhausts has already been obtained (and in new cars, for the first months of their lives, is some 50–70%). It is a matter of forcing the reaction to CO_2 over towards completion. The injection of excess air, and/or the use of catalysts, or the raising of temperature will help. The formation of NO is more difficult to suppress (because spark ignition is bound to occur in the presence of O_2 and N_2).

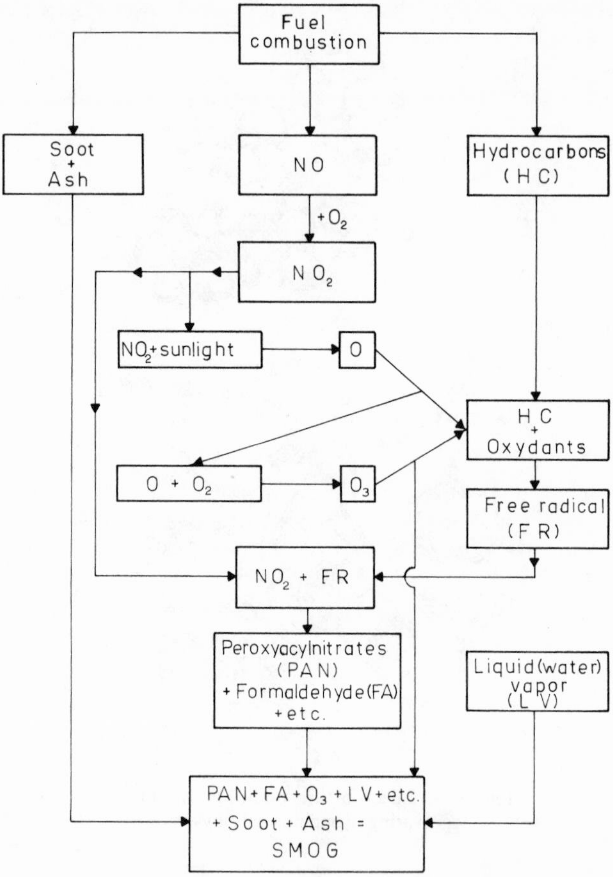

Figure 5. A schematic representation of the formation of smog.

The principles of one method which has been used to bring out this clean-up are shown in Fig. 6. It is the blow-up method, obvious in action from the diagram. Another method is to pass the materials over catalysts. Much research, particularly on this latter method, is in progress.

In 1971, progress has been made in approaching the standards defined by the government, but major aspects are unknown: what will be the eventual (future) level of controls which the government will demand? How long will a catalyst remain effective in in-town driving conditions?

A determinative factor in these matters is the increasing world car population. The car density–time relation cannot continue its present course in the U.S. for more than two to three decades because of overcrowding.

However, continued expansions in the *world* car population can be expected for many decades. *The important result is that even if there is a 95% cleanup of automobile exhausts on all new cars from 1971, the predicted increase in the number of cars will be such that the pollution per day will begin to increase again, towards 1990.*

Vague outlines of some of this knowledge are known to perhaps 0.1% of the populace, although the carcinogenic nature of automobile exhausts, their part in causing emphysema and heart disease, and, particularly, the *inevitability* of a rise in pollutants on even optimistic assumptions of exhaust cleanup did not seem to be realized widely, although admitted by some scientists working in oil companies.

An aspect of pollution caused by automobiles, fossil-fuel driven, which is little discussed, is the long-term climatic effects of the increase in carbon dioxide in the atmosphere. Figure 7 shows aspects of the balance of carbon dioxide production–consumption. The equilibrium between the production and consumption has been disturbed by the evolution of CO_2 resulting from our fossil-fuel economy, part of which is the evolution of CO_2 from automobiles, the rest from technology. In Fig. 8, one sees the parallel between the measured amount of carbon dioxide in the atmosphere and the burning of fossil fuels: the rise indicates the unbalancing of the equilibrium maintained throughout previous history. That the concentration of CO_2 is rising is certain. More doubtful is knowledge of the rise in the temperature of the

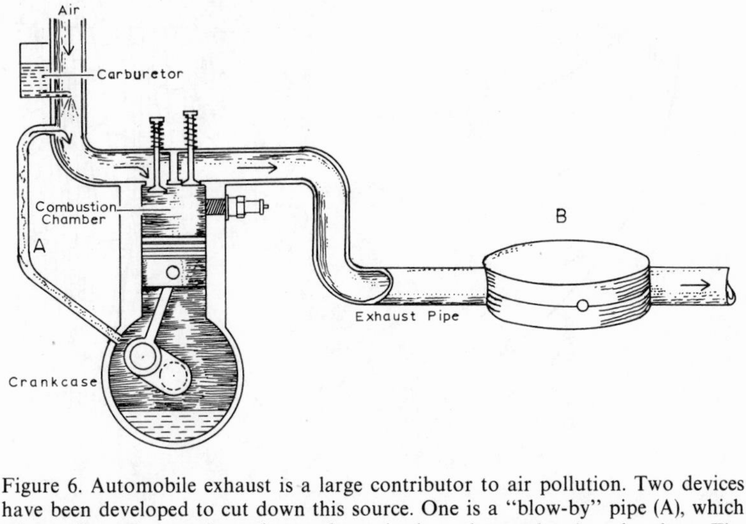

Figure 6. Automobile exhaust is a large contributor to air pollution. Two devices have been developed to cut down this source. One is a "blow-by" pipe (A), which takes unburned gases from the crankcase back to the combustion chambers. The other is an "after-burner" (B), a special muffler that oxidizes carbon monoxide and unburned fuel in the exhaust gases through a catalytic process.

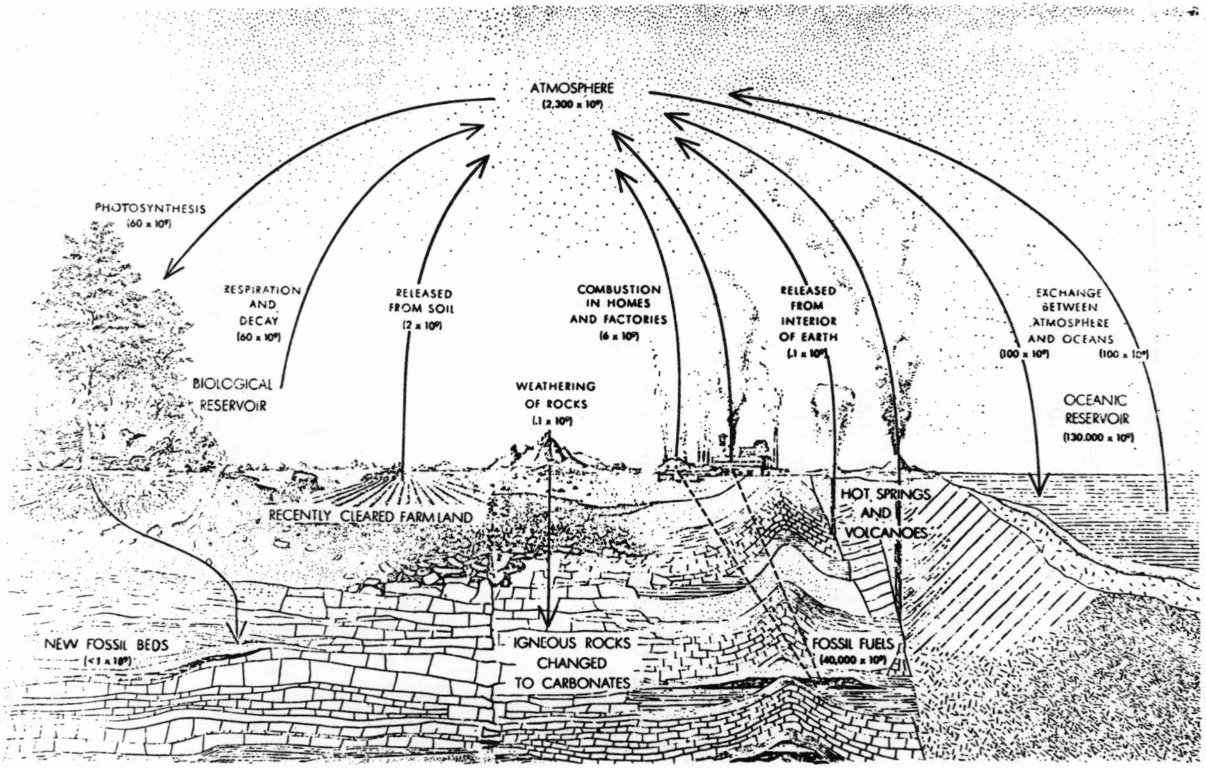

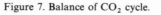

Figure 7. Balance of CO_2 cycle.

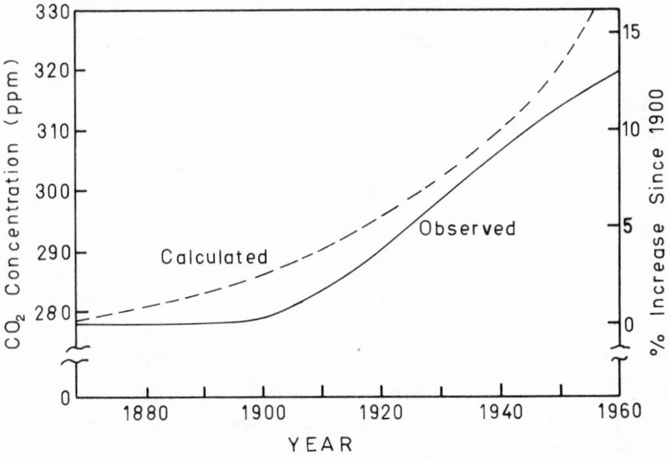

Figure 8. Plots of cumulative amounts of carbon dioxide in the atmosphere vs time. Curves 1 and 2 represent, respectively, the observed and calculated amounts added by burning fossil fuel.

atmosphere expected from the greenhouse effect. The earth's heat balance depends partly upon heat from the sun. The incident radiation is reflected from the earth through the atmosphere, into space. If more carbon dioxide is in the atmosphere, part of the incident and reflected IR radiation is absorbed by the CO_2, and the excited vibrational levels degrade to heat. Figure 9 shows the result of a calculation in which feedback effects, e.g., the presence of an increasing degree of particulate matter in the atmosphere (which would decrease the heat reaching the earth), are neglected. This (certainly simple minded) calculation indicates that by the end of the century the rise in the mean world temperature could be about 2°C. Such a rise would increase the amount of ice melted from Greenland and the Antarctic continent (the northern ice floats). This increased amount of water would increase the world sea level. A calculation, plotted in Fig. 9, suggests that this rise could be as much as 3 ft by the end of the century, increasing rapidly thereafter.*

However, the situation with the climatic effects of carbon dioxide is more complicated than that predicted with these calculations which neglect feedback. Thus, a fossil-fuel economy produces not only CO_2 but suspended particulate matter: this may blot out sufficient sunlight to cause a *decrease* in temperature. In Fig. 10, the rise of dust level is shown along with the rise of the temperature, with a suggestion, based upon the apparent turndown

*The best available calculation as of 1971 (see Chapter II) suggests a rise of about 1 foot in world sea level due to CO_2 by 2000 A.D.

in the temperature graph, that the temperature-decreasing effect of dust has overcome the temperature-increasing effect of the carbon dioxide. We await inundation from the seas but could be saved by a darkening of the sky.

The following are clear conclusions.

1. No cleanup of automotive exhausts which meets the government standards for 1972 had been achieved for an automobile driven by normal drivers for more than 10,000 miles by December 1970. A number of states in the U.S. have passed legislation which would ban internal combustion engines by 1975. It is possible, therefore, that a crisis in pollution from automobile exhausts may occur before 1985, the date at which, if *95% removal of pollutants can be achieved*, assuming continued growth of the car population, there will be an increase again in the pollution rate.

2. If a 95% cleanup of exhausts *is* achieved, under acceptable economic conditions, a change to nonfossil fuel powered transportation will be necessary after 1985.

3. The climatic effects of the increasing carbon dioxide situation or the concommitant effects of solid effluents will force us in this direction in any case, as early as 1990, when the rise in sea level is expected to become significant, or significant sky-darkening will occur.

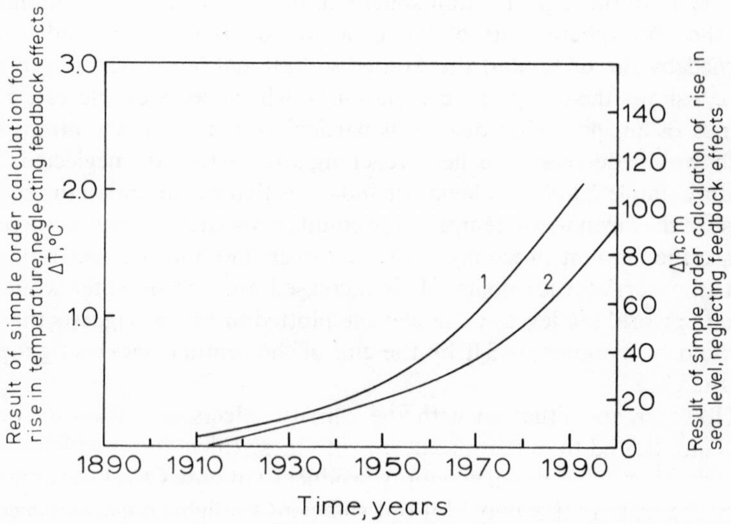

Figure 9. Plots of increase in atmospheric temperature and the consequent rise in sea level vs time. Curve 1 represents the rise in atmospheric temperature and curve 2 the rise in sea level as a result of simple order calculations, feedback effects being neglected.

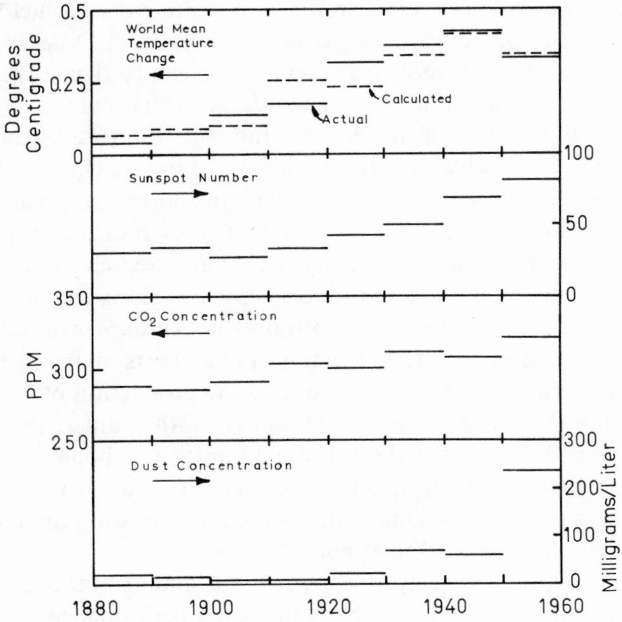

Figure 10. Changes in dust CO_2/time world temperature.

Two other points follow.

1. The performance of the electric cars which will have to replace gasoline-driven cars need not be inferior to those of gasoline-driven cars of the 1970's. The XEP 1970 zinc–air (nickel–cadmium) powered electric car, developed at the General Motors Technical Center in 1970, could reach 60 mph and had a range of 70 miles. Thus, even with the relatively little research in electrochemical storers carried out hitherto, and with the only novel battery system engineered in the last 20 years, a car with a performance near to that of a gasoline-driven car has already been achieved. Much could be achieved in electrochemically powered transportation if research funds were spent on it (i.e., on electrochemical research in batteries and fuel cells) at a rate equal to the rate of expenditure on internal combustion engine research, or if the money (which in 1971 is ca. $250 million) being spent in attempts to clean up the internal combustion engine were directed towards research in the development of electrochemical power sources.

2. The time at which the internal combustion engine will be replaced by electrochemically powered cars is affected by political and psychological considerations. Were controls not to prove effective beyond a 50% cleanup, the pressure towards electrochemically based transportation might be clear within the community in less than a decade.

A fuel-cell-driven car might indeed use hydrazine as a fuel.* To some extent the oil companies can compensate for the Great Disaster which the decrease of sales for automotive gasoline will mean to them. They can buy geothermal springs and uranium mines, i.e., then can become energy sources companies. However, such moves have the sort of future as that of large shipping companies which try to buy aircraft. Great technological revolutions, such as the change from a fossil fuel to an entirely electrically operated economy cannot be stopped by the influence of pressure groups of the economic interests of the old era, anymore than research can alter the fact that gasoline burned in oxygen produces CO_2. But, delay can be introduced. Further, automotive companies will certainly not *encourage* the introduction of an electric automobile. The change of model costs about $1 billion per year: what would the cost be to change to the production of a totally new vehicle, running on a different power source, with a different transducer? Electric motors have lives of about 1 million miles. Car bodies on vibrationless frames will wear out less quickly: selling electric cars may never be such good business as was the selling of the internal-combustion-driven cars with engines limited to about 100,000 miles.

A less obvious difficulty is conservatism among the public. Even if electric cars can be built which have the same performance as gas-driven cars, they will not vibrate, roar, and belch in the manner to which the consumer is accustomed in his present car—which may give some psychological satisfaction. Within a decade of the abandonment of live, steam locomotives, the American railways (where the use is decided by public taste) were dying fast.†

III. ELECTROCHEMICAL POWERING OF TRANSPORTATION

Figure 11 shows the distribution of the sources from which we get our energy. It is still overwhelmingly from burning fossil fuels to create heat, the expansion of gas caused by that heat, the movement of a piston caused by the expansion of this gas, and the working of a generator by the movement

*In 1971, the fraction of electricity which is atomic in origin is 6%; by 2000 A.D., the plan is to have it at 27%, i.e., 73% fossil fuels. Thus, American government plans seem at present inconsistent with the speed with which an atomic-electrochemical economy has to be introduced in order to avoid air-pollutional problems before the end of the century. In fact, the situation of underbuilding of atomic power stations is worse than it seems, because the present plans (and the percentage quoted above for 2000 A.D.) correspond to the expected growth curve without allowing for a massive introduction of battery-driven cars. Either the rate of building of atomic sources of electricity must be greatly increased or fuel cells, rather than batteries, must power the first few generations of electric cars (see Chapter 3).

†The comparison is not merely fanciful. American trains are not dying because they were suppressed by alternative forms of transportation. In technologically advanced countries where the railways are above vascillations of public taste (e.g., Japan), they bloom. Was the decay of the American railroad partly caused by a psychological dissatisfaction with "lifeless" trains?

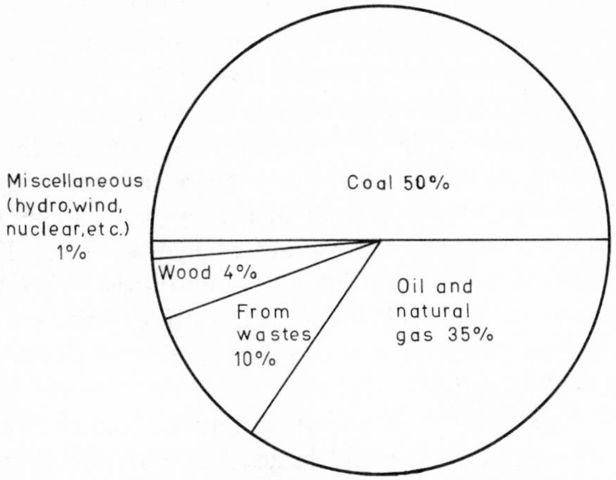

Figure 11. A schematic representation of the origin of total world energy among various sources.

of the piston to produce electricity with a Carnot efficiency loss of 60–90%. The hydrocarbon store is a once only store. In respect to natural gas and oil, it will last a few decades. In respect to coal, it would last for hundreds of years (in the U.S.).

However, there are other uses to which this fossil-fuel capital can be put. It can be converted to chemicals, plastics, and textiles—and even food. That it should be burned to give us energy, obtained in an inefficient way, and with lethal side effects on the atmosphere, is not rational.

Other means of energy conversion than the usual indirect method, may be mentioned. One in which research support is well supplied, for it

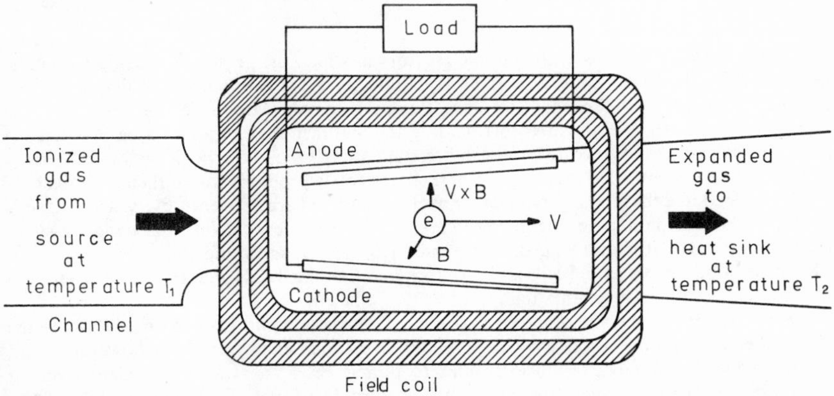

Figure 12. The elements of a magnetohydrodynamic engine. Here, V represents the flow velocity of the ionized gas, B is the strength of the magnetic field (at right angles to the plane of the paper).

uses oil, is magnetohydrodynamics, the principle of which is shown in Fig. 12. This method of energy conversion is not relevant for powering transports because it can be developed (if at all in practice) only in megawatt size. The gas turbine will probably be used for a decade or two in powering large trucks where long distance runs, without much change in speed or shifting of gears, make the performance satisfactory. Both steam and gas turbines are less polluting than a noncleaned up internal combustion engine; they do not give rise to the same degree of unsaturated hydrocarbons. However, for cleaned up sources, there is no advantage, pollution-wise, in gas turbines and steam. Both steam and gas turbines pollute the atmosphere with unsaturates, with the same amount of CO_2, and with (for gas turbines) more noise than internal combustion engines.

Electrochemical power cells are nonpolluting,* and electric cars would make little noise. Fossil fuels could be used as a source of energy to run electric transports, by conversion to electricity in fuel cells on board the cars; the Carnot cycle efficiency loss is avoided,† the pollution would be zero and the CO_2 cut to, say, one-half that for the same mileage by an internal combustion engine. The weight of electrochemical power plants is greater than that of conventional ones. However, theory suggests that electrochemical power cells could be developed to give about 1 kW/kg (i.e., equal to that of the internal combustion engine). The best attained hitherto, is about 0.1 kW/kg.

A most needed area for effort is in research in fuel cells and new electrochemical storage devices. As of July 1971 no U.S. government funding for progress in these areas (or in most other electrochemical areas) has been voted, nor was there any interest in developing electrochemical devices shown by any of some 20 government agencies, some specifically associated with environmental pollution.‡ Significant work in the engineering of

*The nonpolluting character of the atomic reactors in which the power would originate must be stressed. Their only pollution is thermal. This difficulty can be eliminated by building reactors on platforms floating on the sea, or on lakes. Pumping up cold water from deep water near the reactor platforms to cool the area surrounding the platform may mean an area of enhanced nutriment which would provide good fish breeding areas. One of the bogies of atomic energy is the possible uncontrolled runaway. Even then, the waste products would then be dangerous only if trapped in a thermal inversion layer over a populated area. Twenty-five years of atomic energy without a damaging uncontrolled event gives confidence. Were the danger ever to become important, it would be possible to build reactors underground.

†The corresponding loss in fuel cells is the overpotential. But fuel cells tend to be $>50\%$ efficient and internal combustion engines tend to be $<30\%$ efficient.

‡Funding of research in some other areas of energy conversion and storage (in addition to the electrochemical) should of course be made, above all for solar energy. It seems, however, inconsistent in view of ecological considerations to pursue research at a development level for methods (e.g., magneto-hydrodynamical) which burn fossil fuels, and hence produce CO_2. Energy Conversion institutes should be, effectively, institutes for study of advances in fusion, and in electrochemical and photovoltaics methods.

Table 1. It Could Already Be Slightly Cheaper to Run Cars Electrochemically on Hydrogen than Conventionally on Gasoline

Fuel	Price	Cost in cents per hp·h				Cost in cents per mile for typical (20 mpg) car at efficiency shown		
		$\varepsilon = 100\%$	$\varepsilon = 15\%$	$\varepsilon = 50\%$	$\varepsilon = 65\%$	$\varepsilon = 15\%$	$\varepsilon = 50\%$	$\varepsilon = 65\%$
n-octane	25 ¢ gal^{-1} [a]	0.55	3.5			1.37		
hydrogen	50 ¢ lb^{-1} [b]	1.9		3.8	3.0		1.34	1.08
hydrogen	5 ¢ lb^{-1} [c]	0.2		0.4	0.3		0.14	0.11

[a] Excluding tax.
[b] Present bulk price (liquid or gas).
[c] Possible price in future.

electrochemical energy storers and converters was being carried out only by the two leading automotive concerns.

Producing a satisfactory electric car depends largely upon electrochemical research in fuel cells and batteries, together with research on the underlying fundamental electrochemistry. Another aspect of electric transportation, which needs research, is lightweight dc motors. Electric motors have been little developed for situations in which light weight is to be stressed. The weight per unit of power quoted in electrical engineering textbooks is 10 lb/hp. However, this can be greatly reduced (for motors of several hundred hp, to about $\frac{1}{2}$ lb/hp). Another gap which needs funding is the education of dc electrical engineers. For years, little development of dc electrical engineering has been made. Electrochemical power will be dc. It will be difficult to proceed if there is no *government-planned* training of such engineers.

A comparison of some cost factors may be found in Table 1. The example shows the cost of driving 1 mile on hydrogen at two different prices, the price available now, and the probable price in a one-to-two-decade future. *Even now the cost of driving electric cars on hydrogen, using fuel cells, would be less than that of using tax-free present gasoline.*

Our continued use of fossil fuels to drive transports arises partly from a lack of information flow. Established industries naturally have inertia towards radical change.* Steam engines and gas turbines are not likely to

*In a competitive economy this should be satisfactory because, if the growth is desired by consumers, other interests should come forward and provide research money. One difficulty of this idea of capitalist thinking is that the consumers usually do not know what to desire unless they have their desires suitably stimulated. In the present instance, most of them do not know the situation—the inevitability of eventual increased pollution, the absence of independent work to develop electrochemical power sources—to know what desire to exhibit. However, another difficulty is that the utility companies (i.e., those who would profit by research in electrochemistry) can only just meet the demand for electricity now. The introduction of battery-powered transportation too suddenly might overstrain available electricity.

be important for use in passenger cars. The only pollutively viable source for transports* is the electric battery, or the fuel cell, in combination with lightweight electric motors.

The massive electrically oriented revolution in technology which pollutional considerations will bring about during the next two or three decades would not have been necessary had an understanding of *overpotential* allowed Ostwald's suggestions of the 1890's to have been more gradually developed, and the present state of pollution hence avoided.

IV. ILLUSIONS IN RESPECT TO THE ELECTRIC AUTOMOBILE

Several illusions concerning electrochemically powered cars are widespread and may be briefly discussed.

1. *Recharging would be slow.* This difficulty may be overcome by exchanging batteries at "charging stations." One would rent kilowatt hours. "Fast charging" (say 10 min) has been shown to be possible, but it seems a less good solution than renting charged batteries.

2. *Electrochemically powered transportation is also dirty, because it needs atomic power stations.* This difficulty is easily dispelled by study of the relevant literature in atomic energy technology. A difficulty is thermal pollution, but this could be overcome, as indicated above.

3. *Electrochemical transportation would be expensive.* It might be cheaper. The initial cost of electric vehicles, produced in the same quantity as those with internal combustion engines, should be about the same as that of present cars. Running costs would be less.

4. *Much research is being done on new types of batteries and fuel cells.* This is not true. Apart from the hardware developments at NASA, and the molten salt work at the Argonne National Laboratory, *effectively no fuel-cell or battery research on new systems is being carried out under American government funding.* Support of electrochemical work at Universities (1971) has fallen during the last few years (as the relation of pollution to fossil fuels becomes clear!) to a negligible level ($1 million per year). New battery development is not being pursued by American battery companies. Fuel-cell research† is government supported at about 15% of that of 1966.

5. *Electric cars will have a lesser performance than gasoline-driven cars.* That the supposition may not be true in the future is exemplified by the experimental electric car of General Motors, 1970, mentioned above (60 mph and a range of 70 miles). Progressive developments, particularly if extensive

*Including ships. Trains can pick up current. Planes have been little considered for electrochemical power sources. The prospects look poor.

†The dangers of pollution came into focus in the public press from 1969 onwards, the funding of electrochemical research in electric power sources was reduced (between 1968 and 1970, by 75%).

research were funded, seem certain. In some respects—acceleration from low speed ranges—an electric car is superior to a corresponding gasoline-powered vehicle.

V. METALLURGY

Many substances, typified by copper, occur in nature as sulfide ores. One heats them, forms the oxide, reduces this, and, for some metals (e.g., copper), electrodissolves and redeposits for purification. This is illogical because the metallic oxides are more stable than the sulfides. One should have an energetically better situation if one electrolyzed directly, or after dissolution in an electrolyte. In this approach, no S or SO_2 need reach the atmosphere. An alternative electrochemical approach is reviewed in Chapter 9.

Ferrous metallurgy may also be improved in respect to pollution. It may be possible to obtain iron by direct chemical reduction of iron oxides with hydrogen. An approach to this is through electrolysis, made increasingly cheap by the reduction of overpotential, and through the eventual availability of cheaper electricity from atomic sources.

VI. ELECTROCHEMICAL SYNTHESIS

Were it possible to produce hydrazine in large quantities, sufficiently cheaply, it would be an excellent hydrogen carrier, and hydrazine–air fuel cells would become practical for cars. The needed price is 10 cents per pound.

The electrochemical oxidation of SO_2 with O_2 in a fuel cell to give SO_3 and hence sulfuric acid appears to be a possible development, as does the production of HNO_3 from the electrochemical oxidation of NO.

VII. DIRTY LIQUIDS

Dirty liquids can be purified electrochemically by ion exchange, or by direct anodic oxidation to CO_2. Factory wastes could be regenerated locally and the purified water recirculated in factories (see Chapter 4).

VIII. SEWAGE AND RUBBISH

1. Sewage

Thor Hyerdahl has described floating rubbish, even at 1000 miles from land in the Pacific Ocean. He noted, in 1970, deterioration of the ocean, in respect to rubbish, compared with the situation on his first Pacific voyage in 1947. Continued jettisoning of rubbish into the sea is not feasible.

It is possible to oxidize cellulose at 100% current efficiency to carbon dioxide. The solid material in sewage is 60% cellulose, and this can be converted into CO_2 by electrolysis. Chlorination might be used to convert the rest to odorless products. Units could be in-house. The cathode could be of air, i.e., a fuel-cell principle might be usable.

What should one do with the carbon dioxide produced? Formaldehyde is a result of CO_2 reduction at mercury electrode; at a cadmium electrode it is methanol. Formaldehyde, together with air nitrogen and an enzyme, could be made into protein.

2. Rubbish

The present method is objectionable because the emission contains particulate matter, to a degree dependent upon the efficiency of electrostatic precipitation (at ca. $1 million per chimney).

An electrochemical approach would envisage a slurry made from trash in strong acid (at the home site) passing between layers of anodes, or perhaps stirred by a rotating anode. A central cylindrical cathode would act as an air cathode. The method would deal with organic materials; metals in the rubbish would have to be separated, collected, and perhaps electrochemically refined at a central area.

IX. ELECTROEXTRACTIONS (SEE CHAPTER 4)

Many "mines" of the future could be made from the junk heaps of today by the use of electroseparation. Metals can be recycled electrochemically. The electrodissolution of alloys (e.g., brass), and the redeposition of the components involves negligible electricity costs. On a larger scale would be the refining of junked cars. Cars would undergo removal of paint by burning; and then be placed in an electroextraction bath. A redeposition of individual components from the car might be achieved under potentiostatic control on cathodes. One major product would be powdered iron. Table 2* lists the constituents of a car and an analysis of the cost of the materials recoverable.

X. DETECTION DEVICES (SEE CHAPTER 6)

Electrochemical monitors give electrical signals directly. Substances to be detected are often in water, and hence in a suitable form to be sensed electrochemically.

A challenge to electroanalytical chemists is the rapid analysis of the products of the exhaust pipe. Potentiodynamic sweep devices might be successfully applied. By differential techniques the characteristic current

*Prepared by J. MacHardy.

Table 2. Breakdown of Metals in a Typical Car

Metal	Pounds	Price/lb	Value
Iron	3000	50.00/lb[b]	$150.00
Copper	32	0.55/lb	17.60
Zinc	54	0.15/lb	8.10
Aluminum[a]	51	(0.80/lb)	(5.10)
Lead	20	0.16/lb	3.20
Nickel	5	1.28/lb	6.40
Chromium	5	3.00/lb[c]	15.00
		total	$200.30

[a] Not recoverable in aqueous solution. If not removed prior to electrolysis, it will build up in the solution. Price is quoted for scrap, but excluded from total.
[b] Price for top grade scrap. However, the current price of 99.4% iron powder is $0.32 per pound. Less pure grades of iron powder sell for $0.074–0.094 per pound. Even the lowest price would correspond to $220, however, raising the total value of the car to $270.30. The gross return on 1.6×10^4 cars per year would thus be between $\$3.2 \times 10^6$ and $\$4.3 \times 10^6$.
[c] Not quoted on the metal market as such (usually sold as ferrochrome alloys). Price estimated from chemical catalog price in comparison to that of nickel.

peaks might be turned into a rapid measure of the concentration of the constituents.

XI. THE HYDROGEN ECONOMY (SEE CHAPTER 8)

In the concept of the hydrogen economy, abundant electrical energy is assumed to be available from atomic or geothermal sources. At these sources electricity is used to produce hydrogen from sea water, and this is the medium in which the energy is delivered. The hydrogen is then burned in homes and factories to produce heat, or converted to electricity by fuel cells, fresh water being the byproduct.

Because of the low viscosity of hydrogen gas, it is cheaper (at distances above a few hundred miles) to push it in pipes and to reconvert it at site by means of fuel cells than to push electricity through wires (unless the wires are surrounded by liquid hydrogen and made into superconductors).

The hydrogen would provide a single fuel for all purposes, and fresh water. The concept would give rise to an entirely nonpolluting economy which would not seem to have any limitation on its life.

People are afraid of hydrogen. However, one can throw liquid hydrogen on the ground, and flame it. It burns, like gasoline, but does not explode, in an open space. Liquid hydrogen is transported in railway cars. The present fear in the use of hydrogen as a fuel is likely to be seen from a future vantage point as similar to fears expressed by people coming out of the horse and buggy era in respect to the use of gasoline to drive cars.

Jet aircraft could be powered by electrochemically produced liquid hydrogen. The weight needed for the same energy content is less than one-third that for gasoline fuels. The consequences for the range and cost of running aircraft would be considerable. Jet power via hydrogen combustion should be feasible. The result would be the end of pollution from aircraft in addition to a reduction of the cost of fuel for air transportation (cf. Table 1).

Hydrogen is cheap enough now to make these advances economically feasible. Cheaper hydrogen would be obtainable if the electrocatalysis of oxygen evolution could be improved.

XII. THE AGRO- (AND URBAN) NUCLEAR COMPLEX (SEE CHAPTER 7)

In the agronuclear complex, a nuclear reactor is situated near, or on, the sea. It distills sea water, produces steam, electricity, and fresh water. The fresh water irrigates, the electricity drives the machines to work the complex, and excess electricity is exported.

Another concept is the urban-nuclear complex : the atomic reactor is the center of the city. Water would be produced by electrodialysis from brackish water, usually available from wells. Sewage would be processed electrochemically, and *in situ*. With available abundant electricity, materials could be produced within the city, and used materials recycled electrochemically.

XIII. THE ELECTROCHEMICAL WORLD

The immediate research choice is whether the main thrust goes toward the fuel cell, whereupon the major efforts should be in the synthesis of cheap hydrazine and lessening the cost of hydrogen, or towards the high-energy–density battery. The essence of a viable nonpollutive technology is that everything will be run *electrically*. It is not of vital* consequence as to whether the electric energy for transportation comes from chemical energy via fuel cells, or from geothermal sources and atomic reactors, whereupon it will be

*If we burn methanol to provide fuel-cell power for cars, CO_2 will result. Fuel cells therefore could only be a temporary solution—until sufficient atomic reactors have been built to make electricity for batteries to run cars. In any case, the methanol would come from oil which will be becoming an uneconomic source by the end of the century. One possibility would be to recover CO_2 from the atmosphere and reduce it electrochemically to methanol for use in fuel cells for cars. Very much depends on the rate at which atomic reactors are built.

stored electrochemically and cars will be run by means of high-energy–density batteries.

Radiochemically polluted liquids produced as products of atomic fission can be subjected to electrochemical separation. If the liquids are passed over electrodes held at suitable potentials, one could selectively deposit the radioactive metals. Storage of radioactive materials in compressed form (in mine shafts) would be satisfactory for many decades. Eventually, if one's view into the future is beyond several decades, it seems reasonable to assume that we shall have atomic powered space vehicles, and that the cost per pound of putting something into space may have sunk sufficiently so that sending compressed radioactive junk back into the sun would be economically feasible. An atomic-electric economy does not seem to have a time after which it, too, would have to be abandoned.

To sum up, man caused atmospheric pollution by burning fossil fuels to obtain his mechanical and electrical energy, because of a hiatus in the development of electrochemical science—the lack of comprehension by electrochemists in the early years of the century of the electrochemical concept of *overpotential.* No lasting cleanup from fossil-fuel pollution of the atmosphere is possible. The only long-term path is through atomically-produced electricity as the medium of energy, electrochemistry as the medium between electricity, and materials, energy, and technology.

Electrochemical technology would lead an electrochemical transportation system, and to ways which will serve indefinitely in dealing with sewage, rubbish, metal recycling, junked cars, and nonpollutive synthesis of compounds. Large-scale chemistry, in general, would be converted to electrochemistry, with improvement in the antipollutive sense. Metallurgical processes which now pollute could be made electrochemical.

The future of electrochemical technology will be enhanced insofar as atomic energy gives rise to abundant electricity. The agronuclear and urban-nuclear complex are central concepts in which electrochemical processes will play big parts. In the hydrogen economy, hydrogen becomes the medium of electrical energy, it is produced at the site of the reactor and converted to heat and electricity in home and plant.

XIV. RESEARCH FUNDING

Some comparisons in funding are given in Table 3. The major funding of electrochemical research in recent times arose from NASA. When NASA found its electrochemical auxiliary power for space to be successful, it reduced its support of electrochemical research to ca. 20 % of its 1968 amount —almost all for hardware. This catastrophic reduction was not compensated by increase in support from any other agency.

Table 3. Some Peculiar Priorities (1969, Millions of Dollars)

Battery research—all U.S.	5
Fundamental electrochemistry research	<1
G.M. antipollution research	8
N.A.P.C.A. total budget	112
G.M. advertising	240
Yearly model change, car cost	2000

One astounding aspect of the origins of research funding is that the committees called together by the government to consider electrochemical problems usually contain no electrochemists, or at the most one or two of classical* vintage. For example, the U.S. Department of Commerce in December 1967 published a book entitled *The Automobile and Air Pollution.* The panel on electrically powered vehicles consisted of one School of Management consultant, one director of an energy conversion institute known for his work on heat transfer, one professor of environmental engineering, the Director of the Economics Office of the Ford Company, the manager of a process research division, a section head of a government research laboratory in an oil company, a vice-president of a corporation, a manager of a technical program at a company, the Technical Director of the Styling Staff at General Motors (*sic*), the Vice-President of Quality and Reliability of Chrysler, an engineer with knowledge of electric cars from General Dynamics, the vice-president of a battery company, the retired president of an electric car company, one (!) electrochemical engineer, and a thermodynamician. The electrochemical engineer was an experienced and respected man but strongly associated with an organization concerned with traditional electrochemistry. This person on this committee had the *knowledge* (to say nothing of the *vision* and *enthusiasm*) necessary to understand the result of investing in research in modern electrochemical science. However, five persons were executives of companies which might suffer economic retardation were electrochemically powered transportation to be introduced. The focus of the solution to the automobile and air pollution as being connected to research in only electrochemistry and the development of the electrochemical engineering of batteries and fuel cells was not pointed out by the members of the committee.

*Roughly speaking, electrochemists who were trained before about 1950 can be said to be classical, i.e., their orientation is largely thermodynamic and not mechanistic and quantum mechanical.

Table 4. More Peculiar Priorities (Rough estimate of Ph.D.'s per year)

All chemistry	c[a].	2000 (U.S.)
Electrochemistry (1969)	c[a].	4 (U.S.)
		12 (U.K.)
		23 (Germany)
		50 (U.S.S.R.)[a]

[a]Kandidat degrees are equivalent to U.S. Ph.D.

XV. EDUCATION

There are a few universities in the U.S. in which some largely 19th century electrochemistry is taught. There are hardly any in which the new, modern electrochemistry (electrochemistry looked at in terms of solid-state physics and quantum mechanics) is taught. The contrast with other countries is shown in Table 4.

Chemistry departments in some U.S. universities are turning increasingly *away* from research on subjects of national relevance. There is a preoccupation with chemical physics, without the parallel development into fundamental research into areas such as surface chemistry, chemical aspects of the properties of materials, etc., which influence our future in a more direct way.

It is clear that, from the point of view of future, pollution-free technology, electrochemistry in its modern version should be taught in high schools and in every university. The connection between it, atomic energy, and a non-pollutive future should be in a teenager's education just as much as education in the part air plays in his existence. They are, after all, somewhat connected.

Chapter 2

THE INFLUENCE OF THE COMBUSTION OF FOSSIL FUELS ON THE CLIMATE

Gilbert N. Plass

Department of Physics, Texas A&M University
College Station, Texas

I. INTRODUCTION

The average temperature of the surface of the earth depends critically on the presence of three relatively rare gases in our atmosphere: carbon dioxide, water vapor, and ozone. The most numerous molecules in our atmosphere, oxygen, nitrogen, and argon, have almost no direct influence on the climate. They have negligible absorption of electromagnetic radiation at visible and infrared frequencies. Carbon dioxide (0.032% by volume), water vapor (variable, usually less than 2% by volume), and ozone (variable, usually less than 10^{-7} ppm) have strong absorption bands, particularly at certain wavelengths in the infrared, and in the case of ozone also in the ultraviolet. Thus, these relatively rare atmospheric molecules determine how much solar energy reaches the earth's surface and how much infrared radiation is radiated back to space. The proportion of all three of these gases in the atmosphere varies with time. A sufficiently large change in the concentration of any one of these gases changes the temperature of the earth's surface. An increase in the concentration of carbon dioxide, for example, increases the temperature of the earth's surface.

The carbon dioxide and water-vapor theory of climatic change was first proposed by the physicist John Tyndall[1] in 1861. He wrote:

> If, as the above experiments indicated, the chief influence be exercised by the aqueous vapour, every variation of this constituent must produce a change of climate. Similar remarks would apply to the carbonic acid diffused through the air.... It is not, therefore, necessary to assume alterations in the density and height of the atmosphere to account for different amounts of heat being preserved to the earth at different times; a slight change in its variable constituents would suffice for this. Such changes in fact may have produced all the mutations of climate which the researches of geologists reveal. However this may be, the facts above cited remain: they constitute true cases, the *extent* alone of the operation remaining doubtful.

More than a century of scientific work has been necessary in order to calculate with any certainty the extent of the influence of these gases. There are two problems, both of great complexity, which must be solved. The first is the calculation of the influence of changes in the concentration of one of these gases on the radiation flux at all frequencies of importance in the visible, infrared, and ultraviolet, assuming nothing else changes in the atmosphere. Each of these gases has characteristic absorption lines; up to several thousand of these lines are grouped in each absorption band. Each line has a different intensity which varies with temperature and a half-width which varies with pressure and temperature. The effect of an entire absorption band is not the sum of the effects due to each absorption line in the band, since the lines overlap because of their finite half-widths. It is only recently that one has been able to calculate the intensities and frequencies of the spectral lines of triatomic molecules such as these with some degree of assurance by taking account of the many different interactions in such a molecule. However, because of the complexity of the molecular interactions, various approximations must be used. More recent work hopefully gives a more reliable answer than earlier studies.

The second complex problem concerns the effect of changes in the concentration of one of these gases on the climate when the assumption is removed that other factors in the atmosphere remain constant. For example, a change in carbon dioxide amount may change the circulation pattern in the atmosphere which in turn may influence the mean cloud amount. This in turn alters the radiation balance. The change in carbon dioxide concentration may also change the extent of the ice fields which in turn influences the amount of solar radiation reflected back to space. It is only in very recent years that models have been proposed which attempt to simulate conditions of this sort. The interactions are very complex and involve many different kinds of physical processes.

The calculation by Arrhenius[2] of the influence of carbon dioxide on the temperature was the most extensive made during the nineteenth century. However, these are extremely approximate calculations by modern standards. Chamberlin[3] presented in detail the geologic implications of the carbon dioxide theory with such persuasion that the carbon dioxide theory of climatic change was probably the most widely held theory at the beginning of the century. The problem was again considered by Callendar in several articles[4] in which he first pointed out that the atmospheric carbon dioxide amount was increasing due to the burning of fossil fuel. The first extensive calculation of the temperature changes to be expected from given variations in carbon dioxide amount using the modern theory of band structure were made by Plass,[5,6] who also reviewed the status of the carbon dioxide theory of climatic change. Kaplan[7] and Moller[8] have also made calculations for the static model. More recently, Manabe and Strickler[9] and Manabe and Wetherald[10] have attempted to take into account the interaction between the various atmospheric factors.

II. CALCULATION OF TEMPERATURE CHANGES

Carbon dioxide, water vapor, and ozone all have special absorption bands in the region of the infrared spectrum where an appreciable flux of energy is radiated by the earth's surface and atmosphere. When there are clear skies an important part of the downward flux of infrared radiation which reaches the earth's surface arises from emission by the molecules of these three gases. When the concentration of any one of these gases increases, the downward radiation increases. This in turn causes the surface temperature to rise. The final equilibrium temperature is determined by an adjustment in all the factors which determine the outgoing flux of radiation to space. This flux must equal the incoming energy from the solar radiation when averaged over the entire earth for a reasonable length of time.

The calculations of Plass[5] started from laboratory measurements of absorption. These were reduced to the variable pressure and temperature found in the earth's atmosphere by a technique that properly took account of the variation in both intensity and spacing between lines in a spectral band, any degree of overlapping of the spectral lines, and the actual variation in the atmosphere of the half-width of the spectral lines with pressure. A correction was introduced for the fact that the spectral lines have the Doppler line shape (due to the distribution in molecular velocities) at the highest altitudes.

Plass[5,6] found that the equilibrium temperature at the earth's surface would rise 3.6°C if the carbon dioxide concentration is doubled and would fall 3.8°C if the carbon dioxide concentration is halved. These results assume clear-sky conditions and that no other factors in the atmosphere change as a result of the carbon dioxide changes. Plass[5,6] also discussed how these temperature changes would be affected if the average cloudiness in the earth's atmosphere were taken into account. His calculations show that for a reasonable average cloud distribution over the earth that the average surface temperature increases 2.5°C or decreases 2.7°C when the carbon dioxide amount in the atmosphere is doubled or halved. Later calculations by Kaplan[7] and Moller[8] obtained results in essential agreement with these, considering the different methods used and the complexity of the calculations. Under some conditions Moller obtained extremely large temperature changes. These results do not make physical sense and arise only because of Moller's arbitrary assumption of constant absolute humidity in the atmosphere.[10,11] All of these calculations assume that nothing else changes in the atmosphere when the concentration of one of these gases changes.

Variations in the ozone amount also influence the radiation balance. The total ozone amount and its vertical distribution show considerable variation with time at a given location on the earth's surface. The total ozone amount in a vertical air column is usually between 0.15 and 0.40 cm at STP (the height of a column with the same number of ozone molecules all at standard temperature and pressure). The maximum ozone concentration (as measured in centimeters of ozone at STP per km in the atmosphere) may occur anywhere between 10 and 30 km. The maximum value is usually about 10^{-2} cm of ozone per km. The ozone in the lower stratosphere is not in photochemical equilibrium. The amount present is largely determined by the circulation in the stratosphere, which brings new supplies of ozone down from the higher altitudes. This ozone is not destroyed rapidly by the sun's ultraviolet radiation, since very little radiation of the appropriate frequencies penetrates this far down in the atmosphere.

On the other hand, in the upper stratosphere the ozone is in photochemical equilibrium. At these levels the amount of ozone present is determined by the ultraviolet flux from the sun. The ultraviolet radiation in the Herzberg and Schumann–Runge bands causes the ozone to be formed by first dissociating the oxygen molecules. Longer wavelength ultraviolet radiation in the Hartley band around 2500 Å dissociates the ozone molecule. The equilibrium amount of ozone is proportional to the square root of the ratio of the intensity of sunlight in these two frequency intervals. Thus, changes in the spectral emission of the sun necessarily cause changes in the equilibrium amount of ozone.

The influence of ozone variations on the climate has been discussed by Willet,[12] Plass,[13] and Kraus.[14] Plass[13] made quantitative calculations of the effect of ozone variations on the infrared flux. As an example, when the total ozone amount changes from 0.213 to 0.267 cm and the ozone distribution shifts from one with a maximum at 28 km to one with a maximum at 10 km, the corresponding change in the equilibrium temperature at the surface of the earth is 2.1°C, when there are clear skies. Probably more than half of this temperature change is due to the difference in the altitude where the maximum ozone amount occurs; somewhat less than half of the difference is due to the variation in the total ozone amount. This calculated temperature change results from a rather moderate variation in the ozone distribution; larger temperature changes would result from larger variations.

The atmospheric water vapor amount also varies over a wide range. As discussed by Plass,[5,6] the average amount of water vapor in the atmosphere increases rapidly with temperature. This increase is largely the result of the increased ability of the atmosphere to hold larger water vapor amounts as the temperature increases. Although the absolute humidity increases rapidly with temperature, the relative humidity (essentially the ratio of water vapor amount present to amount present in saturated air at the temperature in question) in the atmosphere tends to remain constant as the temperature varies. This increased amount of water vapor present when the temperature increases tends to further increase the surface temperature through the interaction of its infrared bands with the infrared radiation. Wexler[15] found that for a particular water-vapor distribution that the surface temperature is increased 6°C when the relative humidity increases from 50 to 100%.

Plass[6] has emphasized that the effect of the water vapor in the atmosphere is to reinforce temperature changes. These changes may be caused by variations in either the ozone or carbon dioxide amounts in the atmosphere. For example, if the ozone amount increases following a change in the ultraviolet radiation from the sun, the average temperature tends to rise because of increased downward infrared radiation reaching the earth's surface. As a result the average surface temperature increases still more. Even small temperature changes can be magnified appreciably by the various factors discussed here. Temperature changes caused by carbon dioxide variations are reinforced by the water vapor in exactly the same manner as for ozone. Water vapor can also reinforce temperature changes caused initially by other factors, such as changes in the amount of volcanic dust in the air, in the average height of the continents, and in the elements of the earth's orbit around the sun.

All of the previously discussed calculations assume that no other factors will change in the atmosphere when one of the atmospheric constituents changes. Obviously, many different factors are interrelated in a complex manner in our atmosphere, and this assumption may be very far from the truth. In two important papers, Manabe and Strickler[9] and Manabe and Wetherald[10] have postulated a model which takes account of some of these interrelations. Although some interrelations are not included in their model and various approximations are necessarily made, their results should provide a reasonable first approximation for the atmospheric reaction to many of these changes. Their model essentially incorporates radiative transfer into a general circulation model.

Manabe and Wetherald[10] assume that the atmosphere maintains the same distribution of relative humidity as exists today as conditions change. The moisture content of the atmosphere increases rapidly as the temperature changes, but there is some reason to believe that the atmospheric processes readjust so that the relative humidity remains approximately constant. This is certainly a much better assumption than assuming constant absolute humidity as was done by Moller[8] and Manabe and Strickler.[9] This was the reason that Moller obtained unreasonably large temperature changes in some cases. Manabe and Wetherald also assume in their calculations that the lapse rate never exceeds the critical value of 6.5°C km^{-1}, that no temperature discontinuity should exist, that the net incoming solar radiation should equal the net outgoing long-wave radiation at the top of the atmosphere, that local radiative equilibrium occurs whenever the lapse rate is subcritical, and that the heat capacity of the earth's surface is zero. The remainder of the results given in this section are based on this model, which is certainly the most realistic model yet presented.

Calculations of the equilibrium temperature in the atmosphere from this model show the following interesting features: (1) the larger the mixing ratio of carbon dioxide, the warmer the equilibrium temperature of the earth's surface and troposphere; (2) the larger the mixing ratio of carbon dioxide, the colder the equilibrium temperature of the stratosphere; and (3) relatively speaking, the dependence of the equilibrium temperature of the stratosphere on carbon dioxide content is much larger than that of tropospheric temperature. When the carbon dioxide content of the atmosphere is reduced by 50%, the equilibrium temperature of the earth's surface decreases 2.80°C for clear-sky conditions and 2.28°C for average cloudiness conditions. When the carbon dioxide content of the atmosphere is increased by 100%, the equilibrium temperature of the earth's surface increases 2.92°C for clear-sky conditions and 2.36°C for average cloudiness conditions. These results, particularly for average cloudiness conditions, are remarkably close to those obtained earlier by Plass,[5,6] especially when one considers the completely different method used in each case together with the more reliable laboratory data now available. Plass used a more accurate method of estimating the overlap between the spectral lines and the variation of the line intensities and half-width with temperature and pressure. Manabe and Wetherald have used a more complete model of the atmosphere as a whole.

Manabe and Wetherald find that the equilibrium temperature above 20 km decreases as the carbon dioxide content increases. When the carbon dioxide content doubles, the temperature above 30 km decreases by 10°C or more.

One theory of climatic change is based on changes in the solar constant. Obviously, a change in the total amount of energy received from the sun must have effects on the earth's climate. Manabe and Wetherald find that the temperature increases at all heights in the atmosphere as the solar constant increases. The solar constant is approximately 2.0 ly·min^{-1} at the present time. The equilibrium temperature calculated by this model for this value of the solar constant is 287°K, while the value is 270 and 308°K for values of the solar constant of 1.75 and 2.25 ly·min^{-1}, respectively. The equilibrium surface temperature increases very rapidly above 300°K as the solar constant increases and the relative humidity remains fixed. The reason is that the water-holding capacity of the atmosphere increases very rapidly with temperature in this range. The result clearly demonstrates the self-amplification effect of water vapor on the equilibrium temperature of the atmosphere first discussed by Plass.[5,6]

The effect of changes in the concentration of various atmospheric absorbers is also shown by the calculations of Manabe and Wetherald. If the relative humidity in the troposphere increases, the equilibrium temperature of the troposphere increases, whereas there is little change in the equilibrium temperature of the stratosphere. For example, if the relative humidity at the earth's surface has the values 0.2, 0.6, and 1.0, then the surface equilibrium temperature is 278, 285, and 290°K, respectively, with very little change at higher altitudes. On the other hand, if the relative humidity in the stratosphere only is increased, it is found that the troposphere becomes warmer and the stratosphere colder. The dependence of the equilibrium temperature in the stratosphere upon the stratospheric water-vapor mixing ratio is much larger than that in the troposphere. If the stratospheric water-vapor mixing ratio decreases fivefold, the surface equilibrium temperature decreases by 2.0°C; if it increases fivefold, the corresponding temperature increase is 5.6°C.

The surface albedo of the earth is defined as the fraction of the radiation incident on the surface which is reflected back; the remainder of the incident radiation is absorbed. The albedo of an ice sheet or snow-covered ground may be of the order of 0.9, while the albedo of soil or rocks may be in the range 0.1–0.3. Thus, it is important when considering climatic variations to understand how the albedo influences atmospheric temperatures. Again the calculations of Manabe and Wetherald show that the larger the value of the albedo of the earth's surface, the colder the temperature of the atmosphere. The influence of the surface albedo decreases with increasing altitude and is almost negligible above a 30-km altitude. A change in the average surface albedo of the entire earth of 0.1 changes the surface equilibrium temperature by about 10°C.

The calculations of Manabe and Wetherald were also made for three different ozone distributions. The results show that the larger the amount of ozone, the warmer the temperature of the troposphere and the lower stratosphere and the colder the temperature of the upper stratosphere (above

25 km). The influence of the ozone distribution upon the equilibrium temperature was found to be significant in the stratosphere, but to be small in the troposphere. Total ozone amounts of 0.260, 0.341, and 0.435 cm at STP give surface equilibrium temperatures of 287.9, 288.8, and 290.3°K, respectively. However, equilibrium temperature variations of 10–20°K for these different distributions were obtained above 10 km.

The last and perhaps most important atmospheric variable that can influence the equilibrium temperature is the amount of cloudiness. It should be emphasized that all of the values quoted previously for temperature change assumed an average cloud distribution unless stated otherwise. The temperature of the earth's surface would be 307.8°K if there were no clouds but, instead, clear skies according to the model of Manabe and Wetherald. The amount of clouds clearly have a decisive influence on the surface temperature. Their calculations performed for a variety of cloud distributions show that the influence of clouds depends on their heights and the cloud albedo. If the height of a cirrus cloud is greater than 9 km and its blackness for infrared radiation is larger than 50 %, cirrus has a heating effect on the temperature of the earth's surface. However, middle and low clouds have a cooling effect on the temperature of the earth's surface.

Generally speaking, the larger the cloud amount, the colder the equilibrium temperature of the earth's surface, although this tendency decreases with increasing cloud height and does not always hold for cirrus. The equilibrium temperature of the atmosphere with average cloudiness is about 20.7°C colder than that for a clear atmosphere. The following temperature changes are found when there is a unit percentage increase in cloudiness: −0.82°C for low clouds; −0.39°C for middle clouds; +0.38°C for high clouds with zero albedo; +0.04°C for high clouds with 0.5 albedo. The influence of cloudiness is more pronounced in the troposphere than in the stratosphere.

III. CARBON DIOXIDE BALANCE

Carbon dioxide is stored in many different forms on the earth and there is a continuous exchange between the atmosphere, oceans, biosphere, and geosphere. Tables 1 and 2 show the present carbon content of the atmosphere, hydrosphere, geosphere, and biosphere and the exchange activity adopted from Lieth[16] and Bolin.[17] By far the largest amount of carbon is stored in the geosphere, mostly in carbonate rocks and other noncarbonate compounds. The next largest reservoir (but 1000 times smaller) is the hydrosphere. The atmosphere contains a still smaller amount of carbon dioxide; the biosphere is the smallest reservoir.

Table 1. Carbon Dioxide Content

		Carbon dioxide content (in units of 10^9 metric tons of carbon)	Totals
Atmosphere			700
Hydrosphere:	sea surface layers	500	
	deep sea	34,500	
			35,000
Geosphere:	inorganic carbonates	18,000,000	
	inorganic noncarbonates	7,000,000	
	older organic compounds: coal, oil, gas	10,000	
	dead organic matter in ocean	3,000	
	dead organic matter on land	700	
			~25,000,000
Biosphere:	phytoplankton, zooplankton, fish in ocean	10	
	plants on land	450	
			460

There is a continuous exchange activity between these reservoirs, tabulated in Table 2. Many of these numbers are rather uncertain. Nevertheless, this table indicates the major processes involved in the carbon circulation together with an indication of their relative importance. Many of the factors which control the exchange of carbon dioxide between the various reservoirs have changed by many orders of magnitude during the geologic history of the earth. Now the activities of man have introduced serious perturbations in some of these factors as well.

In recent years the burning of fossil fuels has added 10^{10} tons per year of carbon dioxide to the atmosphere. It is predicted that this figure will increase to 5×10^{10} tons per year by the year 2000. The amount of carbon dioxide added to the atmosphere each year can be calculated fairly accurately from the statistics on the use of coal, oil, and gas. At the present time (1970), there is a carbon dioxide concentration of 320 ppm in the atmosphere. The carbon dioxide added each year is now equivalent to 2.0 ppm·year^{-1}. Not all of this carbon dioxide stays in the atmosphere; important amounts are absorbed by the oceans and by plants. Measurements by Keeling[18] suggest that about 40% of the carbon dioxide from fossil fuel is remaining in the atmosphere. His long-term measurements in Hawaii and Antarctica also clearly show the yearly increase in the atmospheric amount.

In addition to burning fossil fuels, man's activities are adding additional quantities of carbon dioxide to the atmosphere by the clearance of forests, the drainage and cultivation of lands, and industrial processes such as lime burning and fermentation. If all of this carbon dioxide remains in the atmosphere, the carbon dioxide concentration would be 437 ppm by the year 2000.

If 40% stays in the atmosphere the carbon dioxide concentration would be 352 ppm by the year 2000, which would cause a temperature rise of approximately 0.5°C. If the burning of fossil fuel is continued at this rate into the twenty-first century, there will be a profound effect on the earth's climate.

Until man recently disturbed his environment, it is believed that the many factors which enter into the carbon dioxide balance were approximately in equilibrium. Approximately 3.5×10^{10} tons of carbon are used each year in photosynthesis by land plants. In a steady state precisely the same amount must be returned to the atmosphere each year by all of the processes of respiration and decay of plants and animals, provided that none of this carbon dioxide is permanently lost in forming new coal, oil, and other organic deposits. At the present time, at least, the amount lost in this manner is very small ($<1 \times 10^9$ tons per year) compared to the carbon dioxide used in photosynthesis and can be neglected in a discussion of the balance of factors from the organic world.

Let us suppose that the steady-state absorption and emission of carbon dioxide by the organic world is disturbed by a sudden increase in the amount of carbon dioxide in the atmosphere. The amount of carbon dioxide used in

Table 2. Exchange Activity

	Exchange activity (in units of 10^9 metric tons of carbon per year)
Withdrawn from atmosphere in photosynthesis by plants on land	35
Returned to atmosphere in respiration by plants on land	10
Plants and animals become dead organic matter on land	25
Released to atmosphere by decay of dead organic matter on land	25
Withdrawn from atmosphere by exchange with surface layers of ocean	100
Returned to atmosphere by exchange with surface layers of ocean	97
Withdrawn from surface layers of ocean in photosynthesis by plants in ocean	40
Plants and fish become dead organic matter in ocean	40
Released to surface layers of ocean by decay of dead organic matter in ocean	35
Released to deep oceanic layers by decay of dead organic matter in ocean	5
Withdrawn from deep ocean and added to surface layers by upwelling water	45
Withdrawn from surface layers and added to deep ocean by descending currents	40
Dead organic matter forming sediments	<1
Returned to atmosphere by combustion of fossil fuel	5
Removed from atmosphere by weathering of rocks	0.05
Returned to atmosphere by volcanic emissions and hot springs	0.05

photosynthesis would increase since plants grow more rapidly in a carbon dioxide-rich atmosphere. However, in a very short time interval the processes of decay and respiration would also have increased. Since an average carbon atom that has been used in photosynthesis returns to the atmosphere from the biosphere in a relatively few years, it follows that the factors from the organic world would once again be in balance a few years after a change in the carbon dioxide concentration in the atmosphere.

The clearing of virgin land for cultivation has released large quantities of carbon dioxide to the atmosphere. Very much less carbon is contained in crops grown on the cleared land than were in the luxuriant growth of the virgin forests. The carbon in the cleared vegetation quickly returns to the atmosphere through either decay or by being burned. Also, large forest fires have probably increased as a result of man's activities. These have also added large amounts of carbon dioxide to the atmosphere that would otherwise have been locked up in the vegetation for centuries or more.

Measurements by Keeling[19] and his coworkers have shown conclusively that the carbon dioxide amount varies in cyclic manner with a yearly period. It has a minimum value in late summer. The photosynthetic activity of land plants reaches a maximum in the summer season and depletes the atmospheric carbon dioxide. To a lesser extent these variations are also the result of biologic activity in surface ocean waters. The variation produced by photosynthesis is far greater in the northern hemisphere than in the southern, because of the much greater proportion of land to ocean area in the northern hemisphere. The yearly variation in the carbon dioxide concentration caused by photosynthesis is 10 ppm on the north coast of Alaska, 6 ppm on the island of Hawaii, and only 1.5 ppm at the South Pole. These measurements also show that the carbon dioxide is very uniformly mixed in the troposphere. Previously reported variations in the carbon dioxide content of different types of air masses have not been confirmed by more recent measurements.

Throughout geologic history, the amount of carbon dioxide lost from the atmosphere by the formation of new coal beds and other organic deposits and by the weathering of igneous rocks and deposition of carbonates has varied widely, as has the carbon dioxide added to the atmosphere by such factors as the evolution of carbon dioxide from hot springs, volcanic vents, gas wells, and other sources. Since variations in each of these factors occur independently, the net gain or loss of carbon dioxide by the atmosphere has changed frequently on a geologic time scale. Although it is not always easy to deduce the net result of these many independent variations for a given epoch, the mere knowledge that these factors have undergone significant fluctuations many times in the past has important implications for studies of the climate.

IV. CARBON DIOXIDE EXCHANGE BETWEEN OCEANS AND ATMOSPHERE

The oceans contain an equivalent amount of carbon dioxide which is 50 times greater than that in the atmosphere. For this reason it is important to understand the various factors which control the exchange of carbon dioxide between the oceans and atmosphere. If the amount of carbon dioxide in the atmosphere is changed, carbon dioxide must be released or absorbed by the oceans before a new equilibrium point is reached.

The equations which determine the equilibrium between the various ions in sea water have been extensively studied,[20,21] and the appropriate dissociation constants are known, although in some cases with only fair accuracy. Carbon dioxide exists in sea water not only as a dissolved gas, but in the equivalent forms of $CO_3^=$, HCO_3^-, and H_2CO_3. The equilibrium amount is influenced by the hydrogen ion concentration, the amount of excess base, and the amount of boric acid present. Rubey[20] also emphasized that the oceans must reach equilibrium with calcium carbonate after a sufficient period of time following a change in the total amount of carbon dioxide in the atmosphere–ocean system. If there is an excess amount of calcium carbonate, it precipitates; if there is too little calcium carbonate, it dissolves and at the same time accumulates from the rivers that flow into the oceans until the solubility product is reached.

The dissociation constants are functions of the temperature. Plass[6] first pointed out that the variations in oceanic temperature which occur as the climate changes must be taken into account. A warm ocean holds much more carbon dioxide equivalent than does a cold ocean. This effect tends to amplify temperature changes caused by variations in the atmospheric carbon dioxide amount. This same article[6] also emphasized that the equilibrium depends in an important manner on the volume of the oceans, which is appreciably reduced during a period of glaciation by the large volume of water locked up in the ice sheets.

These calculations[6] show that if the total amount of carbon dioxide equivalent in the atmosphere–ocean system remains constant, then the atmospheric carbon dioxide amount doubles if the oceanic volume decreases by 7%, if there is insufficient time for calcium carbonate equilibrium, or it doubles if the oceanic volume decreases by 20%, if there is sufficient time for calcium carbonate equilibrium. If the amount of carbon dioxide equivalent in the atmosphere–ocean system increases by 10%, then the amount of carbon dioxide in the atmosphere increases from 3 to 5 ppm or 8 ppm depending on whether there is or is not sufficient time for calcium carbonate equilibrium, respectively.

More recently, Eriksson[21] repeated and extended these calculations. He linearized the equations so that his results are valid only for small departures from the equilibrium situation. Nevertheless, they provide an important indication of the variation of the atmospheric carbon dioxide amount as various factors are changed. He found that when the surface water temperature increased 1°C the atmospheric carbon dioxide pressure increased 5.7 and 4.2%, with and without calcium carbonate equilibrium, respectively. The atmospheric carbon dioxide pressure increases about 3% when the oceanic volume decreases 1%. At the same time about 0.1% of carbon dioxide is precipitated as calcium carbonate. He also points out that any change in the total phosphate content of the ocean causes changes in the equilibrium amount of atmospheric carbon dioxide. However, this is buffered by hydro-oxyapatite.

Keeling and Waterman[22] made extensive measurements of the carbon dioxide content of surface ocean waters. Their results beautifully show the major processes which control the carbon dioxide content of the ocean. Plants withdraw carbon dioxide from surface water during photosynthesis. After their life cycle these organisms sink below the thermocline and decay, thus releasing carbon dioxide to the subsurface waters. The surface waters are thus persistently depleted of carbon dioxide, whereas the deep waters are correspondingly enriched. At those locations where upwelling or vertical mixing promotes upward transport of carbon dioxide from deep waters, the surface concentration of carbon dioxide is higher than would be expected for equilibrium. On the other hand, in those localities where there is little vertical transport of ocean water the surface waters are continually depleted of carbon dioxide, and the carbon dioxide concentration of these surface waters is less than would be expected for equilibrium with the atmosphere.

The north–south currents in the oceans also have an important influence on the amount of carbon dioxide in the surface waters. Oceanic water is cooled when it is carried northward from the equator by a current. This cooler water has less carbon dioxide because of the temperature equilibrium factor controls. These conclusions are confirmed by the measurements of Keeling and Waterman.[22] In general, low concentrations of carbon dioxide are expected in surface waters when rapid cooling occurs and when withdrawal by marine organisms exceeds replenishment by physical transport. A high concentration occurs when there is rapid warming of the waters and when upwelling or vertical mixing supplies more carbon dioxide than can be withdrawn by plants or can escape into the atmosphere.

How rapidly does the atmosphere–ocean system come to equilibrium? There are two parts to this question. First we must consider how rapidly carbon dioxide is exchanged between the surface waters and the atmosphere. Then we must determine the rate at which the surface waters exchange with the deep-ocean waters. Keeling and Waterman[22] state that the absorption rate into the oceans is 7 moles·cm^{-2}·atm^{-1}·yr^{-1}, if the atmospheric residence time of radiocarbon is 5 years. The rate of exchange between the atmosphere and ocean depends critically on the wind velocity. It has been estimated by Kanwisher[24] that the air–ocean exchange is at least 20 times faster at wind velocities of 10 m·sec^{-1} than at 2 m·sec^{-1} and that the exchange is approximately proportional to the square of the wind velocity. Berger and Libby[23] have raised the possibility that some enzyme such as carbonic anhydrase may control the rate at which carbon dioxide is exchanged at the ocean surface. Such an enzyme could greatly hasten the time required for carbon dioxide to dissolve in some areas of the ocean. It also raises the question whether worldwide marine pollution may control the atmospheric carbon dioxide concentration by affecting the organisms that produce such enzymes.

From 1000 to 10,000 years[18,20] is required for the waters of the deep oceans to be renewed by exchange with the surface water. Thus, it would seem that it would require a time of this magnitude for the ocean–atmosphere system to return to equilibrium after some factor has changed the carbon dioxide balance in one part of the system.

V. EVIDENCE FROM PAST CLIMATES

In order to evaluate the possibility of an increase in the atmospheric carbon dioxide amount influencing the climate in the future, it is appropriate to first consider evidence from past climates. Many different theories of climatic change have been advanced. Among the most important are those based on changes in the solar constant, in the continentality (average height of the continents), in the amount of volcanic dust and aerosols in the atmosphere, in the elements of the earth's orbit around the sun and the angle of the earth's axis, in the average albedo of the earth, and in the amount of infrared absorbing gases in the atmosphere, including carbon dioxide, water vapor, and ozone. There have certainly been changes in all of these factors during the geologic history of the earth. Thus, each of these must have had at least some influence on the climate during some particular time period. These various factors are so interrelated that it is most unlikely that any one of them can explain all of the past fluctuations in the climate. It is important to review the knowledge of past climates and see which factors can be explained simply by variations in one of these parameters.

Quantitative calculations are important to decide whether reasonable variations of a given factor can be expected to produce important variations in the climate. Such calculations now predict the result of variations in the amount of the infrared absorbing gases, the earth's albedo, and the solar constant. We should emphasize, though, that even the most recent calculations of Manabe *et al.*[9,10] do not take into account all the complex interactions in the atmosphere. In particular, the coupling between the mean cloud amount and the fraction of the earth's surface covered with ice and snow with variations in the infrared absorbing gases has not been taken into account. At the present time, we cannot answer the question: how much will the mean cloud amount increase and compensate for a temperature rise induced by an increase in carbon dioxide amount? Nevertheless, there seems to be no doubt that a sufficiently great increase in carbon dioxide amount will cause an increase in the average temperature of the earth's surface, an increase that should be of the order of magnitude of that calculated. There is more doubt as to whether a small increase in the carbon dioxide amount will produce a temperature increase of the magnitude calculated.

With the reservations expressed here let us consider the facts about past climates which can be explained naturally by the carbon dioxide theory of climate change and related theories. There is considerable geologic evidence that extensive outbursts of mountain building preceded each of the last two major glacial epochs by at least several million years. The carbon dioxide theory provides probably the only reasonable explanation for this time lag. At the onset of a period of mountain building increased amounts of carbon dioxide are released from the interior of the earth by new volcanoes and hot springs. The warming effect of this extra carbon dioxide offsets the cooling effect of the volcanic dust in the atmosphere and of the increase in the height of the mountains, as shown by the fact that extensive glaciers did not form during this first stage of mountain building. In fact, there have been periods of mountain building during the earth's history which have not been followed by any extensive glaciation. This could be explained by assuming that the carbon dioxide released from the interior of the earth was sufficient to prevent the formation of glaciers.

When glaciers do follow periods of mountain building, the carbon dioxide theory requires a time lag of several million years before the extensive ice sheets form. During this period the carbon dioxide is being removed from the atmosphere by the process of weathering. The most active zone for the decomposition of rock occurs between the surface and the level of the permanent underground water. In mountainous country this level is farther below the surface than in flat country. Thus, in mountainous country there is a considerably larger volume in which active weathering of the rocks takes place. The dominant reaction for the weathering of igneous rocks involves the formation of carbonates, thus removing carbon dioxide from the atmosphere. After a period of mountain building this increased rock weathering, acting over a period of millions of years, sufficiently depletes the atmosphere of carbon dioxide so that a period of glaciation begins. Thus, the carbon dioxide theory clearly predicts an appreciable time lag between a period of mountain building and the following glacial epoch.

If the carbon dioxide amount is suddenly increased in one hemisphere, it will be disturbed uniformly over both hemispheres in a relatively short time, probably less than a few decades. Thus, any large-scale climatic variations due to carbon dioxide must occur simultaneously in both hemispheres. The geologic evidence, supported by radiocarbon dating, supports this view. Epstein *et al.*[25] have made an analysis of the oxygen and hydrogen isotopes in ice cores in Antarctica and the Northern Hemisphere. They provide a time scale for the Wisconsin cold interval from 75,000 to 11,000 years ago. It was warmer both before and after this period. The post-Wisconsin has averaged 8°C warmer than the Wisconsin cold interval. From a study of glaciers Denton and Porter[26] show that the maximum of the Wisconsin glaciation was from 20,000 to 14,000 years ago. At that time sea level was 130 m lower than today and glacier ice covered 27 % of the earth's land surface, compared to 10 % today. They point out that a number of dates for the advance and recession of glaciers are essentially contemporaneous for the different continents.

It is most important that this work has established that glaciation has occurred contemporaneously in both hemispheres. The theory of climatic change based upon variations in the earth's orbit and axis predicts that these changes must be out of phase in the Northern and Southern Hemispheres. Since recent calculations have shown that the temperature changes to be expected from variations in the earth's orbit and axis are very much less than were originally claimed, there no longer seems to be any reason to give important consideration to this theory as an explanation of the most recent periods of glaciation.

It should be pointed out that there might be special situations when glaciation would not occur in both hemispheres simultaneously. For example, there might be large mountainous regions in one hemisphere where glaciers could form. If the other hemisphere had no comparable mountain ranges, the surface temperature might not be low enough to form large ice sheets at the lower elevations.

Another factor in the carbon dioxide balance that has varied widely during the geologic history of the earth is the amount of organic material trapped in new coal and oil deposits and other organic sediments. During the Carboniferous period the amount thus trapped must have been much larger than at the present time. Most of our coal deposits were formed then. The proper conditions for the formation of these deposits were provided by the relatively flat land and the large number of marshes. After a relatively long period of time this accumulation of organic deposits appreciably reduced the amount of carbon dioxide available to the atmosphere–ocean system. It is perhaps significant that the glaciation at the end of the Carboniferous may have been the most severe in the earth's history.

Four distinct periods of glaciation during the last glacial epoch have been known to geologists for many years. Recent work has shown that there have been a dozen or so minima in the oceanic temperature in the last 600,000 years. A characteristic property of a glacial epoch is a constantly changing climate. The carbon dioxide theory can explain such changes. The essential element which prevents the system from approaching equilibrium is the large value of the heat of fusion of water. The extra heat energy required to melt the ice sheets keeps the temperature below the average value appropriate for the amount of carbon dioxide in the atmosphere for many thousands of years.

In order to understand these fluctuations, consider the sequence of events when the total carbon dioxide amount available to the atmosphere–ocean system decreases slightly, perhaps from an increased rate of rock weathering following a period of volcanism. If the reduced atmospheric carbon dioxide amount lowers the average surface temperature sufficiently, glaciers start to form. After thousands of years these glaciers grow into continental ice sheets. A considerable amount of heat energy is released to the atmosphere by the freezing of the water; this energy is eventually lost to space by radiation from atmospheric molecules. Once the glaciers have reached a certain critical size, they seem to grow more readily, probably because of the increased albedo of the surface. The volume of water held in the ice sheets may be from 5 to 10% of the volume of the oceans. However, since the reservoir of carbon dioxide in the oceans is many times the amount in the atmosphere, even this relatively small change in the oceanic volume has a large influence on the atmospheric amount. Since the ice sheets can hold only a very small amount of carbonates compared to the same volume of ocean water, the remaining water in the oceans must release carbon dioxide to the atmosphere in order to return to equilibrium. It probably takes an average parcel of ocean water several thousand years to make a complete circuit from the top to the bottom of the oceans and back again. Thus, there is a time lag of at least this amount before the atmospheric carbon dioxide increases to a new equilibrium value appropriate to the smaller volume of water in the oceans.

As the atmospheric carbon dioxide increases, the average surface temperature rises. The ice sheets begin to melt. However, the very considerable energy required to melt the ice sheets must be supplied by the increased downward flux of radiation from the larger number of atmospheric carbon dioxide molecules. In a typical case, where the carbon dioxide amount increases from approximately one-half of its present value to its present value, the increase in the downward flux is approximately 0.01 $cal \cdot cm^{-2} \cdot min^{-1}$. If one-tenth of this energy is absorbed by the ice sheet and causes melting, it takes 15,000 years to melt an ice sheet 1 km in thickness. During this period the average temperature near the ice sheets does not increase appreciably, since the energy is being used to melt the ice. The vast quantity of cold water from the ice sheets keeps the oceanic temperature low for a considerable period of time. Plastic ice flow and crustal warping may also have an influence during the melting of the ice sheets. Plastic ice flow increases the area of the glaciers without increasing their volume. The result is to increase further the albedo. The weight of the ice sheets causes crustal warping, which decreases the average height of the land. This tends to prevent the reformation of the ice sheets when they are melting. Possibly the most important of these various time-lag mechanisms is absorption of the energy required for melting.

Finally, when a major fraction of the ice sheets has melted, the temperature of the oceans begins to rise. The increased oceanic volume no longer holds enough carbonates to be in equilibrium with the atmosphere. Slowly the oceans withdraw carbon dioxide from the atmosphere. As the atmospheric carbon dioxide amount decreases, the temperature falls again and the cycle starts over. When some other factor raises the total carbon dioxide amount to such an extent that glaciers cannot form to an appreciable extent at the beginning of one of these cycles, the glacial epoch is at an end.

Thus, many of the climatic variations of the past can be explained simply by variations in the atmospheric carbon dioxide amount of reasonable magnitude. Most of the climatic variations mentioned above have not been satisfactorily explained by any other theory. This is not to say that variations in the solar constant, earth's albedo, and volcanic dust in the atmosphere may not have been important at certain periods of the earth's history. We do maintain that sufficiently large variations in the carbon dioxide amount must have an important influence on the climate and that indeed this has happened during many different geologic epochs.

Many authors have emphasized that not only lower temperature but also increased or at least undiminished rainfall is necessary for the formation of extensive ice sheets. It seems reasonable to expect increased precipitation when the atmospheric carbon dioxide amount decreases. When the atmospheric carbon dioxide amount is halved, the average temperature at the upper surface of a cloud is lowered by an amount of 1.3 to 2.2°C for clouds whose upper surfaces are at 4 and 9 km, respectively. Thus, when the atmospheric carbon dioxide amount is decreased, the cloud top can cool more rapidly by increased radiation to space. This in turn increases the convection inside the cloud and allows it to grow in the vertical direction. The time is hastened when the cloud becomes large enough and has sufficiently strong convective currents to form precipitation. In addition, the meridional circulation is increased so that the atmosphere absorbs more water vapor from the oceans. It is possible that an increased cloud cover reflects more of the sun's radiation back to space so that the temperature falls still further. If the relative humidity of the atmosphere stays constant, as seems likely,[10] there is less total amount of water vapor in the atmosphere at the reduced temperature. This also tends to increase the original cooling effect because of the decreased absorption and emission from the infrared bands of water vapor. Although the complex interactions between all of these factors are imperfectly understood, it seems likely that an initial change in carbon dioxide content causes a temperature change that is reinforced by changes in the water vapor amount in the atmosphere and in the mean cloud amount. It also seems likely that cold and wet climates tend to occur together, as do warm and dry climates.

This review of past climates of the earth indicates that variations in the amount of infrared-active gases in the atmosphere is one of the important factors which should be considered in any explanation of climatic variations. Large variations in the atmospheric carbon dioxide amount must have occurred many times in the geologic history of the earth. Sufficiently large variations in this variable must influence the average temperature at the earth's surface.

VI. CARBON DIOXIDE FROM FOSSIL-FUEL COMBUSTION

Throughout most of the million or more years of man's existence his fuels consisted of wood or the remains of other plants which had grown only a few years before they were burned. The effect of this activity on the atmospheric carbon dioxide was negligible, since it only sped up slightly the natural decay processes which return carbon from plants to the atmosphere. However, in the last 100 years the combustion of fossil fuel has returned to the atmosphere appreciable quantities of carbon dioxide that had been slowly extracted by the plants during a time span of half a billion years and buried in the sedimentary rocks. The present rate of production is about 100 times the average rate of release of carbon dioxide from the interior of the earth by hot springs and volcanoes and is also about 100 times the rate at which carbon dioxide is removed from the atmosphere by the weathering of rocks.

Callendar[4] first pointed out the climatic effects of the burning of fossil fuels, although he was probably mistaken in connecting the slight temperature rise during the first part of this century to this cause. Thus, mankind is now carrying on a unique geophysical experiment which cannot be repeated in the future. Within a few centuries we are returning to the atmosphere large quantities of organic carbon that has been stored in the sedimentary rocks. This experiment at the very least should provide us with a significant new understanding of the factors which determine our climate and of the factors which determine the division of carbon dioxide between the atmosphere, hydrosphere, geosphere, and biosphere.

Table 3. Carbon Dioxide Added to Atmosphere by Consumption of Fossil Fuels

Decade	CO_2 added, ppm/decade	Cumulative total, ppm	Atmospheric concentration, ppm	Predicted temperature rise, °C
1860–69	0.6	0.6	295	0.00
1870–79	1.0	1.6	296	0.00
1880–89	1.4	3.0	296	0.01
1890–99	2.1	5.1	297	0.02
1900–09	3.4	8.5	298	0.03
1910–19	4.6	13.1	300	0.04
1920–29	5.3	18.4	302	0.06
1930–39	5.6	24.0	305	0.08
1940–49	7.2	31.2	307	0.10
1950–59	10.3	41.5	312	0.13
1960–69	14.5	56.0	317	0.18
1970–79	19.9	75.9	325	0.24
1980–89	28.0	104.0	336	0.33
1990–99	38.5	142.0	352	0.46
2000–09	53.4	196.0	373	0.63
2010–19	78.0	274.0	405	0.88
2020–29	110.0	384.0	449	1.2
2030–39	154.0	538.0	510	1.7
2040–49	216.0	754.0	597	2.4
2050–59	304.0	1058.0	718	3.4
2060–69	428.0	1486.0	889	4.8
2070–79	602.0	2088.0	1130	6.7

The past consumption of fossil fuels can be calculated with considerable accuracy. Tables of the coal, oil, and gas consumption of the world are found in most almanacs and encyclopedias. The amount of carbon dioxide released from the use of each type of fuel can be estimated and the carbon dioxide added to the atmosphere in each decade calculated. The results are given in Table 3. The second column gives the amount of carbon dioxide added to the atmosphere in each decade in ppm (parts per million), while the cumulative totals are given in the third column.

The predicted atmospheric carbon dioxide concentration in ppm is given in the fourth column. This is calculated using 295 ppm as the atmospheric concentration at the beginning of this period. This is approximately the mean value of measurements made in the latter part of the nineteenth century. Furthermore, this calculation assumes that 40% of the carbon dioxide released into the atmosphere stays there. This is the figure which applies over the past decade according to the measurements of Keeling.[18] The remainder of the carbon dioxide is absorbed by the oceans and other natural processes. The assumption that 40 % of the carbon dioxide remains in the atmosphere is only an approximation and may not apply when larger quantities are released into the atmosphere.

The fifth column in Table 3 gives the predicted temperature rise according to the calculations of Manabe and Wetherald.[10] Their result predicts a temperature rise of 2.36°C for average cloudiness conditions when the carbon dioxide content doubled. This result assumes a constant relative humidity in the atmosphere as the temperature changes. As already discussed, this result is uncertain in that this effect is reinforced by such factors as the increasing temperature of the ocean waters and the decreased albedo of the earth's surface as a whole because of the diminished area of the ice sheets. Changes in the average cloudiness also have an important effect on the average temperature of the earth's surface that unfortunately cannot be calculated at the present time. Nevertheless, there seems to be little doubt that a sufficiently large change in carbon dioxide concentration must produce climatic changes of the order of magnitude predicted here.

The values of the atmospheric concentration up to the present time (1970) as given in Table 3 are in excellent agreement with the measurements of Keeling[18,19] and earlier workers in this field. The data for the carbon dioxide added to the earth's atmosphere for future years is based on the expected increase in demand for energy sources, assuming that nuclear energy is used to generate only a minor fraction of this total requirement. If energy requirements continue to increase at the present rate, in the decade 2070–2079 the consumption of fossil fuels will add twice the present atmospheric amount of carbon dioxide to the atmosphere in one decade! The temperature at this time is predicted to average 6.7°C warmer than today with an atmospheric carbon dioxide concentration of 1130 ppm. These figures only emphasize that mankind cannot allow the use of fossil fuels to continue to increase at an exponential rate. Catastrophy will certainly occur at some point in the future not much more than 100 years from now. A temperature change of 6.7°C in the average temperature of the earth's surface would have the most profound effects. All of the ice sheets would melt raising the sea level sufficiently to place most of the world's largest cities under water. The warmer climate would undoubtedly completely change the present pattern of rainfall over the earth's surface.

It seems almost certain that the consumption of fossil fuels will increase at nearly the predicted rate up to the year 2000. If mankind decides to limit the use of fossil fuels to the amount being used in the year 2000, the atmospheric concentration and predicted temperature increase in later years will be as given in Table 4. This assumption of constant use after the year 2000 is probably too conservative. However, even with this assumption the atmospheric concentration is predicted to be 496 ppm by the year 2080

Table 4. Carbon Dioxide Added to Atmosphere by Consumption of Fossil Fuels (Assuming Fossil-Fuel Consumption Remains Constant at Estimated Rate for Year 2000)

Decade	CO_2 added, ppm/decade	Cumulative total, ppm	Atmospheric concentration, ppm	Predicted temperature rise, °C
2000–09	45	187	370	0.60
2010–19	45	232	387	0.74
2020–29	45	277	406	0.89
2030–39	45	322	424	1.0
2040–49	45	367	445	1.2
2050–59	45	412	460	1.3
2060–69	45	457	478	1.5
2070–79	45	502	496	1.6

with an increase in the average temperature of 1.6°C. Although this may not appear to be a large temperature increase, even this change in the average temperature of the earth's surface would cause important changes in the climate.

Other activities of man also add carbon dioxide to the atmosphere and have not been taken into account in this tabulation. The clearance of forests, the drainage and cultivation of lands, and industrial processes such as lime burning and fermentation all add carbon dioxide to the atmosphere as already discussed.

How long will the known reserves of fossil fuels last? The use of all the coal, oil, and gas beneath the earth's surface is estimated to produce 4×10^{13} tons of carbon dioxide. These reserves at the estimated rate of use will be completely exhausted before the year 3000, with their use peaking some time in the next few hundred years. When all of the fossil fuels are exhausted, man will have increased the total carbon dioxide content of the atmosphere–ocean system from 1.32×10^{14} tons to 1.72×10^{14} tons. If there were sufficient time for the atmosphere–ocean system to come to equilibrium, calculations by Plass[6] indicate that there would be 10 times the present amount of carbon dioxide in the atmosphere. This amount would be even greater if there were insufficient time for complete equilibrium. A simple extrapolation of Manabe and Wetherald's result[10] (which is not justified) would indicate a temperature increase of the order of 24°C. A more realistic value is probably given by the calculations of Plass[6] which indicate an increase in the average temperature of 12°C in this situation. There presumably would not be time for the atmosphere–ocean system to come to equilibrium with the calcium carbonate in the oceans. When this finally occurs, presumably after several thousand more years, the atmospheric carbon dioxide amount will be four times its present value, according to Plass,[6] and the average temperature will be 7°C warmer than today.

In summary, significant amounts of carbon dioxide are being released into the atmosphere by the combustion of fossil fuels. Conservative predictions of the use of fossil fuels show that the atmospheric carbon dioxide amount will increase significantly in coming decades. If fossil fuel use continues to increase at the present rate, the atmospheric carbon dioxide concentration will be almost four times the present amount by the year 2080. An increase of this magnitude would certainly cause important changes in our climate. It would be significantly warmer, probably drier on the whole, together with an appreciable increase in the mean level of the oceans. When the reserves of fossil fuel are depleted, the resulting climatic changes would be even larger and probably intolerable, if not fatal, to our civilization. It is obvious that civilization cannot tolerate the continued use of fossil fuels in ever greater amounts.

REFERENCES

[1]J. Tyndall, *Phil. Mag.* **22** (series 4) (1861) 169, 273.
[2]S. Arrhenius, *Phil. Mag.* **41** (1896) 237.
[3]T. C. Chamberlin, *J. Geol.* **5** (1897) 653; **6** (1898) 609; **7** (1899) 545, 667, 751.
[4]G. S. Callendar, *Quart. J. Roy. Meteorol. Soc.* **64** (1938) 233; **66** (1940) 395; *Weather* **4** (1949) 310; *Tellus* **10** (1958) 243.
[5]G. N. Plass, *Quart. J. Roy. Meteorol. Soc.* **82** (1956) 310.
[6]G. N. Plass, *Tellus* **8** (1956) 140.
[7]L. D. Kaplan, *Tellus* **12** (1960) 204; see also, G. N. Plass, *Tellus* **13** (1961) 296.
[8]F. Moller, *J. Geophys. Res.* **68** (1963) 3877.
[9]S. Manabe and R. F. Strickler, *J. Atm. Sci.* **21** (1964) 361.
[10]S. Manabe and R. T. Wetherald, *J. Atm. Sci.* **24** (1967) 241.
[11]G. N. Plass, *J. Geophys. Res.* **69** (1964) 1663.
[12]H. C. Willet, *Climatic Change: Evidence, Causes, and Effects*, Ed., H. Shapley, Harvard Univ. Press, Cambridge, Mass., 1953.
[13]G. N. Plass, *Quart. J. Roy. Meteorol. Soc.* **82** (1956) 30.
[14]E. B. Kraus, *Quart. J. Roy. Meteorol. Soc.* **86** (1960) 1.
[15]H. Wexler, *Climatic Change: Evidence, Causes, and Effects*, Ed., H. Shapley, Harvard Univ. Press, Cambridge, Mass., 1953.
[16]H. Lieth, *J. Geophys. Res.* **68** (1963) 3887.
[17]B. Bolin, *Scientific Amer.* **223** (1970) 124.
[18]C. D. Keeling, *Proc. Am. Phil. Soc.* **114** (1970) 10.
[19]C. D. Keeling, T. B. Harris, and E. M. Wilkins, *J. Geophys. Res.* **73** (1968) 4511.
[20]W. W. Rubey, *Geol. Soc. Amer. Bull.* **62** (1951) 1111.
[21]E. Eriksson, *J. Geophys. Res.* **68** (1963) 3871.
[22]C. D. Keeling and L. S. Waterman, *J. Geophys. Res.* **73** (1968) 4529.
[23]R. Berger and W. F. Libby, *Science* **164** (1969) 1345.
[24]J. Kanwisher, *J. Geophys. Res.* **68** (1963) 3921.
[25]S. Epstein, R. P. Sharp, and A. J. Gow, *Science* **168** (1970) 1570.
[26]G. H. Denton and S. C. Porter, *Scientific Amer.* **223** (1970) 101.

Chapter 3

ELECTROCHEMICAL POWER SOURCES FOR VEHICLE PROPULSION

E. H. Hietbrink, J. McBreen, S. M. Selis,* S. B. Tricklebank, and R. R. Witherspoon

Electrochemistry Department
Research Laboratories, General Motors Corporation
Warren, Michigan

I. INTRODUCTION

In recent years considerable attention has been given to investigations leading to a possible alternative to the internal combustion engine as the primary power source for vehicle propulsion. Electric power plants, one of the possible alternatives, is the subject of this review.

The first part of the chapter examines some of the incentives for electric vehicles. Following this, there is a discussion of the obstacles to their development. Power-plant requirements to propel various vehicles are then discussed. The energy and power density of power sources, to propel five representative type vehicles, were established using assumptions of top speed, range, and acceleration.

The main part of the chapter discusses batteries and fuel cells. Included are batteries and fuel cells that have been developed or are undergoing refinements, as well as some of those that are in the research stage. This discussion is focused on systems for which possible vehicle use can be foreseen. No attempt has been made to predict when some of these emerging batteries may be developed, but some of the problems that are obstacles to their development are discussed.

* Deceased, May 13, 1971.

The conclusion of the chapter is a plausible projection of the steps in which electric-vehicle technology might evolve.

II. INCENTIVES FOR ELECTRIC VEHICLES

The basic incentive for research and development of electric vehicles is the increased need and concern to mitigate and hopefully eliminate environmental pollution. The battery- or fuel-cell-powered vehicle is generally considered to be pollution-free; however, it is not in itself a complete solution in that the energy source for charging batteries and/or manufacturing a fuel must also be considered. This more general consideration leads to a second incentive in that the electric vehicle provides a greater flexibility for the efficient utilization of more abundant energy resources associated with future advances in nuclear technology.

It is logical to assume that the evolution of practical electric vehicles will primarily be a function of (1) their ability to meet specific transportation requirements more economically than competing concepts within the framework of restraints associated with utilization of current fossil-fuel energy sources and environmental pollution; (2) their ability to be more economically compatible with projected changing sources of energy in the future, such as the evolution of nuclear power plants which quite likely will result in a predominantly electric economy; and (3) their ability to be more compatible with new transportation-system concepts of the future, particularly in the large metropolitan areas or new cities and industrial complexes that could be inherent to and necessarily integrated with the evolution of large nuclear power plants.

1. Conservation and Utilization of Fossil-Fuel Supplies

In recent years concern over the conservation and associated utilization of fossil-fuel supplies has been mounting at a rapid rate. This is an incentive for the development of electric vehicles, particularly a fuel-cell-powered vehicle since a high efficiency is inherent to the fuel cell as it is not limited by the Carnot efficiency.

The fossil-fuel resources of the U.S. are of the order of 25% of the total fossil-fuel resources of the world. Provisional estimates of fossil-fuel resources by Duncan and MacKelvey of the U.S. Geological Survey,[1] show that the U.S. has known recoverable reserves of about $5.5Q$ ($Q = 10^{18}$ Btu) and marginal and undiscovered reserves of approximately $124Q$. Based on the U.S. Geological Survey of 1961 and other sources,[2,3] fossil-fuel resources recoverable at reasonable costs are estimated at about $28Q$. It is interesting to note that in terms of specific resources this $28Q$ consists of coal ($21Q$), crude oil and shale oil ($6Q$), and natural gas ($1Q$). An optimistic estimate of $130Q$ by the Department of the Interior[3] includes both reasonable- and high-cost resources, but for economic reasons, the exploitation of the ultimate reserve may not be attained.

Along with this prognosis of the potential resources of fossil fuel, estimates have also been made of the total energy demand and cumulative energy consumption in the U.S.[2,3] The current energy demand is estimated to be about $0.064Q$ per year and is projected to be $0.14Q$ per year by the year 2000 and about $0.64Q$ per year by 2100. It has also been estimated that the resulting cumulative energy consumption to the present is approximately $2Q$ and is projected to be $5Q$ by the year 2000 and $28Q$ by approximately 2080. Thus, if the total energy demand is met entirely with fossil fuel, the estimated recoverable resources of $28Q$ would be depleted within another century; hence, the concern over conservation and utilization of fossil fuel is not unjustified.

In considering incentives for electric vehicles, it is appropriate to examine the petroleum resources in more detail. Transportation accounted for 20% of the total energy consumption in 1960, and based on projections of miles traveled by the year 2000,[4] it is estimated that at that time transportation will account for approximately 25% of the total energy consumption. For all practical purposes, nearly all of the energy consumed by transportation comes from petroleum, which is a relatively small part of fossil fuel resources, not only in the U.S. but in the world.

There are two types of data which apply to petroleum reserves. These are the proved reserves and the ultimate recovery of crude oil present within the reservoir. Proved reserves represent the amount of oil known to be recoverable commercially from the producing strata on the basis of current technology and economic conditions for short-term future production. Future production is calculated on current production and recovery techniques and normally does not include more efficient recovery methods that may develop with advancing technology. Most of the statistics in the past were based on a 20% recovery level, which left 80% of the crude oil in place. In the past decade the reserve figures were primarily computed on the basis of conventional recovery methods and a recovery of 29% of the original oil in place. More recently, pilot-plant operations of thermal recovery methods have indicated that there will be a greatly improved recovery capability available within this decade.[5,6] Thus, in most reserves, it now appears possible to recover nearly 60% of the oil in place.

The history of crude oil production in the U.S. shows a growth rate of 7.9% per year in the time period from 1900 to 1930 and 4.2% per year during 1930–1958. During the time period 1960–1965, the growth rate was reported in the order of 3%, but the imports from Canada and foreign areas complicates the projection beyond 1960. Figure 1 shows a projection of the crude oil resources available for ultimate recovery. These projections are made on

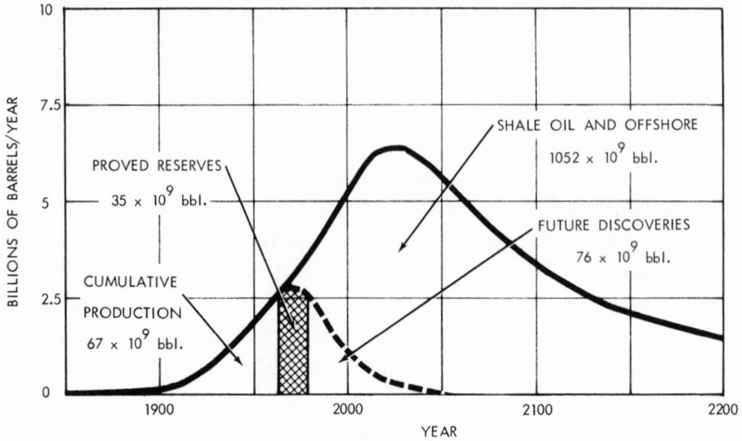

Figure 1. Projection of the ultimate crude oil resources available in the U.S. based on 1964 data.

the basis of crude oil production history, and it is assumed that there will be no imports and all U.S. oil consumption will be taken from our own resources, including offshore and shale-oil sources.[2,5,7]

The industry-proven reserve quoted at the end of 1964 is 35 billion barrels, or $0.2Q$. However, this does not include the offshore and oil-shale reserves. When these sources are included, the total reaches a sum of over 1200×10^9 barrels, or $7Q$, as indicated in Fig. 1. The shale-oil and offshore ultimate production is projected in Fig. 1 on the basis of an assumed growth rate in the order of 3% per year, and the area below the curve represents an approximation of the ultimate reserves.

It must be emphasized that any projection, such as that shown in Fig. 1, must be used carefully in drawing any conclusions since the effect of imports has not been included and estimates of unproven reserves can only be made with a relatively low confidence level. However, it is evident that within the relatively near future the proven reserves will not meet the demand and a transition must be begun from crude oil to the more abundant shale-oil and offshore reserves and ultimately to coal for supplies of liquid and gaseous fuels. Long before the exhaustion of supply is imminent, sufficient reserves must be assured for special high-priority needs of the future. In addition to gasoline and other petrochemical products for transportation, petroleum and coal resources provide the basic ingredients for a rapidly growing chemical industry and fuel oil for residential use. It is projected that a transition to shale-oil and offshore reserves and ultimately to coal will increase the cost of gasoline and other fuels.[8] Also, transportation will be competing with other users for petroleum, and it would be reasonable to expect increased concern and ultimate control of the allocation of petroleum among the industries that depend on this less abundant resource.[9]

These factors provide incentives for development of the electric vehicle, particularly a fuel-cell power plant, since the fuel cell has the potential of utilizing the fuel more efficiently than a combustion engine thus stretching the fuel supply over a longer time period. The increased efficiency of a fuel cell would also be an economic advantage in that it would tend to counter increases in fuel costs.

If changes in the uses of fuel result in a greater utilization of coal, the most abundant fossil-fuel resource, it would appear that this would also favor the electric vehicle with either a fuel-cell or battery power plant. At the present time coal is primarily used by the utility industry to generate electricity, which is the obvious ingredient for a battery-powered vehicle. However, the use of coal in this manner will require the development of appropriate technology to control SO_2 and other pollutants of the combustion process so as to meet rigid standards that are anticipated. The other approach would be to use coal as the basic ingredient for the manufacture of gasoline or other fuels which could be used in fuel cells. The ability of the fuel cell to use a variety of fuels derived from coal would appear to place the fuel cell at an advantage.

2. Future Changes in Energy Sources

Nuclear energy is projected to become a major contributor in the energy market near the end of the century, and since this energy will be primarily generated as electricity, the use of such energy sources will be a major incentive for electric vehicles. Thus, the evolution of battery-powered vehicles could be directly related to the use of nuclear power. Nuclear energy sources are anticipated in the long-range future because of the necessity for an energy source that conserves fossil fuels for other specific needs.

The history of the sources of energy as well as projected future changes in energy sources to the year 2000 have been investigated by Landsburg and Schurr and are shown in Fig. 2.[10] One can see that in 1850 fuelwood was the principal source of energy, with coal accounting for a little more than 10%. By 1910 the picture was completely reversed with coal being the predominant source. By 1920 the use of oil and gas was rapidly increasing and they surpassed coal by 1946, and by 1960 oil and gas accounted for 70% of the total energy consumption. Projections beyond 1960 show that by 2000, coal will be back to the point of supplying about 10% of the energy and nuclear power about 11% with an anticipated continuous growth rate. Other sources project a more rapid growth of nuclear energy. Projections by Sporn and the Atomic Energy Commission show that by 2000 approximately 50% of the production of electricity will be with nuclear power plants and that nuclear

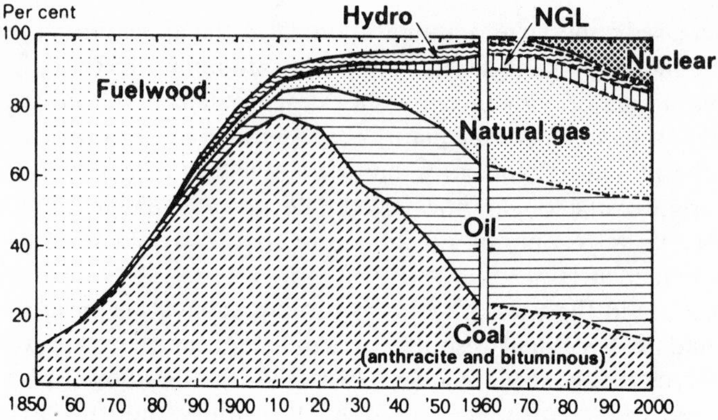

Figure 2. Changing sources of energy, 1850–1960, with projections to 2000.[10]

energy will account for about 21–23% of the total energy requirements.[11] Other estimates are that nuclear power plants will account for 25% of the production of electricity by 1980 and approximately 69% by 2000.[12]

Although there is a considerable variation in the projected growth rate of nuclear power, it does appear that there is general agreement that it will evolve as the major energy source; and the real question is related to the time period in which its impact will really be felt. Some reports[13] indicate that nuclear power by the fission process is already economical in certain geographic areas, and indications are that utility companies are favoring nuclear power rather than fossil fuel in planned construction of new plants after 1980. However, apart from economics there are other problems associated with the use of nuclear power. These are primarily the concern over disposal of radioactive waste, radioactive leakage, and thermal pollution of cooling water. It is also noted that current fission reactor technology is not the ultimate answer because the basic fuel, uranium 235, is in relatively short supply. It is estimated that the real gain in nuclear energy will occur near the end of the century when there is a possibility that breeder reactors will be developed to use to a much greater degree the natural resources of the available fertile isotopes of uranium and thorium. The other possibility, and possibly the greatest hope, is that fusion reactors will be developed, thus providing a practically inexhaustible supply of electrical energy. This hope of ultimately having an unlimited supply of electrical energy is one of the major incentives for electric vehicles. Automobiles could utilize fuel cells operating on synthetic noncarbonaceous fuels, such as ammonia or hydrazine; or they could use the electricity directly to charge batteries.

3. Environmental Pollution

The relatively recent concern over environmental pollution, specifically atmospheric pollution resulting from combustion of fossil fuels, has created a great interest in research and development of electric vehicles. One would be naïve to think that the electric vehicle, as a single entity, would eliminate pollution of the biosphere. As long as the energy demands, including the generation of electricity, are predominately met by the combustion of oil, coal, and natural gas, the concentrations of carbon dioxide, carbon monoxide, sulfur oxides, hydrocarbons, nitrogen oxides, and particulate matter in the biosphere will be affected. In the future when nuclear power plants are projected to meet a continuously increasing portion of the energy demand, the picture will completely change, and the major concerns are expected to be associated with thermal and radioactive pollution. Thus, it follows that in the long term, fuel cells or battery-powered vehicles in the transportation industry associated with nuclear power plants in the utility industry could possibly be the answer to pollution of the biosphere. However, in the interim when fossil fuel is still the major source of energy, the emphasis is on obtaining a better understanding of the accumulated effects of inherent pollutants and limiting their concentration from all sources of fossil-fuel combustion, as necessary, by establishing suitable standards and controls. Within this framework, one can cite the following advantages and incentives for development of electric transportation vehicles.

1. Two basic aspects of the air-pollution problem associated with excessive concentrations of pollutants are currently of primary concern. One is the gross accumulation of pollutants in the total biosphere, and the other is the high concentration of pollutants in specific, highly congested areas, such as large cities and industrial complexes. Since the electric vehicle does not contribute to pollution at its point of usage, it would, therefore, help to alleviate the immediate problem of high concentration of pollutants in specific localized areas.

2. In controlling the pollutants from the combustion of fossil fuels to meet specified standards, it appears that the "control effectiveness" of electric vehicles could be greater relative to combustion engine vehicles. In the case of an electric vehicle, the responsibility of controlling pollutants is delegated to the utility industry rather than each individual vehicle owner. In addition, the pollutants generated by the utility industry are more amenable to being monitored and controlled, assuming, of course, that the technology required to limit pollutants from fossil-fuel electric power plants to required levels of concentrations is evolved. One would project that it would be more practical, economical, and efficient to control pollutants from a relatively small number of large stationary combustion units as compared to millions of small, mobile combustion engines.

3. One final consideration is the growing concern over increased concentrations of carbon dioxide in the biosphere resulting from the combustion of fossil fuel. The effect of this is the so-called "greenhouse effect," which

results in a change in climatic conditions (see Chapter 2). The fact that the carbon dioxide content of the atmosphere is increasing with time has been firmly established by precise measurements.[14] Although the effects of increased concentrations are greatly uncertain, one should not dismiss the possibility that restrictions could be imposed in the future on the amount of fossil fuel that could be consumed in combustion processes over a specific time period. This would add to the incentives for developing nuclear power plants as the ultimate energy source, hence an incentive for the development of batteries and fuel cells as compatible power sources for transportation vehicles.

III. OBSTACLES TO ELECTRIC VEHICLES

By far the greatest obstacle to the evolution of electric vehicles is the fact that battery and fuel-cell technology must be greatly advanced in order to satisfy the stringent requirements associated with current vehicles. To emphasize the importance of this obstacle, vehicular propulsion requirements are discussed in a separate section below in order to define the power and energy–density goals that must be met by future batteries and fuel cells. To achieve these technological goals for the general purpose passenger car, a great deal of research effort is required. In addition to achieving the technological goals, there is still the problem of engineering development to obtain a practical unit from an economical and operational point of view.

A second consideration is the development of the nonbattery and fuel-cell components of electric vehicles. Electronic controls, motors, and other drive-line components have already been demonstrated to be technically feasible. However, there is always a need for innovation and new approaches. Thus, the emphasis here is not so much on research as on engineering development to obtain high reliability, low cost, and high efficiency. It would appear that any advances that could be made relative to the economics and efficiency of drive-line components would enhance the initial evolution of electric vehicles in specialized applications using current technology. These improvements would partially counterbalance limited battery performance but would still be inadequate for general passenger-car applications.

One final major obstacle to the evolution of electric vehicles that must be considered is the highly developed transportation system currently in existence. Consequently, it would appear that any significant shift from internal combustion vehicles will take place only if a less expensive or more comfortable and convenient alternative is offered within the framework of rigid pollution standards. The successful development of electric vehicles would certainly cause some changes in the automotive, petroleum, and electric utility industries. However, there is every indication that the evolution of the electric vehicle market would be gradual with a minimum disruption to these industries and their suppliers. The electric utility companies view the electric vehicle as an attractive market for electricity during off-peak load periods; hence, there is a strong incentive for the power industry to develop rechargeable battery systems.[15]

IV. VEHICLE POWER-PLANT REQUIREMENTS

In order to evaluate the application of future electric power plants to vehicle propulsion, it is necessary to know the power requirements of the types of vehicles being considered. For this discussion, five specific types of vehicles which appear to be of greatest interest for electrification have been selected for evaluation. These are listed in Table 1.

The urban car is suggested as a small shopper type of vehicle primarily limited to travel on city streets. A top speed of 50 mph has been selected which on first consideration may seem rather high as its use would be mainly within the city. However, this higher top speed would give the car a limited capability on city expressways. In addition, it will be seen in the next section that most of the power is required for acceleration (approximately 66 %) and a higher than necessary top speed does not require a large increase in power to overcome road-load.

The family car is the familiar six passenger automobile; whereas, the commuter car is somewhat smaller and would be used primarily for commuter travel. Thus, the range requirement for the commuter car could be less than the family car; however, the acceleration capability of both would be approximately the same. The metro truck can be regarded as a delivery truck for metropolitan use with a maximum payload of about two tons. The urban coach is considered as a small to medium sized city bus for travel only on city streets with a carrying capacity of approximately 40 passengers.

The weight allocated to the battery power source is somewhat arbitrary and involves some assumptions, but the premise that approximately 35 % of the passenger-vehicle curb weight and 25 % of the city-bus curb weight could

Table 1. Characteristics of Selected Vehicles

	Gross vehicle weight (gvw), lb	Approximate curb weight, lb	Suggested max. cruise speed, mph	App. weight available for battery power source, lb
Urban car	2000	1650	50	450
Commuter car	3000	2600	70	650
Family car	4000	3500	80	850
Metro truck	10000	6000	60	2175
Urban coach	20000	14000	45	3100

be allocated to the complete propulsion system including motors, controls, gear box, and power source has been established.[16] Approximately the same percentage values were used here in evolving the weights allocated to the battery power source listed in Table 1. The weight allocations for motor, controls, and miscellaneous drive-line components were based on experience in electric drive applications.

1. Power Requirements

The total power requirement to propel a vehicle is the summation of the power required for acceleration, aerodynamic drag, rolling resistance, and grade requirements. In addition, auxiliary power, such as that required for power steering, power brakes, and air conditioning, should be considered in the total vehicle power requirements. In this discussion, primary emphasis is placed on establishing representative power and energy requirements for vehicle *propulsion only*, and as such the auxiliary power requirements have not been included.

The typical acceleration requirements for the vehicles being considered are presented in Fig. 3. The shaded area for the passenger vehicles is bounded by what is typical of a conventional family car and what is considered to be a minimum acceleration requirement from the standpoint of mixing with traffic and safety considerations. A minimum performance requirement is necessary as it must be assumed that there will be a gradual introduction of electric cars which, apart from some possible special applications, will be required to share the roads with existing vehicles with conventional internal-combustion-engine power plants. The requirements in Fig. 3 are not intended to be precise values for detailed vehicle design, but rather to be representative values for purposes of making first-order projections of the required battery-power-source characteristics. The maximum power required from the power source is determined primarily by the acceleration requirement. Table 2 gives the power densities required for acceleration of the five vehicles. These were determined by taking into account tractive effort considerations

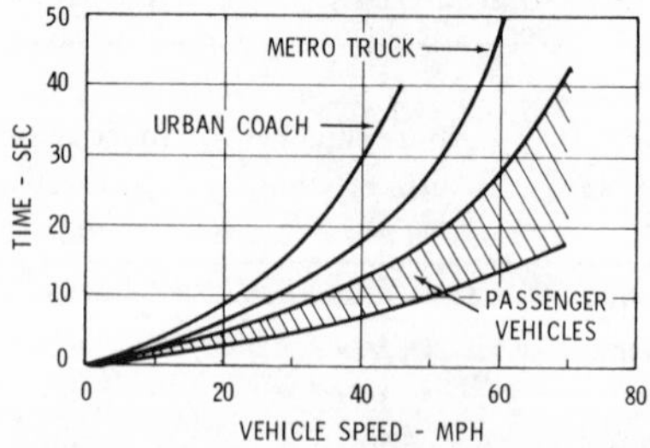

Figure 3. Typical acceleration requirements for selected vehicles.

Table 2. Energy- and Power-Density Requirements for Various Vehicles

	Constant speed cruise				Acceleration
	Range, miles	Velocity, mph	Energy density, Wh/lb	Power density, W/lb	Power density, W/lb
Urban car	50	40	25	20	65
Commuter car	100	60	55	33	70–103
Family car	200	70	122	43	70–110
Metro truck	100	40	33	13	40
Urban coach	125	30	42	11	35

and dividing by an appropriate efficiency of motor, controls, driveline, and the weight allocated to the battery power sources in Table 1.[17] For the commuter car and the family car, the lower value is required to maintain a 70-mph speed on a 3% grade; and the upper value is required for the maximum acceleration characteristic of Fig. 3. For the urban car, the power density is that required to meet a minimum acceleration of about 15 sec to 45 mph.

2. Energy Requirements

The energy density (Wh/lb) of the power source is primarily a function of the vehicle-range requirement at specified vehicle speeds. The curves in Fig. 4 present a first-order projection of vehicle range as a function of the energy density of the power source and vehicle speed. With a fixed weight allocated to the power source, the range is directly proportional to the energy density of the power source. Thus, it is convenient to plot the range in miles normalized to the energy density of the power source in Wh/lb. The associated W/lb of the power source required to meet the zero-grade road-load power as a function of vehicle speed is also shown. The important parameters in the calculations for the curves in Fig. 4 are the zero-grade road-load requirements, the weight allocated to the power source, and the efficiency of the propulsion system from the battery output to the wheels. For these calculations, a combined efficiency for motor, electronic controls, and drive line in the range 45–78% was used. Experience has shown that the combined motor and control efficiency is lower at low speed when the motor rpm and power level are low, and increases substantially as the motor rpm and power level is increased. The drive-line efficiency from the output of the motor to the wheels is normally in the order of 90%.

Table 2 gives a summary of the energy density and associated power density for each of the vehicles being considered at selected values of range and velocity. The data were obtained from the curves in Fig. 4. These

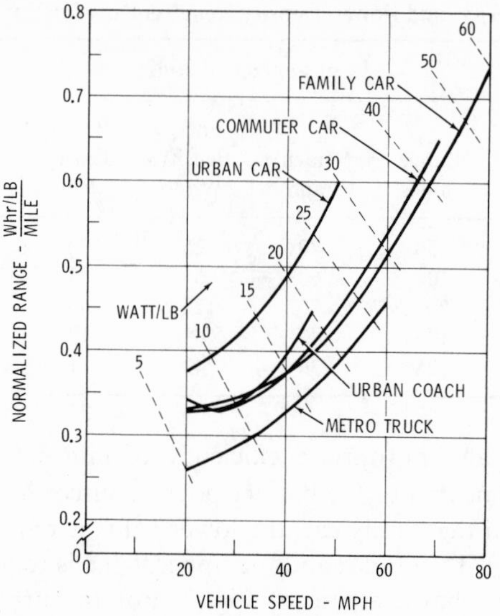

Figure 4. Power- and energy-density requirements as a function of vehicle range and speed for selected vehicles.

road-load energy values used in conjunction with the accelerative power requirements (also given in Table 2) give an estimate of the power source needed for each of the vehicle types.

3. Electric Energy Requirements from the Utilities

Since electrification of vehicles would require substantial amounts of electric power, it is of interest to consider the demand on the power-distribution system. A first-order approximation can be obtained by factoring the charge–discharge efficiency of battery power sources into the results from Fig. 4. Thus, the energy input to the vehicle from the electric utility can be estimated. These results are shown in Fig. 5 in terms of kWh/mile from the electric utility as a function of vehicle cruise speed at zero-grade road-load.

With the results shown in Fig. 5, it is now possible to estimate the impact on the U.S. electric utility industry if substantial electrification of automobiles were to occur by 2000. One must start with the premise that the projected car registrations[18] and the National Power Survey and the AEC estimates of future electrical generation figures[19] are valid. Other necessary assumptions are (1) an average electric vehicle use of 8000 miles/yr; (2) 1 kWh/mile required from the generating plant (this value would allow for inefficiencies in the transmission of power from the generating source); (3) the present diurnal

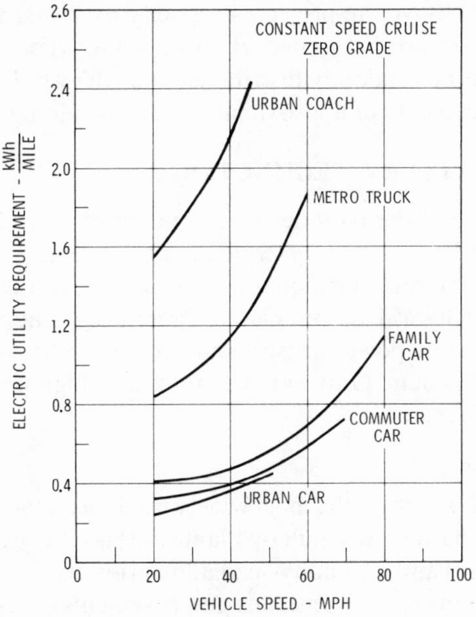

Figure 5. Energy requirement from the electric utility source as a function of vehicle range and speed.

load pattern of the utilities will be the same; and (4) slow-rate battery charging, distributed over the nightly off-peak hours.

Using these assumptions, the data in Table 3 were calculated. As on the average only 52% of the generating capacity is in use all of the time,[15] it is very reasonable to expect that an additional 20% power could be generated by the utilities during off-peak hours. If this is the case, Table 3 shows that off-peak power generation could satisfy the required load for complete automobile electrification in 2000. This means that additional generating

Table 3. Projected Electrical Energy Requirements for Electric Passenger Cars

Year	Total No. passenger cars, million	No. electric passenger cars, million	Estimated electric energy required for electric cars, billion kWh	Estimated electric power generation U.S.A., billion kWh	Estimated power available by leveling diurnal load (20% increase), billion kWh[a]
1980	110	1.1	9	2830	566
	110	11.0	88	2830	566
2000	150	15.0	120	8250	
	150	150.0	1200	8250	1650

[a] This assumes that charging will occur during off-peak hours.

capacity, over and above the projected capacity by 2000 (based on present growth rates), would not be needed. If an opposing view is taken that the charge pattern will be randomly distributed over the whole day, then additional generating capacity of approximately 15% would be sufficient.

V. CURRENT BATTERY TECHNOLOGY

Although none of the batteries that are currently available fulfill both the high energy- and power-density requirements for vehicle propulsion, some of these batteries exhibit exceptional power density and could find use as power batteries in hybrid power plants. Present-day fuel cells have high energy density and low power density and thus could be used as an energy source in a hybrid system. Only batteries that have high power density will be discussed in this section.

1. Lead–Acid Battery

The lead–acid battery is the most widely used secondary battery system. The first lead–acid battery was built by Planté in 1860.[20] Further refinements, such as pasted plates and antimony–lead alloy grids, were made in the latter part of the last century.[21] Numerous improvements, such as the use of synthetic separators, uncalcined oxides, plastic battery cases, and organic expanders, have been incorporated in the battery over the past thirty years.[22]

The overall cell reaction is

$$Pb + PbO_2 + 2H_2SO_4 \leftrightarrows 2PbSO_4 + 2H_2O$$

and the open-circuit voltage is 2.1 V.

The electrodes of a lead–acid cell are made by casting a reticulated grid from a lead alloy and incorporating a basic lead sulfate paste in the grid structure. The paste for the positive plates is made by adding dilute sulfuric acid to a finely comminuted litharge–lead mixture. The negative plate pastes contain, in addition, expanders, such as lignin sulfonic acid, barium sulfate, and carbon black. The cell pack consists of alternative positive and negative plates interspersed with separators. These separators are usually microporous rubber membranes. The positives and negatives are welded together to their respective terminals. The cells are filled with dilute sulfuric acid and slowly charged to form the plates. This acid is removed from the cells, and the cells are then filled with sulfuric acid (sp. gr. 1.2–1.3). The cells are then ready for use.

The performance of typical lead–acid batteries is shown in Fig. 14 and Table 4. Although the energy density of the system is extremely low, the power density is relatively high. The low energy density at high rates of discharge is due to poor utilization of the negative-plate active material. The sponge lead in the negative plate is rapidly passivated by a film of lead

Table 4. Data for Ambient Temperature Batteries

System	Thermodynamic potential, V	Energy density		
		Theoretical, Wh/lb	Actual, Wh/lb	Nominal, Wh/in.^{3}
Pb–Acid	2.10	87	2–15	1.5
Ni–Cd	1.29	107	8–25	1.2
AgO–Zn	1.86, 1.61	220	25–50	3.5
Ni–Zn	1.70	170	12–27	2.0

sulfate when the battery is discharged at high rates; hence, the decrease in capacity and energy density.

If the battery electrolyte is not allowed to become too concentrated, if the cells are not electrically abused by gross overcharging or reversal, and if the battery is not left on stand for prolonged periods, battery failure is usually due to shedding of the positive-plate active material. Antimony, which is added as a casting aid to the positive-plate grid, fortuitously mitigates the shedding problem. However, some of the antimony is transferred to the negative as cycle life progresses. Contamination of the negative plate with antimony lowers the hydrogen overvoltage on the lead negative. This buildup of antimony on the negative plate results in the deterioration of the charge rentention and the rechargeability of the plate.

Before the lead–acid battery can be considered as a good power battery for vehicle propulsion, the following problems have to be resolved:

(1) improvement in the negative-plate active-material utilization at high rates;
(2) elimination of antimony from the positive plate with the development of another method for elimination of positive plate shedding;
(3) the development of new cell configurations to eliminate extraneous resistance drops in the cell pack.

2. Nickel–Cadmium Battery

The early work on the nickel–cadmium battery was carried out by Jungner in Sweden.[23] The original battery was a low energy–low power device. In the early 1930's, the introduction of sintered-plate structures for the battery electrodes converted the nickel–cadmium battery into a high-power system.

The overall cell reaction is

$$2NiOOH + Cd + 2H_2O \longrightarrow 2Ni(OH)_2 + Cd(OH)_2$$

and the open-circuit voltage is 1.29 V.

The electrode plaques are made by sintering carbonyl nickel powder to a nickel-screen current collector. The resultant plaques are about 80–85%

porous. The positives are made by incorporating nickel hydroxide into the porous structure. Cadmium is incorporated on the negative electrode plaques in a similar fashion.[24] The separators are nonwoven polypropylene, and the electrolyte is KOH ($\simeq 30\%$).

The performance of the nickel–cadmium battery is given in Fig. 14 and Table 4. The nickel–cadmium cell is fully developed, has long life, is mechanically rugged, and offers a quick recharge. When the nickel–cadmium cell is charged at high rates, the possibility of thermal runaway exists. Oxygen evolved at the positive can react with the cadmium in the negative with considerable heat generation. This heat generation lowers the cell resistance, decreasing the charge efficiency of the positive, and facilitates rapid diffusion of oxygen to the negative plate and the further generation of heat. This situation can easily occur when the battery is charged at high rates at constant voltage. Controlled charging and proper separator design mitigates this problem.

The major drawback to the nickel–cadmium battery is the expense and scarcity of the active materials; this precludes the large-scale use of this battery for applications such as vehicle propulsion.

3. Silver–Zinc Battery

The silver–zinc battery has the highest energy and power density of any commercially available aqueous-electrolyte secondary battery. The theoretical possibilities of the system had long been realized. However, it was not until André[25] demonstrated in the early 1930's that the active anode and cathode materials could be kept separated during the charge and discharge processes that a practical silver–zinc battery was realized. This was accomplished by the use of a regenerated unplasticized cellophane membrane as a semipermeable separator.

The overall reaction of the silver–zinc cell is

$$AgO + Zn \rightarrow ZnO + Ag$$

The electrolyte is 40–45% KOH.

The discharge of the silver electrode takes place in two steps, which are

$$2AgO + Zn \rightarrow Ag_2O + ZnO \qquad E^0 = 1.86\ V$$

and

$$Ag_2O + Zn \rightarrow 2Ag + ZnO \qquad E^0 = 1.61\ V$$

In the positive-limiting silver–zinc cell, a reversible discharge should give half the capacity at 1.86 V and the remainder of the capacity at 1.61 V. In actual practice, the upper level voltage is about 1.7 V and the lower level voltage is 1.5 V. Even at low rates of discharge, no more than 30 % of the capacity is delivered at the upper level voltage; and at extremely high rates of discharge, the upper level voltage is completely depressed.

Silver electrodes consist of a structure of finely divided silver and have a porosity of 50–60 %. Zinc electrodes are about 70 % porous and are made by pressing zinc oxide with a binder onto an expanded metal current collector. The cell pack is constructed by wrapping the positive electrodes with an inter-separator of woven nylon or nonwoven polypropylene. This assembly is then wrapped in the main separator system, and then the positive-plate separator assembly is stacked together with the negative plates to form the cell pack.

Performance data for the silver–zinc battery are given in Fig. 14 and Table 4.

The following, in order of importance, are the major problems associated with the silver–zinc cell:

(1) zinc shape change;
(2) zinc penetration of the separator;
(3) silver attack of the separators;
(4) oxidation of the separators.

When a silver–zinc cell is cycled, zinc is removed from the plate edges and agglomerates toward the plate center. This phenomenon is known as *shape change*. The loss in geometric plate area and the concomitant densification of the plate center results in a loss in capacity of the cell. This problem has been mitigated somewhat by the extension of the edge of the zinc electrode beyond the positive,[26] the use of Teflon[27] as a binder, and the use of dish-shaped negative plates.[28] However, no complete solution of this problem has yet been found.

Overcharging promotes the growth of zinc into the separator. Trails of zinc can grow through the separator and short out the cell. Silver monoxide, Ag_2O, is soluble in the electrolyte to the extent of $10^{-3}M$.[29] This soluble silver reacts with cellulosic separators to yield metallic silver and oxidized products of cellophane. As a result of this, the separator is slowly attacked and is replaced by a conductive film of metallic silver powder. Since this silver attack occurs on discharge, on charge stand, and on discharge stand, it limits the calendar life of the cell.

Small amounts of oxygen are evolved at the silver positive during charge, and as a result perhydroxyl ions are formed in the electrolyte. These perhydroxyl ions attack the cellulosic separators with the generation of oxalates and carbonates. The resultant weakening of the separator folds promotes zinc penetration.

The most serious drawback of the silver–zinc battery is the excessive cost and scarcity of silver.

4. Fuel Cells

The principle of the fuel cell was first demonstrated by Grove in 1839. Further pioneering work was carried out by Baur, Jacques, Davtyan, and Bacon. Excellent reviews of these early studies are given elsewhere.[30–32]

In its simplest form, the fuel cell consists of two porous, catalytic electrodes (Fig. 6) which are suitably catalyzed for the particular fuel (anode) and oxidant (cathode). The two catalyzed electrodes are separated by an electrolyte which may consist of strong acids (H_2SO_4 or H_3PO_4) or alkalies (KOH), and, in some cases, ion-exchange membranes. Both organic (e.g., polystyrene sulfonic acid) and inorganic membranes have been used. During cell operation, fuel and oxidant are fed to the appropriate electrodes, and power can be drawn from the unit as long as fuel and oxidant are supplied and the reaction products removed. Thus, no electrical recharge is necessary; and if air is used as the oxidant, a fuel-cell power source can be quickly refueled in a manner similar to that now used with the internal combustion engine. Furthermore, the chemical energy of the fuel-cell reactants can be converted to electrical energy with a high degree of efficiency since fuel cells do not suffer from the Carnot limitation.

In practice, the fuel cell is of course more complicated than that described above. To maintain continuous operation of a fuel-cell system, a number of auxiliary devices have to be included with the cell stacks. A schematic diagram of an air fuel-cell system is shown in Fig. 7. The auxiliary equipment, such as pumps, fans, and radiators, consume power from the fuel-cell stacks, and reduce efficiency and reliability. The necessity for all these devices is evident when one considers that a 1-kW hydrogen–oxygen

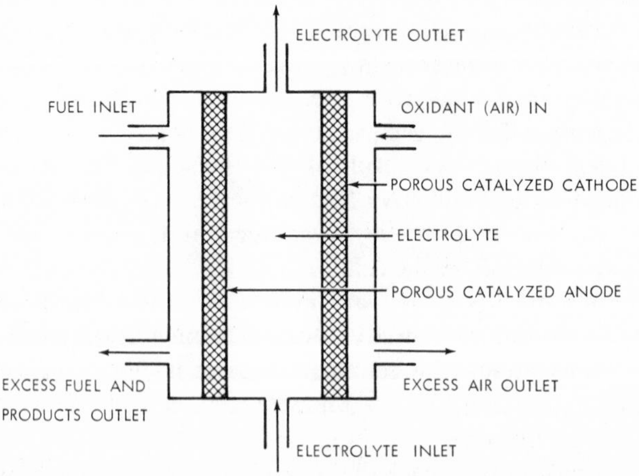

Figure 6. Simplified fuel cell.

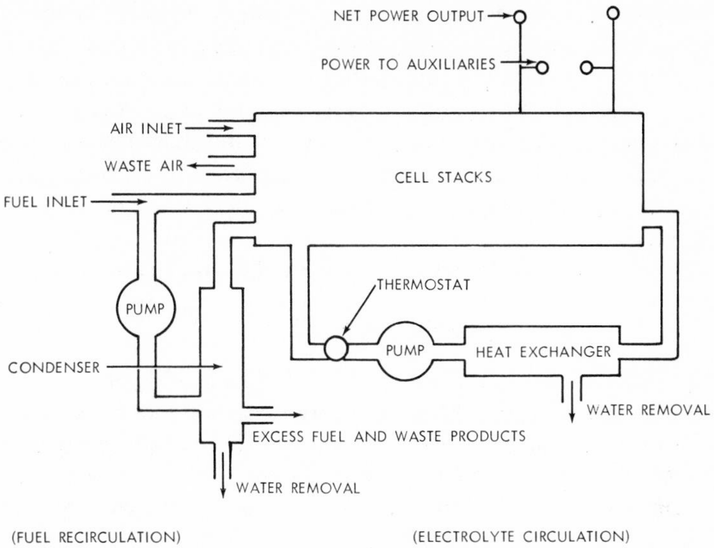

Figure 7. Simplified fuel-cell system.

fuel-cell battery, operating at 0.7 V/cell, has to also reject 1 kW of heat through a radiator and dispose of 1 pint of water per kWh of power. Furthermore, when reformed hydrocarbons ($CO_2 + H_2$) or cracked ammonia ($N_2 + H_2$) are used as fuels, extra pumps are needed, since not more than about 80% of the available fuel can be removed in a single pass through the anode compartment without a severe reduction in anode performance. If steam reformers (for carbonaceous fuels and hydrocarbons) or crackers (for ammonia) are included, the overall system complexity is increased. Thus, fuel cells are complicated chemical power plants for producing electricity, and require a higher degree of design sophistication than that found in conventional storage batteries.

The greatest obstacle to fuel-cell development, for general use, has been the lack of an inexpensive electrode catalyst. Practically all of the fuel-cell systems used to date employ platinum-metal catalysts of 0.50–10 mg/cm^2 or 0.50–10 g/ft^2. The fact that platinum metals (palladium excluded) cost an average of \$5.00/g results in an electrode cost of \$2.50–50.00/ft^2. If expensive catalysts are only used in the anode, and single cells of 1 ft^2 deliver 80 W at 100 mA/cm^2, then the cost for catalyst alone would be \$30.00–500.00/kW. Since a minimum of 20 kW is required for even small vehicles, the catalyst cost alone for such a battery would be \$600.00–10,000.00.

The low catalyst loadings that are required for hydrogen oxidation make this fuel attractive. On the other hand, high catalyst loadings are required for hydrocarbon oxidation, and direct oxidation of ammonia is not feasible even with high catalyst loadings and temperatures of 150°C. In the case of hydrocarbon fuels, oxidation proceeds at a low rate and catalysts are easily poisoned. Storage problems militate against the use of hydrogen. Ammonia, methanol, and hydrocarbons are easily stored, and these chemicals can be used as a source of hydrogen fuel. This is accomplished by cracking ammonia, steam reforming methanol or hydrocarbons, or by partial oxidation of hydrocarbons.

Catalysts such as Cr_2O_3 or Fe_2O_3 are used to crack ammonia.

$$2NH_3 \xrightarrow{\Delta} N_2 + 3H_2$$

Thus, the gas delivered to the battery is 75% H_2 and 25% N_2.

In the steam reforming process, the hydrocarbon fuel is reformed with steam, over a nickel-based catalyst at about 800°C, to produce a gas mixture consisting of 45% H_2, 11% CO, with the balance as CO_2 and water. Since this reaction is highly endothermic, large amounts of heat are required. Methanol can be reformed at 250°C.

The partial oxidation process is an exothermic partial burning of the fuel with less than the stoichiometric amount of air. No catalyst is required and the reaction occurs at 1100°C. Although the products contain only 17% H_2, this can be increased to 22% H_2 by passing the product gas over an iron oxide catalyst at 600°C to produce the shift reaction

$$CO + H_2O \rightarrow CO_2 + H_2$$

Another fuel that has been considered is hydrazine, which now costs about $1.00/lb in the U.S. and $0.45/lb in Japan. It has a distinct advantage over most other fuel-cell systems in that low-cost catalysts are usable at the anode, and the fuel is soluble in the electrolyte. Nickel boride and iron catalysts have been found suitable and provide low anodic polarization at 300 mA/cm^2 or higher, so the power density of a hydrazine–air fuel cell should be higher than that for the hydrogen or ammonia systems. Nickel boride, even in small amounts, costs about $0.10/g, so catalyst costs are only about $0.05–1.00/$ft^2$ or $0.60–12.00/kW or less.

Some of the various fuel-cell systems which have received considerable attention are listed in Table 5. The approximate energy storage in British thermal units per gallon (Btu/gal) and the approximate energy storage capability in horsepower-hour per gallon (hp-h/gal) are also shown. A fuel-cell efficiency of 60% is assumed along with a 70% efficiency for motors and controls.

The approximate status and relative cost and complexity of the various fuel-cell systems are summarized in Table 6.

Table 5. Energy Storage Capability for Various Fuel-Cell Systems[33]

Fuel	Heat of combustion, Btu/gal	hp-h/gal
Hydrogen	30600	5.1
Ammonia	36400	6.0
Hydrazine	58500	9.7
Hydrocarbons	56100 (methanol) to	9. 3 to
(includes Methanol)	120000 (diesel fuel)	19.8

There have been relatively few instances of fuel cells being used for vehicular propulsion. The hydrogen fuel cell is the best developed of all fuel-cell systems, and in the past this system has been used to power experimental vehicles; the two main examples are (1) the Electrovan by General Motors Corporation[34] and (2) the Union Carbide (H_2–air)–lead–acid hybrid-powered Austin.[35]

In the Electrovan there were 32 fuel-cell modules of a 1-kW nominal rating which weighed 53 lb each yielding a total weight of 1700 lb for the modules alone. The total system weight was about 2640 lb and the volume was 61 ft^3. The peak power of 100 kW was usable only for short acceleration purposes.

In the case of the fuel-cell–battery hybrid, the fuel cell delivered about 6 kW from a system using lightweight gas bottles and air. The energy storage was about 85 Wh/lb, thus giving the vehicle a range of about 200 miles.

The main problems restricting the use of fuel cells for vehicular applications at the present time are low power density per unit weight (Fig. 14) and volume, excessive cost, limited life, and the impracticability of the direct oxidation of hydrocarbons. The direct oxidation of hydrocarbons has been hitherto impractical[36] because of the excessive catalyst loadings that are required (10–40 mg/cm^2). Furthermore, the necessity of using acid electrolytes

Table 6. Status and Relative Cost of Various Fuel-Cell Systems

System	Catalyst cost	Development status	Fuel cost	Power density, kW/ft^3
H_2–Air	moderate	highest	high	1–2
NH_3–(cracked) $N_2 + H_2$	moderate	high	high	1–2
Hydrazine	low	high	very high ($0.45–1.00/lb)	>5
Hydrocarbons (direct)	high	low	very low	>1
Hydrocarbons (indirect)	moderate	high	very low	1–2

and temperatures in excess of 150°C greatly increases the cell materials cost. It is highly unlikely that hydrocarbons will be used to power vehicles via direct-oxidation fuel cells even in the relatively far future, since CO_2 pollution will always be present and hydrocarbons will continue to decrease in availability.

Possible future developments in fuel cells are discussed later in this chapter.

VI. EMERGING BATTERY TECHNOLOGY

1. Ambient Temperature Systems

Most of the recent work in ambient temperature batteries has revolved around the development of inexpensive batteries with high power capabilities and new high-energy systems. The nickel–zinc battery and the rechargeable alkaline zinc–manganese dioxide battery are the most promising new high-power systems. Zinc–air batteries and organic electrolyte batteries are possible contenders as high-energy systems. An ideal energy source for vehicle propulsion is the fuel cell and several types of fuel cells are under investigation.

(i) *Nickel–Zinc Battery*

The nickel–zinc battery was first mentioned by Michalowski in 1901.[37] Since then, development work on the system has proceeded sporadically,[38–48] and a battery has recently been marketed.

The overall reaction of the nickel–zinc cell is

$$2NiOOH + Zn + 2H_2O \leftrightarrows 2Ni(OH)_2 + Zn(OH)_2$$

The open-circuit voltage of this system is 1.72 V, and the electrolyte is 30–40% KOH.

The positive electrodes in the nickel–zinc battery are similar to those in the nickel–cadmium battery; the separator system and negative electrode are similar to that found in the silver–zinc battery.

Performance data for the nickel–zinc battery are given in Fig. 14 and Table 4. The system has high rate capabilities and has higher energy density than the lead–acid battery.

The major problems associated with the nickel–zinc system are the following:

(1) zinc shape change;
(2) oxidation of the separators;
(3) zinc penetration of the separators.

The zinc shape-change problem in nickel–zinc cells is similar to that found in silver–zinc cells. Because overcharging is necessary to maintain cell capacity, considerable oxygen evolution occurs on the positive plate. It is desirable to use lower concentrations of KOH in nickel–zinc cells than in silver–zinc cells, and the solubility of oxygen increases with decrease in KOH concentration. These high concentrations of oxygen in the cell electrolyte promote rapid oxidation of cellulosic separators. Thus, it is necessary to use inorganic separators or nonoxidizable organic separators in the nickel–zinc cell.

Because overcharging is necessary, the reducible zinc species [ZnO and $Zn(OH)_4^=$] are rapidly exhausted in the negative plate. Thus, metal deposition at the negative plate becomes diffusion controlled. Diffusion-controlled deposition results in a dendritic deposit.[49] Whenever the flux of zincate ions (to the growing metal substrates) from the negative-plate ambient electrolyte becomes less than the flux of zincate ions from the separator membrane, zinc deposition will preferentially occur in the pores of the separator. If these metallic trails through the separator are sufficiently small, they are oxidized on reaching the catholyte because of the oxygen evolution on the positive. It is only when the separator is oxidized and massive amounts of zinc penetrate the separator system that catastrophic cell failure by shorting occurs.

Although this system may yet prove to be an attractive replacement for the nickel–cadmium battery in certain applications, the scarcity and cost of nickel preclude extensive use of the nickel–zinc battery for vehicle propulsion.

(*ii*) *Zinc–Manganese Dioxide Battery*

The alkaline zinc–manganese dioxide battery is an outgrowth of the Leclanché cell. The Leclanché cell is a low rate system suitable for intermittent use.[50] In the late 1940's and early 1950's, a number of modifications were introduced to make this system a high-rate battery.[51] These modifications were (1) the introduction of KOH as the electrolyte; (2) the use of a steel jacketed can with proper seals to contain the cell; (3) the modification of the positive electrode to facilitate high rate discharge; and (4) the development of a sponge-type zinc electrode. In the course of this development, it was found that this system was capable of recharge. Further minor modifications, such as a multilayer cellulosic separator system and the introduction of a heavier gauge steel jacket to withstand pressure buildup in the cell, were made to yield a rechargeable cell with limited cycle life.

The open-circuit voltage of the alkaline zinc–manganese dioxide cell is 1.55 V. If one assumes that the first step of the discharge is

$$2MnO_2 + Zn + H_2O \rightarrow 2MnOOH + ZnO$$

the theoretical energy density is 136 Wh/lb. A second step with an open-circuit voltage of 0.95 V,

$$2MnOOH + Zn + H_2O \rightarrow 2Mn(OH)_2 + ZnO$$

gives a total theoretical energy density of 220 Wh/lb.

The first step of the discharge of the manganese dioxide electrode involves the incorporation of protons into the MnO_2 lattice. This reaction is a homogeneous solid-state reaction, and the open circuit of the cell at this stage depends on the ratio of Mn(IV) and Mn(III) in the solid phase. This reaction is a high rate reaction, and, as a result, alkaline zinc–manganese dioxide flashlight cells are capable of short-circuit currents as high as 25 A. The second step of the reaction is not well understood, but it apparently involves soluble intermediates of Mn^{3+} ions and Mn^{2+} ions in the electrolyte.

If a secondary zinc–manganese dioxide cell is repetitively discharged below 0.9 V, the cell rapidly loses capacity. The reason for this irreversibility is not well understood, but a number of explanations have been advanced.

1. A resistive film of Mn_3O_4 is formed at the current collector–MnO_2 mix interface.[52]

2. On charge, $Mn(OH)_2$ is oxidized to a form of manganese oxide (Mn_2O_3 or Mn_3O_4) which cannot be further oxidized to MnO_2.[53]

3. A mixed oxide of zinc and manganese ($ZnO{\cdot}Mn_2O_3$) is formed on discharge and cannot be recharged.[54]

The alkaline zinc–manganese dioxide battery is attractive because of the low cost of the active materials, the inherent high rate capabilities, good low-temperature performance, and excellent shelf life. The problem of irreversibility has to be overcome before the battery can be considered as a possible candidate for a power source for vehicle propulsion.

(iii) Metal–Air Batteries

For near-term electric vehicles of moderate size and range, the ambient temperature metal–air batteries utilizing an alkaline electrolyte appear to be one of the best possibilities.

Metal–air batteries are similar to the more conventional batteries in that a metal is used as the consummable negative electrode to produce electrons. The positive electrode, instead of being manganese dioxide as in the common dry cell or lead dioxide in the lead–acid battery, consists of a porous gas diffusion electrode which accepts electrons via the reduction of oxygen from the air. The two most common examples are the zinc–air and iron–air systems which use an alkaline electrolyte of NaOH or KOH. The overall cell reactions for these two systems are

$$2Zn + O_2 \longrightarrow 2ZnO \qquad E^0 = 1.61\ V$$

$$2Fe + O_2 + 2H_2O \longrightarrow 2Fe(OH)_2 \qquad E^0 = 1.27\ V$$

The products of the discharge are either the metal oxide or hydroxide, depending on the metal used. Both the zinc–air and iron–air systems have relatively high theoretical energy storage in the vicinity of 400 Wh/lb for the

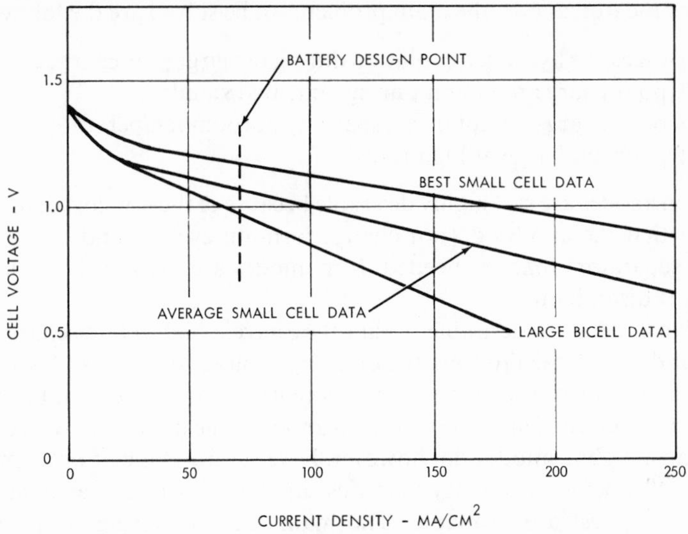

Figure 8. Zinc–air-battery single-cell performance data.

reactants only. If well-developed batteries are considered, a final device normally will have about one-fourth to one-fifth the theoretical value. Metal–air batteries should then be possible which have energy storage at least in the range 80–100 Wh/lb.

In recent years a considerable amount of effort and money has been spent on the zinc–air battery to increase its power output and yet retain the good energy-storage properties of the earlier low-power output units.[55] Some of the newer units can deliver 125–150 Wh/lb at a 15-h rate and as much as 50–60 Wh and 50–60 W/lb at the 2–5-hr rate which is required for electric vehicles. Figure 8 shows typical discharge curves for small laboratory cells as well as larger cells of the size which could be used for vehicular applications. It can be clearly seen that there is considerable loss in performance in going from the small laboratory cells to more practical units. This data, although for *primary-type cells*, does indicate that the performance levels needed for vehicular use are at least possible.

The iron–air system has been studied much less than the zinc–air battery. The discharge rate of the iron–air cell is limited by both the anode and cathode rather than the cathode alone as in the zinc–air battery. The cell voltage of the iron–air cell is about 0.34 V lower than the zinc–air cell, and two voltage plateaus are found rather than one for the zinc anode.

Iron and zinc anodes have both been studied rather extensively in other primary and secondary battery systems, so the problem areas are reasonably well defined.

For the iron anode, the main problems to be solved are the following:

(1) excessive hydrogen evolution during recharge (low charge efficiency);
(2) poor charge retention during activated stand;
(3) poor charge acceptance, especially at room temperature;
(4) poor discharge at high rates.

Iron anodes are still highly desirable because of the low cost of materials, the fact that the anodes do not change form on cycling, and the relatively simple separators that are needed. Iron anodes are used in the nickel–iron battery (Edison battery).

Zinc anodes are desirable in that they have a high rate-discharge capability, and, as opposed to iron, the discharge voltage curve has no secondary plateaus. In spite of the higher anode potential, zinc anodes can be charged faster and more efficiently than the iron anode as the hydrogen overvoltage is high on zinc. Zinc anodes do, however, have several difficult problem areas to be solved before secondary batteries capable of deep and extended cycle life (1000 cycles) are available. The main areas requiring study are the following:

(1) cell shorting due to the growth of dendritic zinc to the cathode;
(2) anode shape change on cycling;
(3) impurity effects which may result in poor zinc deposits under certain conditions;
(4) the rechargeability of the high-capacity anodes (greater than 0.30 Ah/cm^2) that are necessary in the zinc–air cell (this is a problem, especially under limited electrolyte conditions).

For the ambient temperature metal–air systems, the air cathode is the rate-limiting part of the device. The cathode is also the life-determining part of the total system. This is particularly true for the secondary metal–air batteries. The two main approaches to the secondary metal–air batteries which have been considered are shown in Fig. 9. These approaches are the direct-charge method and the third-electrode method. In the direct-charge method, the cathode must be able to withstand the highly corrosive effects of high positive potentials and erosion due to oxygen evolution during charge. The problems of making charge-resistant "secondary-type" cathodes are severe since materials which are resistant to the environment during charge are expensive and their life is still undesirably short. This is because the cathodes consist of an ultrafine pore structure with particles and pores in the range of 0.05 to about 0.50 μ, thus presenting a structure which can be easily corroded or eroded during cell charge.

In the third-electrode approach, primary-type cathodes made of low-cost materials, carbon, etc., may be used since the anode is charged using a

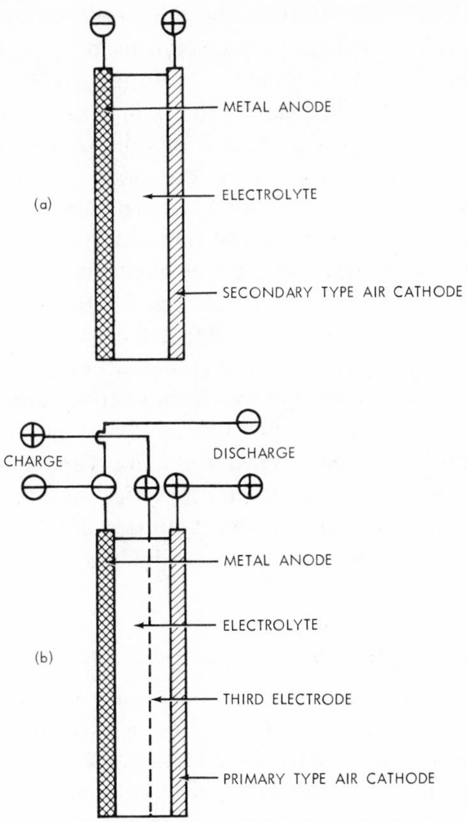

Figure 9. Approaches to rechargeable metal–air batteries: (a) direct charge; (b) third-electrode method.

separate counter electrode. However, this approach necessitates the use of more separators to isolate electrically the third electrode from the cathode, thus increasing cell thickness and resistance. One more disadvantage of this approach is the external switching necessary to go from charge to discharge. Considerable engineering could reduce the disadvantages of the third-electrode approach.

An alternate approach to the problem of secondary metal–air batteries uses a flowing zinc-slurry anode.[56] In this approach, a suspension of zinc particles is circulated through the cell until the electrolyte is saturated with zinc oxide and the capacity decreases. The zinc oxide is recharged externally to dendritic zinc powder which is washed and returned to the cell via a powder-feed mechanism. Present power density is only about 12 W/lb and energy storage is about 30 Wh/lb at the 5-h rate. The present low energy density is due to the large volume of electrolyte used, and the power is limited by the cathode-discharge rate capability. A solution to these drawbacks along with improvements in the efficiency of the pumps and motors is needed to make this system an attractive candidate for vehicle propulsion.

Of the various possible approaches discussed here, it appears that the third-electrode method is the most likely to produce tangible results in the shortest period of time. For batteries of sufficient power to be realized, it is necessary that cathodes be developed which have low cost, are long lived, and which will deliver cell voltages above 1.0 V at 75–100 mA/cm^2 and not less than 0.90 V at the current densities of 200–250 mA/cm^2 that are needed for acceleration purposes. It is quite possible that cathodes similar to the catalyzed active carbon–Teflon type with thin active layers will be suitable with further study and development.

With the solutions to these several problems, it appears quite likely that rechargeable metal–air batteries with properties suitable for the smaller electric vehicles can be developed. Such batteries should be able to deliver up to 80 W/lb and 80 Wh/lb in typical vehicular applications requiring 2–5-h discharge rates.

(iv) Organic Electrolyte High-Energy Batteries

Alkali metals and alkaline earth metals are extremely attractive as battery anodes because these metals have low equivalent weight and highly negative electrode potentials. Since these metals react with water, the electrolyte must be either a molten salt or an electrolyte with an aprotic solvent. Recently, considerable work has been carried out on batteries with lithium anodes and organic electrolytes. The chief impetus for this work is the attractive possibility of developing a high-energy battery that operates at ambient temperatures.[57–59]

The organic electrolyte cells that have been considered employ alkali or alkaline earth metals as the negative-plate active materials. The cathode active materials are metal oxides, halides, sulfides, bromine, or organic compounds. The electrolyte is based on an organic solvent such as propylene carbonate, ethylene carbonate, nitromethane, methyl chlorocarbonate, tetrahydrofuran, acetonitrile, or dimethyl formamide. The most common electrolyte solutes are lithium perchlorate, lithium aluminum chloride, and the hexafluorophosphates of sodium, ammonium, and tetra alkyl ammonium. The conductivities of these electrolytes are of the order of 10^{-3}–10^{-2} $\Omega^{-1} \cdot cm^{-1}$. The considerations which have led to the choice of these materials are as follows:

(1) The active materials in the charged state must be insoluble in the organic solvent.

Table 7. Electrode Couples for Organic Electrolyte Batteries

Electrode couple	Thermodynamic potential $E°$, V	Theoretical energy density, Wh/lb
Ca–CuF_2	3.51	604
Ca–NiF_2	2.82	501
Li–AgCl	2.85	231
Li–CdF_2	2.67	880
Li–$CuCl_2$	3.08	505
Li–CuCl	2.74	312
Li–CuF_2	3.53	750
Li–NiF_2	2.83	617
Li–NiS	1.80	500
Li–CuS	2.24	500
Li–Br_2	4.00	572

(2) The solvent must be aprotic and capable of forming solutions which have reasonably high conductivities.
(3) The reaction products should be soluble in the electrolyte so as to enhance the active-material utilization and the kinetics of the discharge process.

Table 7 lists the thermodynamic potentials and the theoretical energy densities for the various couples that have been proposed for high-energy organic electrolyte batteries. Table 8 lists some complete cells that have been investigated. In all of these cells, the lithium electrode is reversible and cycle life is limited by the positive electrode. With thin lithium layers (0.25 mA-h/cm^2) the anode can be cycled more than 1000 times in $LiAlCl_4$–ClMC electrolyte.[60] AgCl is the most reversible halide cathode. NiS, on discharge, forms insoluble Li_2S and cannot be recharged. The Li–Br_2 battery has a microporous polyethylene separator that permits conduction by transport of

Table 8. Organic Electrolyte High-Energy Cells

Cell No.	
1	Li/$LiClO_4$–PC/NiF_2
2	Li/LiBr–$LiClO_4$–PC ⁝ Separator ⁝ LiBr–Br_3^-–PC/Br_2
3	Li/$LiAlCl_4$–PC/AgCl
4	Li/$LiClO_4$–PC/NiS
5	Li/$LiAlCl_4$–ClMC/CuCl
6	Li/$LiAlCl_4$–PC/CuCl

PC = propylene carbonate: ClMC = methylchlorocarbonate.

lithium ions but prevents the diffusion of tribromide ions to the negative plate.[61,62] The bromine electrode consists of carbon, and unlike most nonaqueous batteries there is little polarization on the positive electrode during discharge. Battery performance is limited by the resistance of the separator. Cycle life of ~1000 cycles has been reported for this system.[59]

No organic electrolyte cell has been developed into a practical battery. The main problem areas are short shelf life, limited cycle life, low cycling efficiency, and the limitation to low discharge rates. In the past, most of the limitations have been due to the poor cathode kinetics and to the solubility of the positive-plate active materials. Whenever the charged cathode species go into solution, they diffuse to the anode and react with the lithium electrode. For instance, soluble copper chloride will be reduced at the lithium anode to form copper dendrites which can short out the cell. Solubility of the positive-plate active material is reflected in short shelf life and poor cycling efficiency and life. In the case of primary organic electrolyte batteries considerable improvement in shelf life has been made through the use of insoluble cathode materials, such as NiS[59] and CdF_2.[63]

Although much work has been done on organic electrolyte batteries, much of the effort, to date, has been devoted to the collection of basic *electrolyte* data. Thus, there is still a great dearth of information about the electrode kinetics of these systems. The exact effects of impurities, such as water, on these cells is yet unknown. The kinetics of these systems and the cycle life have to be markedly improved before any of these batteries can be considered for vehicle propulsion.

(*v*) *Fuel Cells*

In the future, depending on the degree of public pressure and government regulations, electric vehicles will be required with either low- or zero-polluting propulsion systems. If low pollution levels will be permitted, then there is the possibility of using hydrocarbon fuel cells. More stringent pollution controls, such as a limitation on CO_2 emissions, or a shortage of hydrocarbon fuels would necessitate the use of other fuels, such as hydrogen, hydrazine, or ammonia, which produce only water and nitrogen as the final products.

Hydrocarbon fuel could be fed directly to the anode or be reformed prior to introduction into the fuel stack. Complete oxidation of certain hydrocarbon fuels has been demonstrated in the laboratory.[64] The direct use of such fuels in a fuel cell has yet to be demonstrated, and the constraint of low pollution levels would confine direct hydrocarbon cell operation to the "dead-ended" mode. A low-polluting hydrocarbon fuel cell is more likely to be realized through the indirect route of hydrocarbon reforming followed by the electrochemical oxidation of the hydrogen-containing gas in the fuel cell. Such a system has been recently demonstrated.[65] This 12.5-kW unit has a volume of 36 ft^3 and weighs about 1500 lb. The pollution levels from this device are lower than any of the projected 1975 emissions standards for the internal combustion engine.

If hydrocarbons cannot be used, an alternative source of fuel is necessary. In a separate section of this book (see Chapter 7), the probable costs of producing electrolytic hydrogen and ammonia using cheap electrical power from nuclear reactors is discussed. Since the methods for producing these materials are well known and are now used on a smaller scale, it would seem quite likely that these fuels could be easily produced on a larger scale. In the case of hydrazine, however, the picture is somewhat less certain. Although a few methods for producing hydrazine have been reported,[67–70] none of these processes have yet evolved into large-scale production of hydrazine. The obvious advantages of using hydrazine as a fuel are high energy storage, low catalyst cost, low anodic polarization at high current densities, and the fact that hydrazine is an easily handled liquid. These factors should be an impetus to study methods of producing low-cost hydrazine.

If hydrogen is used as a fuel, a suitable means for storing the gas will be needed; and if ammonia is the fuel, an auxiliary cracking unit will have to be included in the fuel-cell system. Hydrogen can be stored either as a liquid or as a compressed gas. Liquid storage, using conventional cryogenic methods, is preferred. Ammonia storage is more efficient than hydrogen storage in terms of watt-hours per unit volume, as the density of liquid hydrogen is only 0.07 g/cm^3 as opposed to 0.77 g/cm^3 for ammonia. Hence, a given volume of ammonia contains about 1.80 times as much energy as the same volume of hydrogen. However, it is obvious that on a weight basis, hydrogen has superior energy-storage capability. The cracking unit for an ammonia fuel cell increases the power-plant weight; thus, the overall system power density may be lower than that of the hydrogen fuel cell. With both types of fuel cell, the anode catalyst would be the same.

Future work on fuel-cell systems must revolve around increasing the system power density, reducing the catalyst cost, and improving the overall system reliability.

For fuel cells to be practical for electric vehicles, the power density must be increased by a factor of two to three times that of average levels in older systems (peak power 2–4 kW/ft^3). This improvement could be achieved either through increased operating current densities or by engineering refinements to reduce cell thickness and weight.

From some of the studies which have been made to date, it is evident that high current-density operating capability in fuel cells is at least probable. Short-term measurements on both hydrogen and air electrodes have been measured at many laboratories which show usable voltages at current densities of 200–500 mA/cm^2 with noble-metal catalysts on the anode and low-cost cathode catalysts when using an alkaline electrolyte.[32] When very high platinum-catalyst loadings are used on both electrodes (40 mg/cm^2), operation at 2.5 A/cm^2 with a cell voltage of about 0.50 V has been achieved in the case of hydrogen–oxygen cells using a matrix electrolyte of potassium hydroxide.[66] In the case of hydrazine, several anode catalysts promote oxidation at 500 mA/cm^2 with little polarization. The effect of increased operating current density is obvious when one considers that the power of the fuel-cell–lead–acid-battery hybrid, discussed previously,[35] could be increased from 6 to 12 kW, if the cathode current density could be increased from 50 to 150 mA/cm^2 while still maintaining the same electrode polarization.

A design approach which greatly reduces the cell-stack volume and weight is the "dynamic mass-transport" method.[71] Ultrathin cells (about 0.020 in. thick) and a bipolar construction are used in this engineering approach. Instead of the usual porous gas-diffusion-electrode construction, the anode and cathode catalyst are bonded on opposite sides of a thin, corrugated metal sheet. These sheets are alternatively stacked along with separators and gaskets to yield a fuel-cell stack with a "filterpress design." Soluble reactants, such as hydrazine or hydrogen peroxide, are fed through ports in the cell gaskets on either side of the separator. Gaseous reactants, such as air, are fed through the ports as an emulsion of gas and electrolyte. Such a design requires two reactant streams with pumps and a suitable cell separator to prevent mixing of the fuel and oxidant. A schematic diagram of the overall system is shown in Fig. 10. With such thin cells, power densities as high as 2.5 kW/liter and less than 8 lb/kW have been obtained with hydrazine and hydrogen peroxide. With the hydrazine–air system, 1 kW/liter and 16 lb/kW are realized. In a recent announcement,[72] Alsthom Company and Jersey Enterprises have made arrangements to run a joint 5-year development program on the methanol–air system using the "dynamic mass-transport" approach. Battery size is to be in the range of 5–15 kW and program cost is stated to be about $10 million.

Since the "dynamic mass-transport" approach should not preclude the use of several other fuels, such as reformed hydrocarbons, cracked ammonia, or hydrogen, several promising fuel-cell systems for electric vehicles may be indicated.

To be economically feasible, the platinum-metals catalysts must be either reduced to very low levels (below 0.50 mg/cm^2), or be replaced by a lower cost material. When low catalyst levels are used, however, the effect of poisons, such as sulfide ion, surfactants, copper, and iron,[73] becomes pronounced and severely affects life and performance. In laboratory investigations it has been possible to exclude impurities from the system and sustain long-term operation at reasonable current densities. In a large battery,

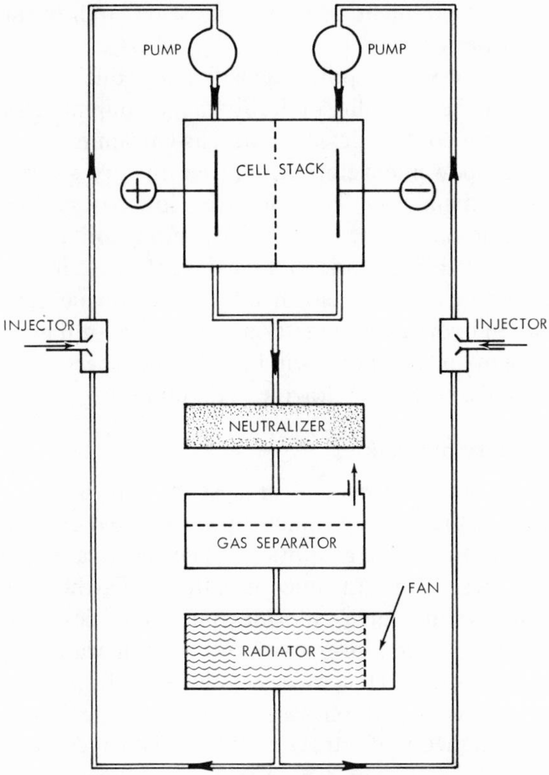

Figure 10. Schematic of the "mass-transport" fuel-cell system.

however, it becomes almost impossible to keep the system free from contaminants. One approach which may mitigate this problem is the use of reverse polarity pulses which force the anode into a condition where impurities are removed. Such pulsing in a battery with large-area electrodes might be difficult.

The possibility of the development of low-cost catalysts to replace the platinum metals should not be overlooked. There have been reported the use of certain oxide catalysts, such as the "tungsten bronzes,"[74] which (when doped with *ca.* 250 ppm Pt) show catalytic activity similar to platinum in certain reactions. Other oxide and sulfide materials with catalytic activity for fuels, such as methanol and formaldehyde, have been made.[75] Some degree of success has been attained using iron phthalocyanine[76] supported on carbon as an oxygen-electrode catalyst in acid or neutral systems. It appears possible that one or more of these materials will eventually reach a practical stage of development. However, most research in this direction is little supported at present.

Along with increasing the power density and reducing the catalyst cost, it is vital to improve the reliability of the auxiliary equipment, such as pumps, blowers, temperature control devices, and gas entrainment separators.

If the nuclear power development programs are successful, low-cost electric power should make available relatively low-cost synthetic fuels, such as hydrogen, ammonia, or hydrazine. Depending on the rate at which pollution controls are being made more stringent, there may be a place for certain types of hydrocarbon or reformed hydrocarbon fuel cells. In the more remote future when hydrocarbon materials are scarce or if no CO_2 products are permitted, the main fuel choice will be either hydrogen gas or hydrogen–nitrogen compounds, such as hydrazine and ammonia.

2. High-Temperature Systems

High-temperature electrochemical systems which have molten salt electrolytes might be the bases of vehicle propulsion batteries. These systems have some important intrinsic advantages. The electrical conductivities are much higher with these than with aqueous batteries. Furthermore, the molten alkali metal and alkaline earth chlorides and bromides that are used as electrolytes in these systems are more stable than water and have high decomposition potentials. Hence, electrodes which provide larger cell voltages can be used without the concomitant electrolyte decomposition that would occur with aqueous electrolytes. Still another consideration is the excellent electrode kinetics that can be realized with molten salts. Some of the electrode processes to be referred to below apparently involve simple electron transfers between atoms and their relatively unsolvated ions, and these reactions are inherently very fast. But an even more important factor is the elevated temperatures that greatly enhance the electrode reaction rates.

Of course, the use of molten salt electrolytes and reactive electrode materials at high temperatures will entail a number of complications not encountered with aqueous systems functioning under ambient conditions. For one thing, there is the matter of activating the battery unit which contains quantities of solid salts at room temperature. There are several reasonable approaches to raising the internal temperature of the battery so as to melt the electrolyte. One of these is electrical resistive heating. Other possibilities involve chemical heating by a pyrotechnic reaction, either external to or within the cells. But with any of these approaches, there must also be provision for reactivating the battery should the latter be cooled either by accident or on purpose.

The need for superior thermal insulation in a reasonable volume represents another complication. If heat must be continuously generated by current passage through the electrolyte or by nongalvanic reaction of electrode materials, then this heat represents a diminished electrochemical efficiency, and if the heat is provided from an external chemical source, then this source must represent additional volume and weight.

With higher and higher operating temperatures, problems with construction materials become increasingly severe. No metal or metal oxide has been demonstrated practically to resist attack by molten alkali metal and alkaline earth halides if these salts contain but trace quantities of water, hydroxide ion, and oxide ion. Moreover, the reactive electrode metals and strong oxidizing agents also attack many materials of construction. Thus, there are formidable materials problems to be solved with regard to cell cases, grids, separators, hermetic seals, and the junctions between electrodes and terminals.

The general approach to choosing an effective electrochemical system is to find a reaction which is accompanied by large negative free-energy changes and which can be carried out as the sum of separate electrode processes. It is also desirable that reacting substances have low equivalent weights. However, these factors which adduce to high energy densities theoretically available are not the only ones to be considered in practice. The systems must be rechargeable; subject to feasible designs as batteries that include cases, grids, separators, and terminals; and simple from the standpoint of reaction mechanism so that the favorable kinetics referred to above can be realized.

There are other necessary qualifications. The reactants must have minimal solubility in the electrolyte because any solubility followed by migration to an opposite electrode represents nongalvanic discharge. This could be a serious problem with negative electrodes, for certain metals dissolve extensively in their molten halides. But the problem is not so severe with alkali and alkaline earth metals, and it is further mitigated by using mixed electrolytes rather than single components.

The alkali metals, lithium and sodium, are used in preference to the alkaline earths, such as beryllium, magnesium, and calcium, even though reactions with the latter provide larger energies per unit weights of reactants. The reason is that lithium and sodium are liquid at usual molten salt electrolyte temperatures, while the alkaline earths are solid and subject to being coated (passivated under certain conditions) with films of oxidation products. Moreover, in charging the system, it may be difficult to achieve a suitable morphology for the solid negative-electrode materials.

The halogen and oxygen groups of the periodic table are the most desirable positive-electrode reactants. Fluorine and oxygen would give systems with the highest energies, but no practical electrodes have been developed for fluorine or oxygen in molten salt electrolytes. Although they have somewhat less capacity, the choice of chlorine or sulfur represents desirable compromises in system selection.

In this section a number of high-temperature battery systems, that after preliminary investigations have shown promise as potential power sources for automotive propulsion, are reviewed in some detail. Some of the principal features of these systems and the problems that have to be faced before the development of practical systems will be described. It is not the intention to imply that these are the only high-temperature systems that have potential as practical batteries; but up to the present these systems have received the most attention.

(i) Lithium–Chlorine Battery

Probably the first molten salt battery that was seriously considered as a possible automotive power source was the lithium–chlorine system.[77] The basic system is conceptually very simple, as seen from the reaction

$$2Li + Cl_2 \underset{\text{charge}}{\overset{\text{discharge}}{\rightleftharpoons}} 2LiCl$$

During discharge Li^+ ions are formed at the molten Li anode, and at the porous carbon cathode externally stored Cl_2 gas is reduced to Cl^- ions. The net result of the electrochemical reaction is the formation of LiCl, which is also the electrolyte. The OCV of about 3.5 V is constant throughout the discharge. Recharge is by simple electrolysis of the LiCl. Counting reactants only, the system has a theoretical energy density of approximately 1000 Wh/lb.

Molten LiCl has the highest electrical conductivity of any simple fused salt ($5.92\ \Omega^{-1}\cdot cm^{-1}$ at 650°C),[78] and this, coupled with excellent electrode kinetics, gives the system a high power-density capability. In a specially designed laboratory primary cell, power densities up to 40 W/cm^2 have been demonstrated,[79] *which is far in excess of the power required in a practical device* All batteries have some internal resistance, and there is a practical limit to the currents that can be passed through the system after which the I^2R heating becomes so large that the battery will overheat. In many cases the advantages of the higher powered battery are more than counterbalanced by the added complexity and weight of the thermal control system.

Extensive investigations into the properties of the electrodes and the electrolyte have been made. These have included the study of both fundamental properties and the development of new electrode concepts.

The reaction at the Li electrode involving a single electron transfer is extremely simple, and within experimental limits no polarization was observed at current densities up to 40 A/cm^2.[77] Lithium is less dense than LiCl, and since both are liquids at cell operating temperatures the Li will float on the top of the electrolyte. To enable the establishment of a stable electrode–

electrolyte interface, a Li wick electrode has been developed.[80] The Li is stored remotely and a porous metal wick transports the Li, on demand of the electrode reaction, to the electrode–electrolyte interface. Capillary-action forces within the wick, which are greater than the hydrostatic head of LiCl, provide the driving force for the Li transport. The theory of the wick electrode has been confirmed experimentally,[80] and the limiting currents observed for this electrode were high enough to satisfy any cell requirements.

The Cl_2 electrode reaction which occurs on porous carbon or graphite is not as simple as the lithium anode, and polarization has been observed at relatively low current densities. Since the porous carbon electrode is not wetted by LiCl, a different model from that used in wetted fuel-cell gas-diffusion electrodes has been proposed to explain the operation of the Cl_2 electrode.[77]

Chlorine gas flows through the porous plug where it dissolves in the electrolyte at the gas–liquid interface. The dissolved Cl_2 then diffuses through the electrolyte to the carbon–electrolyte interface where the Cl_2 dissociates and the charge-transfer step occurs. Since the Cl_2 gas must flow through the porous carbon, the electrode performance is dependent on porosity, pore size and pore-size distribution, pressure, and thickness of the electrode.[81] The mechanism of the Cl_2 reactions on graphite has been studied, and the rate determining steps of both the anodic and the cathodic reactions have been determined.[82] The effect of electroinactive impurities in the Cl_2 is very marked,[83] and even small amounts of impurities reduce the maximum current densities obtainable with any particular carbon material.

In a battery system where mechanical separators are not used, self-discharge is possible when the reactants are soluble in the electrolyte. The dissolved reactants can diffuse through the electrolyte and react nongalvanically with a resultant loss in coulombic efficiency. However, studies of the solubility and diffusion of Li[84] and Cl_2[85] have shown that at normal operating temperatures the values of these parameters are small enough so that self-discharge is not expected to be intolerably large.

The Li–Cl_2 battery, when developed, promises to be a high-powered device. However, the factors that give it such promise—extremely active reactants, good electrode kinetics as a result of the high operating temperatures and low internal losses because of the fused salt electrolyte—give rise to the greatest single obstacle to its successful development, viz., materials. No single, cell-construction material has practically demonstrated compatibility with both Li and Cl_2 at these temperatures, thus complicating the construction of any long-lived device. This is especially true in the area of non-conductive seals.

Figure 11 is a sketch of a possible configuration for an advanced Li–Cl_2 cell. Reference to it will clearly show how the use of an active gaseous reactant

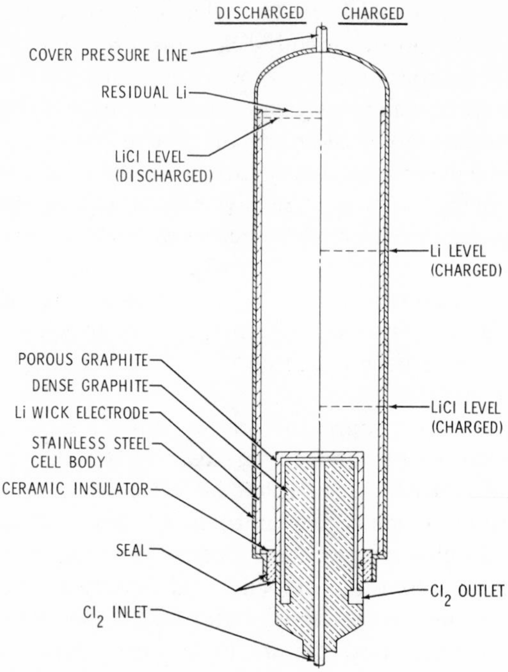

Figure 11. Possible design for an advanced lithium–chlorine cell showing both the charged and discharged fluid levels.

poses additional problems. Chlorine evolved on the cathode during recharge will rise through the electrolyte and react with Li floating in its reservoir unless some means of channeling the evolved Cl_2 away from the Li reservoir is designed into the cell. One way by which this can be accomplished is by the use of a valve electrode,[86] which is a normal porous carbon electrode which has a porous ceramic layer sprayed on its surface. The principle of operation is that while the ceramic is wetted by the LiCl, the porous carbon remains nonwetted. Since the pores of the insulating ceramic are filled with electrolyte, capillary action results in a pressure difference which forces the Cl_2 evolved on the carbon during charge back through the porous carbon to the Cl_2 feed manifold. Even though it is possible to return the Cl_2 generated on charge to the Cl_2 feed manifold, it must still be collected and be made available for the next discharge cycle. This means that the Cl_2 must be cooled and be transferred to a storage tank where it can be condensed. This processing requirement adds to the complexity and to the weight of any total system.

Finally, since this system operates at the highest temperature of any of the high-temperature batteries being considered here, the thermal requirements will be the most stringent. More heat will be required to bring the system from ambient to operating temperatures, and more thermal insulation will also be necessary.

In spite of the weight penalties introduced by the Cl_2 processing equipment and the thermal controls and insulation, energy densities of 135–180 Wh/lb and power densities of 90–180 W/lb have been estimated for the complete system.

(ii) *Lithium-Charge-Storage-in-Carbon Battery*

This system and the lithium–chlorine are essentially the same in several respects. That is to say, the differences are generally quantitative rather than qualitative, and the conditions with the charge-storage system are such as to mitigate some of the difficult technological problems with lithium–chlorine cells. For example, the charge-storage system uses a eutectic electrolyte[87] which is 59 m/o LiCl–41 m/o KCl and melts at 350°C, well below the melting point of pure LiCl (608°C). Consequently, the charge-storage system is usually operated between the eutectic temperature and 500°C, instead of at 650°C as is the case for the Li–Cl_2 cell.

The reduced temperature not only leads to a direct relaxation of some materials problems, but it allows the construction of negative electrodes that can be handled much more conveniently compared with electrodes of pure lithium metal. These negative electrodes are solid at operating temperatures and are made by alloying lithium with aluminum. They have useful capacities of about 12 A h/in^3 and can be charged and discharged at peak current densities of 2 A/cm^2.[87] The solubility of lithium metal in the melt in equilibrium with the alloy is about 1% of the lithium solubility in the melt in equilibrium with pure lithium. If this decrease represents the reduced activity of lithium in alloy, the loss in cell voltage should not be more than about 0.2 V and the open-circuit voltage of a fully charged cell (3.25 V) supports this conclusion.

The positive electrodes are fabricated from porous carbon. Examples of excellent materials are FC-11, FC-13, and FC-50 carbons* and Saran char.† Results with these in single-electrode studies have been reported in some detail.[89]

The mechanism of the positive-electrode function is not clear, but there is evidence that there are contributions to the energy storage from both the cell reactions and the charging and discharging of the double layer.

* These carbons are available from the Pure Carbon Company, St. Marys, Pennsylvania.
† This carbon was prepared by a method similar to that described by Reed and Schwemer.[88]

At large positive potentials, there appears to be high-energy bonding between positively charged carbon atoms and chloride ions; at relatively negative potentials, there are bonds between carbon and cations. It was shown[85] that potassium ions were adsorbed preferentially over lithium ions. On the basis of cyclic voltammetric studies and chemical analyses, the potential at zero charge is -0.75 V vs Cl^-/Cl_2.

It has been found[87] that the chemisorption of electrochemically active additives onto the carbon electrode surface improves the capacity of the electrodes. Also, other materials can be mixed with the carbon during the preparation of the electrode material. For example, appreciable increases in storage capacity have resulted from the incorporation of metallic tungsten into the porous carbon.[90] This additional capacity is attributed to the electrochemical behavior of the tungsten—oxidation of the metal during charge and reduction of tungsten ion during discharge.

There is some difference of opinion concerning aspects of carbon electrodes. Some investigators have determined that capacity is not a function of BET area measured on bulk pieces and that the entire volume of the carbon is used; during charge, the carbon fills with electrolyte with preferential penetration by KCl.[89] Another worker has found that capacity depends very strongly on measured BET area and that the interiors of these pieces do not participate in the electrode process.[91]

The system is unique among the high-temperature batteries being considered in this chapter in that both electrodes are solid at operating temperatures. As a result, a conventional flat-plate design can be utilized to construct cells which leads to ready stacking of cells to form a battery. Data for a flat cell (referred to as a *pancake cell*) have been reported[87] and are given in Table 9. Figure 12 shows the discharge profile for this cell when discharged at a constant 30-A rate. The terminal voltage of the cell continually decreases

Table 9. Performance Characteristics of Lithium-Charge-Storage Cell (Pancake Design)[50]

Cell weight	2.9 lb
Cell resistance	6.7 mΩ
Capacity	50 A·h
Energy delivered	98.7 W·h
Energy delivered—*iR* corrected	109 W·h
Leakage current (at 1.75 V)	0.63 A
Leakage current (at 2.75 V)	0.65 A
Energy density delivered	34 Wh/lb
Energy density—*iR* corrected	37.6 Wh/lb
Cathode capacity	5.5 Wh/in.^{3}

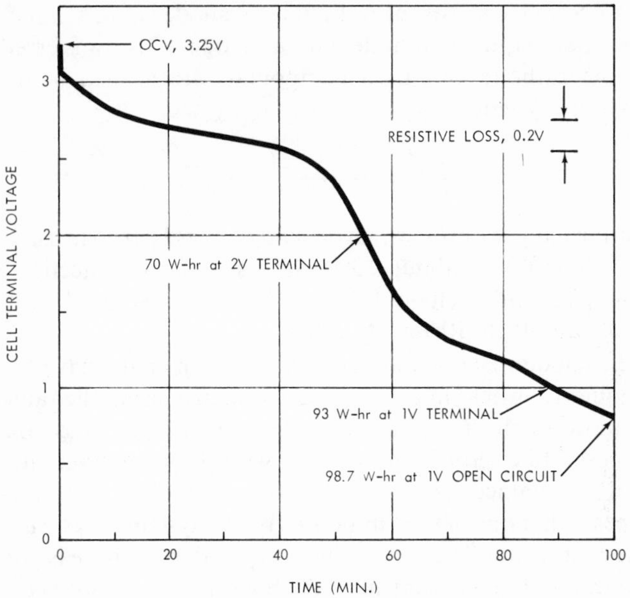

Figure 12. Cell discharge profile for a pancake lithium-charge-storage cell.[87]

over a fairly wide range during discharge. Although this complicates its use as a power source, control systems exist that can overcome this problem.

The energy density of the experimental flat cell was only 38 Wh/lb, but new, improved carbon electrodes are expected to increase the energy density to over 70 Wh/lb. It has also been estimated that power densities of 100 W/lb can be achieved over most of the discharge range.

(iii) Lithium–Sulfur Battery

Another group of closely related systems that has received attention is the lithium–chalcogen group. These utilize Li anodes in conjunction with cathodes of sulfur, selenium, or tellurium. However, Se and Te are rare elements in the earth's crust. Their main source is as a byproduct from the smelting and refining of copper and lead which creates a supply situation which would appear to militate against development of large-scale usage of these elements.[92] For this reason, batteries that use either Se or Te must be rejected for large-scale automotive use unless new sources of these elements are discovered. Preliminary investigations using S (for which there is an adequate supply) have shown that this could be a high-power and high-energy system suitable for vehicle propulsion.[93,94]

The Li–S cell consists of a liquid Li anode, a Li^+ ion-conducting electrolyte, and a liquid S cathode. On discharge, Li is transferred from the anode to the cathode and Li–S compounds are formed in the cathode according to the reaction

$$2Li + xS \underset{\text{charge}}{\overset{\text{discharge}}{\rightleftarrows}} Li_2S_x$$

Cell operating temperatures, which depend on the electrolyte used, are about 400°C and the OCV is about 2.3 V. Several different eutectic electrolytes have been used and include LiF–LiCl–LiI (mp 341°C),[93] LiCl–LiI–KI (mp 264°C)[94] and LiBr–RbBr (mp 278°C).[93]

The Li anode reactions are simple and similar to that of the Li–Cl_2 system. Liquid Li is held in a porous metal matrix using the same capillary forces as those of the Li wick electrode. It is also expected that current densities on the Li electrode comparable with those observed in the Li–Cl_2 battery can be obtained.

It appears that one of the limitations of this system may be the performance of the S electrode. The electrochemistry of the electrode is not as simple as the anode, and the mechanism of the S electrode has not yet been elucidated. Sulfur is a nonconducting liquid and attention must be given to current collection at the cathode. Typical current collectors are porous matrices of metal[93] or carbon[94] in which the S is held by capillarity.

During discharge, lithium sulfides are formed at the electrolyte–current collector–sulfur interface. It has been suggested[93] that these sulfide products dissolve in the S and diffuse away from the reaction site. On recharge the reverse process occurs. Thus, high-surface-area electrodes with small pore diameters would increase the rate of the diffusion step. The electrode performance has been shown to depend on pore size, porosity, and the type of material used for the current collector. For instance, if the material were too porous, the electrode would be limited by its current-collection ability. On the other hand, if the pores were too small, diffusion of reaction products to and from the electrode–electrolyte interface could be hindered. Sulfur, being a liquid, has a potential advantage over gaseous Cl_2 as a cathode reactant. As a liquid it can be stored in the cell itself and there is no need for a gas feed, recovery, or processing system in a complete battery package. Another advantage of the Li–S system is an operating temperature which is lower than that of Li–Cl_2. The operating temperature could possibly be lowered still further by the use of different electrolytes, but such a reduction would reduce the S electrode performance.[95] As with many cell parameters, the choice of operating temperature involves a number of trade-offs, and in this case it is between the increased cell material stability at lower temperatures and the enhanced cell performance at higher temperatures. In spite of the reduced temperatures, Li_2S–S mixtures corrode many materials,[94,95] and care must be taken in the choice of construction materials, particularly for S electrode current collectors. The reactivity of dissolved Li in the electrolyte also limits the choice of insulating seal materials.

Sulfur utilization is another problem in the system and more basic research is needed to fully understand the S electrode. Nearly pure lithium sulfide (Li_2S) has been identified by x-ray diffraction analysis in some locations of a discharged cathode,[95] and it is believed that Li_2S is the end product of the cell reaction. In most cell tests, the maximum S utilization has been only about 50%; however, at very low current densities this has been increased a little.

The theoretical energy density for the system, assuming complete reaction of the S to form Li_2S, is 1200 Wh/lb, based on reactants only. It is not known at present whether the causes of the low S utilization are of a fundamental nature. If they are and if the stoichiometry of the end product is LiS, then the theoretical energy density is reduced to about 700 Wh/lb.

Several analyses have been made of the Li–S system for a variety of vehicle applications.[93,94,96] These analyses have yielded estimates for energy densities of 70–190 Wh/lb and power densities of 35–130 W/lb.

(iv) Sodium–Sulfur Battery

The sodium–sulfur battery system, which has been described in some detail,[97,98] does not use a molten salt electrolyte. However, it is usually grouped with the molten salt batteries because it operates at elevated temperatures (300°C), and its cathode reaction product is a molten salt.

There is a certain unique and elegant basis on which a rechargeable Na–S cell can be made operable. This involves the use of the so-called sodium beta-alumina (β-alumina)[97,98] as a solid electrolyte. The β-alumina is described stoichiometrically as $Na_2O \cdot 11Al_2O_3$. The structure of this ceramic-type material is such that open planes are formed in directions normal to the c axis of the crystal. These open planes are formed by Al–O–Al columns which space apart the spinel blocks of the β-alumina. The plane containing the columnar O is incompletely filled, and hence there is room for alkali metal ions (especially Na^+) to reside and have great mobility within the plane. An apparently more suitable layered phase is formed by the addition of MgO to form the compound $Na_2O \cdot MgO \cdot 5Al_2O_3$, which has been designated β''-alumina.[99] This material has a very similar open-plane structure to β-alumina and exhibits the same enhanced Na^+ mobility in the plane normal to the c axis. Other stabilizing oxides can be added to form similar structures. Thus, there exists a series of Na^+ ion-conducting ceramic membranes which can serve the function of an electrolyte. At 300°C, the resistivity of these materials is approximately 3–5 $\Omega \cdot$cm.

Beta-alumina can be made nonporous, and it is inert to Na and to Na_2S–S mixtures. It therefore functions as both an electrolyte and a separator between negative and positive electrodes. It would be impossible to operate the battery without this separator as Na and S readily react to form a series of polysulfides, and hence self-discharge would be very limiting.

Generally, it has been proposed that the β-alumina be formed in tubular shapes with the Na anode being contained in them. These tubes are dipped into the cathode reactant, a sulfur–sulfide mixture, which is supported on a graphite-felt current collector. During discharge, sodium is oxidized at the sodium–ceramic interface and the Na^+ formed migrates to the sulfide–graphite–ceramic interface. At this interface a series of sodium polysulfides is formed. The overall reaction is

$$2Na + (x - 1)Na_2S_x \rightleftarrows xNa_2S_{x-1} \qquad x = 3, 4, 5$$

A single laboratory cell constructed with an electrolyte thickness of 0.8 mm and a 3-mm layer of cathode reactant gave the performance curves shown in Fig. 13.[97] Using these data, a scaleup of the system to a 2-kW-size unit was proposed, the specifications of which are given in Table 10. The rated energy density of the unit at 147 Wh/lb represents 45% of the theoretical energy density which is an unusually high percentage. Most batteries have practical energy densities that are <20% of their theoretical values.

The system has development problems similar to those of the other high-temperature systems. In addition to the need for an improved ceramic

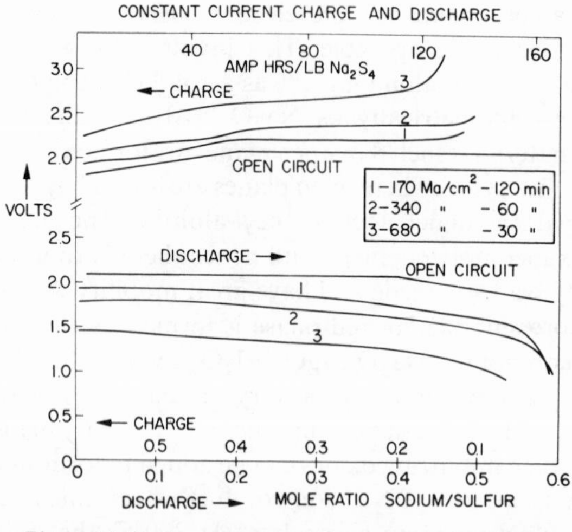

Figure 13. Performance curves for a sodium–sulfur cell with a 0.8-mm-thick β-alumina electrolyte.[97]

Table 10. Specifications of a Proposed 2-kW Sodium–Sulfur Cell[97]

Average power	2 kW
Peak power	4 kW
Open-circuit voltage	2.08 V
Average discharge voltage	1.75 V
Capacity	1850 A·h
Weight	29 lb
Volume	400 in.^{3}

electrolyte, there are material problems associated with cathode current collectors and seals which have to be solved before the sodium–sulfur system can be considered a long-lived battery. Also, the packaging of a battery that has a large number of thin, fragile membranes, will be a challenging engineering task. Even though it does not appear to have a demonstrated advantage, more organizations have announced interest in this battery than any other high-temperature system.

3. Summary of Expected Battery Performances

Table 11 summarizes the achieved or projected energy and power densities of the battery systems and fuel cells discussed in this chapter. The overall data is also pictorially summarized in Fig. 14, which is a Ragone plot of power density as a function of energy density. In Section IV, the power-plant

Table 11. Achieved or Projected Battery Characteristics

Battery	Energy density, Wh/lb	Power density, W/lb
Lead–acid	2–15	80–15
Nickel–cadmium	8–25	250–30
Silver–zinc	25–50	150–20
Nickel–zinc	12–27	200–30
Zinc–manganese dioxide	10–25	100–15
Metal–air	50–80	35–5
Organic electrolyte[a]	90–110	20–5
Lithium–chlorine[a]	135–180	180–90
Lithium–charge-storage[a]	70	100
Lithium–sulfur[a]	70–190	130–35
Sodium–sulfur[a]	80–150	160–90
Fuel cells	Depends on Fuel Tank Size	30–5
Fuel cells[a]	Depends on Fuel Tank Size	up to 100

[a] Projected systems.

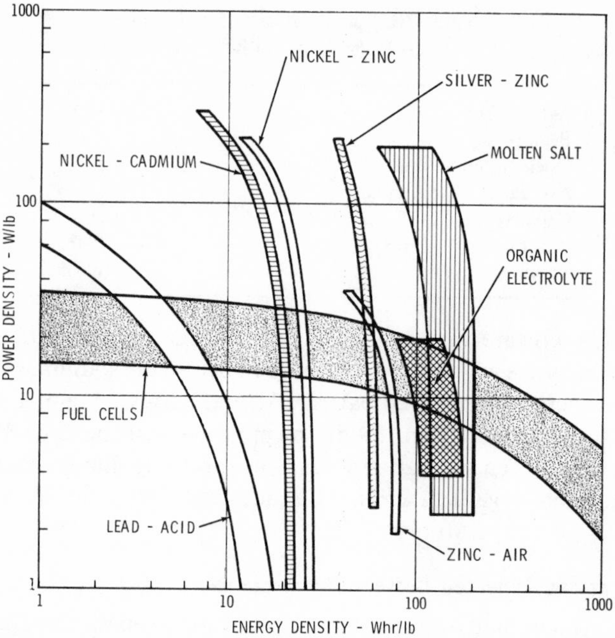

Figure 14. Plot of power density as a function of energy density for most of the systems discussed in this chapter. Taken in part from Ragone.[101]

requirements for several representative types of vehicles were established. If the Ragone plot is used in conjunction with these power-plane requirements (given in Table 2 and Fig. 4), several trends with regard to the status of batteries for vehicular propulsion become obvious.

It has been assumed in this chapter that, apart from special purpose vehicles, electrically powered vehicles will have to have certain minimum performance characteristics consistent with the expectation of the need to intermix with existing transportation modes. There is no doubt that should the vehicle specifications (such as size, range, top speed, or acceleration) be sufficiently altered, then single-battery vehicles could be built as evidenced by some of the experimental units in existence today.[100] However, for the representative vehicle types considered here, there are no batteries presently available which could satisfy both the power and energy requirements for a single-battery power plant.

If the emerging ambient temperature batteries are considered, it can be seen that these would also fail to meet the requirements of a single-battery vehicle. Fuel cells at their present stage of development have energy densities limited only by the energy conversion efficiency and the size of the fuel storage system. On the other hand, power densities of these fuel cells are relatively low and for this reason, apart from such drawbacks as catalyst costs, are not useful as a single power source. Scientific advances in fuel-cell systems, particularly in the area of specific power improvements, could change the situation, and their possible usefulness would be enhanced. One point to be noted about the ambient temperature systems is that some show greater promise as energy batteries while others have power advantages. The significance of this is discussed in the next section.

The *high-temperature* batteries that were discussed are in the research stage and have not yet reached a point in their development where they can be packaged into a self-sustaining test unit. It can be seen from Table 11 that should the projected capabilities of these systems be reached, each, with the exception of the Li-charge-storage, could satisfy the requirements of the most stringent vehicle application—the family car. The Li-charge-storage system could meet the reduced requirements of the other applications.

VII. PROJECTIONS OF ELECTRIC POWER-PLANT DEVELOPMENT

The review of batteries and fuel cells showed that deficiencies in either power or energy densities of available systems limit their use in electric vehicles. However, it also showed that some of the systems being developed offer promise for future use. If there are no existing batteries that could operate as a single-battery power source for an all-electric family car, what is the development path that could be envisaged for this type of vehicle? The *family car* is used here as an example because it has the most stringent requirements on the power source. It is recognized that other vehicle types with different requirements could have alternative development paths and that their requirements could possibly be met by improvements to existing batteries.

It was shown earlier that most of the power in a vehicle power plant is for acceleration and gradeability, and that only about 30–50 % of this power is required for constant speed cruise in the intermediate speed ranges. This obvious division of power requirements, together with the analysis of the capabilities of near-term batteries, points the way to possible future routes for electric vehicle development.

A first possible electric vehicle that could evolve is some form of hybrid vehicle. This would be a heat engine coupled with a battery system; wherein the heat engine provides vehicle range and is sized for average power over a given driving cycle or road-load power for maximum sustained cruise speed. The batteries provide additional power for acceleration and gradeability requirements. With this arrangement, the vehicle can also be driven in an all-electric mode, without the engine, for short distances with limited performance. The Pb–acid battery has sufficient power density for this type of power plant; however, its relatively poor energy density would provide only a very short range in the all-electric mode of operation.

Further possible evolutionary paths for electric vehicles could be the replacement of the heat engine by a battery system. This would require the development of an energy battery such as a rechargeable Zn–air battery or an organic electrolyte system. Since fuel cells are primarily energy systems, these could be used in conjunction with a power battery in a hybrid mode. As was the case for the heat engine–battery hybrid, the energy source, in this instance a battery or fuel cell, is used to recharge the power batteries during certain phases of the vehicle's operation.

The ultimate step in the development would be the replacement of the dual battery or battery–fuel cell hybrid by a single electrochemical power source which would have both the needed energy and high power characteristics. At present, the only battery group that appears capable of meeting the requirements is the high-temperature, molten salt group. However, this group, which appears to have the most potential, also presents the most challenging research problems. The problems, associated with the high operating temperatures and extremely energetic reactants, are principally in the *materials area*. Basic research *in this area* resulting in significant scientific advances is required before the molten salt batteries can be engineered into complete systems.

The other possibility for a single-system power source is the fuel cell. For this to occur, an increase in the power density of fuel cells, both in terms of weight and volume, will be necessary. This could be accomplished by increasing the current densities achievable with practical electrodes and by reducing the thickness and spacing of the electrodes. A fuel-cell system would be the closest approach to our present concept of vehicle transportation, since it would offer rapid refueling and give "tank-full" range of operation.

It has been shown that the incentives for electric vehicles are clear. The principal obstacle to their successful development is the availability of suitable electrochemical power sources. For this reason there is a need and justification for considerable expenditure for research and development efforts in the field of battery and fuel-cell systems.

REFERENCES

[1]A. B. Cambel, *Energy R & D and National Progress*, U.S. Government Printing Office, Washington, 1964, p. 91.

[2]M. K. Hubbert, *Energy Resources*, National Academy of Science, Publication 1000-D, 1962.

[3]*Civilian Nuclear Power—A Report to the President*, U.S. Atomic Energy Commission Report, Nov. 20, 1962.

[4]H. H. Landsberg and S. H. Schurr, *Energy in the United States*, Random House, New York, 1967, pp. 15, 75.
[5]R. Covington, *Oil Gas J.* **62** (1964) 112.
[6]*Bituminous Coal Facts 1958*, National Gas Association, Washington, June 1, 1958.
[7]H. H. Landsberg, L. L. Fischman, and J. L. Fisher, *Resources in America's Future*, John Hopkins Press, Baltimore, 1963.
[8]R. A. Rice, ASME Publication—70—WA/Ener—8, New York, Dec. 1970.
[9]J. F. Weinhold, ASME Publication—70—WA/Ener—9, New York, Dec. 1970.
[10]H. H. Landsberg and S. H. Schurr, *Energy in the United States*, Random House, New York, 1967, p. 30.
[11]A. B. Cambel, *Energy R & D and National Progress*, U.S. Government Printing Office, Washington, 1964, p. 22.
[12]M. M. Yarosh, ASME Publication—70—WA/Ener—12, New York, Dec. 1970.
[13]E. Cony, *Atomic Power Push, Wall Street Journal*, Feb. 15, 1965; see also, *Report on Economic Analysis for Oyster Creek Nuclear Electric Generating Station, Nuclear News*, April 1, 1964.
[14]S. F. Singer, *Scientific Amer.* **223**, No. 3 (1970) 175.
[15]E. Hines, M. S. Mashikian, and L. J. Van Tuyl, SAE Publication—690441, Chicago, May 1969.
[16]J. H. B. George, L. J. Stratton, and R. G. Acton, *Prospects for Electric Vehicles*, A study for Department of Health, Education and Welfare, Contract No. PH 86–67–108, Arthur D. Little, Inc., May 1968.
[17]E. H. Hietbrink and S. B. Tricklebank, ASME Publication—70—WA/Ener—7, New York, Dec. 1970.
[18]*The Automobile and Air Pollution: A Program for Progress—Part II*, Subpanel Reports to the Panel on Electrically Powered Vehicles (R. S. Morse, Chairman), U.S. Department of Commerce, Dec. 1967, p. 103.
[19]*The Automobile and Air Pollution: A Program for Progress—Part II*, Subpanel Reports to the Panel on Electrically Powered Vehicles (R. S. Morse, Chairman), U.S. Department of Commerce, Dec. 1967, p. 101.
[20]G. Planté, *Recherches sur l'Electricite*, Bureaux de la Revue la Lumiere Electrique, Paris, 1883.
[21]G. W. Vinal, *Storage Batteries*, 4th Ed., John Wiley & Sons, New York, 1955.
[22]J. A. Orsino, *Pb–Acid Secondary Cells, Kirk–Othmer Encyclopedia of Chemical Technology*, 2nd Ed., John Wiley & Sons, New York, 1964, Vol. 3, pp. 249–271.
[23]S. U. Falk and A. J. Salkind, *Alkaline Storage Batteries*, John Wiley & Sons, New York, 1969, Chapter 1.
[24]A. Fleischer, *Trans. Electrochem. Soc.* **94** (1948) 289.
[25]H. André, *Bull. Soc. Franc. Electriciens* (6th Series) **1** (1941) 132.
[26]J. McBreen, U.S. Pat. 3,505,115 (1970).
[27]G. A. Dalin and M. J. Sulkes, *Proc. 19th Ann. Power Sources Conf.*, Atlantic City, N.J., PSC Publications Committee, 1965, p. 69.
[28]J. Goodkin, U.S. Pat. 3,493,434 (1969).
[29]R. F. Amlie and P. Ruetschi, *J. Electrochem. Soc.* **108** (1961) 813.
[30]C. Berger, *Handbook of Fuel Cell Technology*, Prentice Hall, Englewood Cliffs, N.J., 1968.
[31]G. J. Young and H. R. Linden, *Fuel Cell Systems*, American Chemical Society, Washington, D.C., 1965.
[32]J. O'M. Bockris and S. Srinivasan, *Fuel Cells: Their Electrochemistry*, McGraw-Hill, New York, 1969.
[33]Condensed from Table 2–5, p. 55, *Fuel Cells: A Review of Government Sponsored Research 1950–1964* NASA SP-120, Scientific and Technical Division, National Aeronautics and Space Administration, Washington, D.C., 1967.
[34]F. M. Wyczalek, D. L. Frank, and G. E. Smith, SAE Publication—670181, Detroit, Mich., Jan. 1967.
[35]K. V. Kordesch, *J. Electrochem. Soc.* **118** (1971) 812.
[36]B. S. Baker, *Hydrocarbon Fuel Cell Technology*, Academic Press, New York, 1967.
[37]T. Michalowski, Russian Pat., Group XI, No. 5100, April 28, 1901.
[38]The Drum Battery Co., German Pat. 659,659 (1937).
[39]J. M. Fay and J. J. Drum, *Railway Engineer* **54** (1933) 258, 280.

[40]Reports on the Drum Battery, *Railway Gazette* **56** (1932) 382, 407, 543, 640.
[41]V. V. Romanov, P. D. Lukovtsev, G. H. Kharchenko, and P. I. Sandler, *Zh. Priklad. Khim.* **33** (1960) 1556.
[42]N. A. Zhulidov and F. I. Yefremov, *The New Nickel–Zinc Storage Battery*, Air Force Systems Command, WPAFB, Translation FTD-TT, 64–605, 1964.
[43]P. Goldberg, *Proc. 21st Ann. Power Sources Conf.*, Atlantic City, N.J., PSC Publications Committee, 1967, p. 70.
[44]E. P. Broglio, *Proc. 21st Ann. Power Sources Conf.*, Atlantic City, N.J., PSC Publications Committee, 1967, p. 73.
[45]P. V. Popat, E. J. Rubin, and R. B. Flanders, *Proc. 21st Ann. Power Sources Conf.*, Atlantic City, N.J., PSC Publications Committee, 1967, p. 76.
[46]M. J. Sulkes, *Proc. 23rd Ann. Power Sources Conf.*, Atlantic City, N.J., PSC Publications Committee, 1969, p. 112.
[47]A. Charkey, *Proc. 23rd Ann. Power Sources Conf.*, Atlantic City, N.J., PSC Publications Committee, 1969, p. 115.
[48]F. P. Kober and A. Charkey, preprint, Paper No. 18, Power Sources Conf., Brighton, England, Sept. 1970.
[49]J. W. Diggle, A. R. Despic, and J. O'M. Bockris, *J. Electrochem. Soc.* **116** (1969) 1503.
[50]G. W. Vinal, *Primary Batteries*, John Wiley & Sons, New York, 1950.
[51]J. L. S. Daley, *Proc. 15th Ann. Power Sources Conf.*, Atlantic City, N.J., PSC Publications Committee, 1961, p. 96.
[52]D. Boden, C. J. Venuto, D. Wisler, and R. B. Wiley, *J. Electrochem. Soc.* **115** (1968) 333.
[53]H. Y. Kang and C. C. Liang, *J. Electrochem. Soc.* **115** (1968) 6.
[54]K. Miyazaki, preprint, Paper No. 35, Power Sources Conf., Brighton, England, Sept. 1970.
[55]A. Fleischer, *Survey on Metal–Air Cells*, Air Force Systems Command, WPAFB, Tech. Rep., AFAPL-TR-68-6, March 1968.
[56]H. Baba, SAE Publication—710237, Detroit, Mich., Jan. 1971.
[57]K. H. M. Braeuer and J. A. Harvey, *Status Report on Organic Electrolyte High Energy Density Batteries*, Tech. Rep. AD 654813 (1967).
[58]R. Jasinski, *Electrochem. Technol.* **6** (1968) 28.
[59]F. Conti and G. Pistoia, *J. Appl. Chem. Biotechnol.* **21** (1971) 77.
[60]J. P. Gabano, G. Gerbier, and J. F. Laurent, *Proc. 23rd Ann. Power Sources Conf.*, Atlantic City, N.J., PSC Publications Committee, 1969, p. 80.
[61]J. R. Weininger and F. P. Holub, *Proc. Symposium on Battery Separators*, Columbus, Ohio, The Columbus Section of the Electrochemical Society, 1970, p. 122.
[62]J. L. Weininger and F. P. Holub, *J. Electrochem. Soc.* **117** (1970) 342.
[63]M. Shaw and R. Shand, *Proc. 23rd Ann. Power Sources Conf.*, Atlantic City, N.J., PSC Publications Committee, 1969, p. 76.
[64]J. O'M. Bockris, H. Wroblova, E. Gileadi, and B. J. Piersma, *Trans. Faraday Soc.* **61** (1965) 2531.
[65]R. W. Benedict, *Little Black Box*, A Discussion of the TARGET Program, *Wall Street Journal*, May 19, 1971.
[66]Interagency Advanced Power Group Project Brief, PIC No. 2053, April 1970.
[67]J. A. Pursley, U.S. Pat. 3,268,425 (1966).
[68]R. E. Lacey, U.S. Pat. 3,281,211 (1966).
[69]J. C. Chu, U.S. Pat. 3,280,015 (1966).
[70]J. H. G. van der Stegen, W. Visscher, and J. G. Hoogland, *Electrochem. Technol.* **4** (1966) 564.
[71]B. Warsawski, *Basis for the Development and Industrialization of the Fuel Cell*, Publication of the Research Division, ALSTHOM, Massy, France, 1968.
[72]*Wall Street Journal*, Dec. 7, 1970.
[73]R. R. Witherspoon and R. L. Adams, Paper No. 344, Electrochemical Society Meeting, Montreal, Canada, Oct. 1968.
[74]B. Broyde, U.S. Pat. 3,502,506 (1970).
[75]V. Mehra and W. R. Wolfe, Jr., U.S. Pat. 3,505,118 (1970); also, W. R. Wolfe, Jr., U.S. Pat. 3,492,164 (1970).
[76]A. Kozawa, V. E. Zilionis, and R. J. Brodd, *J. Electrochem. Soc.* **117** (1970) 1474.

[77]D. A. J. Swinkels, *J. Electrochem. Soc.* **113** (1966) 6.
[78]E. R. Van Artsdalen and I. S. Jaffe, *J. Phys. Chem.* **59** (1955) 118.
[79]E. H. Hietbrink, J. J. Petraits, and G. M. Craig, *Advances in Energy Conversion Engineering*, ASME, 1967, Vol. 1, p. 933.
[80]D. A. J. Swinkels and S. B. Tricklebank, *Electrochem. Technol.* **5** (1967) 327.
[81]D. A. J. Swinkels and R. N. Seefurth, *J. Electrochem. Soc.* **115** (1968) 994.
[82]W. E. Triaca, C. Solomons, and J. O'M. Bockris, *Electrochim. Acta.* **13** (1968) 1949.
[83]D. A. J. Swinkels, *J. Electrochem. Soc.* **114** (1967) 812.
[84]R. A. Foust, Jr., and D. A. J. Swinkels, Paper No. 188, Electrochemical Society Meeting, New York, N.Y., May 1969.
[85]S. B. Tricklebank, Paper No. 48, Electrochemical Society Meeting, Detroit, Mich., Oct. 1969.
[86]D. A. J. Swinkels, *Electrochem. Technol.* **5** (1967) 396.
[87]R. A. Rightmire, J. W. Sprague, W. N. Sorensen, T. H. Hacha, and J. E. Metcalfe, SAE Publication—690206, Detroit, Mich., Jan. 1969.
[88]M. W. Reed and W. C. Schwemer, *J. Electrochem. Soc.* **114** (1967) 582.
[89]J. L. Benak, J. E. Metcalfe III, and J. W. Sprague, Final Report, Contract DAAKO2-68-C-0253, USAMERDC, The Standard Oil Company (Ohio), Cleveland, Ohio, April 1, 1969.
[90]W. K. Behl, D. D. Beals, and G. R. Frysinger, *J. Electrochem. Soc. Japan* **37** (1969) 215.
[91]S. M. Selis, Paper No. 294, Electrochemical Society Meeting, Los Angeles, Calif., May 1970.
[92]A. M. Lansche, *Selenium and Tellurium*; *A Materials Survey*, U.S. Dept. of the Interior, Bureau of Mines, Information Circular 8340, 1967, p. 1.
[93]H. Shimotake, M. L. Kyle, V. A. Maroni, and E. J. Cairns, *Proceedings of the 1st International Electric Vehicle Symposium*, Phoenix, Ariz., The Electric Vehicle Council, 1969, p. 392.
[94]L. A. Heredy, N. P. Yao, and R. C. Saunders, *ibid.*, p. 375.
[95]E. J. Cairns, M. L. Kyle, V. A. Maroni, H. Shimotake, R. K. Steunenberg, and A. D. Tevebaugh, *Report to NAPCA*, from the Chemical Engineering Division, Argonne National Laboratory, Argonne, Ill., July 1970.
[96]N. P. Yao, L. A. Heredy, and R. C. Saunders, Paper No. 60, Electrochemical Society Meeting, Atlantic City, N.J., Oct. 1970.
[97]J. T. Kummer and N. Weber, SAE Publication—670179, Detroit, Mich., Jan. 1967.
[98]N. Weber and J. T. Kummer, *Proc. 21st Ann. Power Sources Conf.*, Atlantic City, N.J., PSC Publications Committee, 1967, p. 37.
[99]M. Bettman and C. R. Peters, *J. Phys. Chem.* **73** (1969) 1774.
[100]See, for example, *Proc. 1st International Electric Vehicle Symposium*, Phoenix, Ariz., The Electric Vehicle Council, 1969.
[101]D. V. Ragone, SAE Publication—680453, Detroit, Mich., May 1968.

Chapter 4

THE ELECTROCHEMICAL TREATMENT OF AQUEOUS EFFLUENT STREAMS

Anselm T. Kuhn

Department of Chemistry
University of Salford
Salford 5, Lancashire

I. INTRODUCTION

For virtually every type of effluent problem, an electrochemical solution can be envisaged, either by the armchair chemist or his laboratory counterpart. Such approaches have been comprehensively reviewed elsewhere[1] and are beyond the scope of this chapter, which concerns itself with those efforts which have been, or which promise to be, successful.

Electrochemical solutions to problems in effluent treatment are not exempt from those criteria which all processes must satisfy, namely, their cost efficiency as expressed in terms of capital investment costs, and the operation expenses. The would-be practical scientist loses sight of these facts at his peril.

The power consumption of an electrochemical plant and, hence, its main operation cost is, broadly speaking, $V \times A$, where V is the voltage of the system, and A is the current passing through it. But in all processes, and especially in effluent treatment processes, the relevant factor is not merely the current A but the current efficiency, which we may designate $Ax/100$, where x is the percentage of the current which is expended on the useful and desired reaction, as opposed to undesirable and wasteful side reactions. In processes such as aluminum manufacture or chlorine electrolysis x can be

90–98%. But with the more dilute solutions found in effluent treatment x can fall way below this figure. When such factors as water saving, saved dump costs, and reclaimed products are taken into account, it is often the case that efficiency values of x as low as 10% are still acceptable. But when x falls to 1% or so, the economics of electrochemical treatments cease to be attractive. The spearhead of current development work in electrochemical engineering is the effort to design novel types of cells which will perform more efficiently with dilute solutions, and these cells, and their performance, will be fully discussed.

In fundamental terms, the electrochemist has four very powerful tools at his disposal—all will be fully examined here. In the electrodialysis process, charged species (ions) can be removed from an effluent stream, and may be concentrated into a smaller stream. The technique, then, is one which though in no way solves the effluent problem recasts it in a form which may be more amenable to treatment. The electroflotation process can be used to separate dilute suspensions into slurries and clear liquid. It is a variation of the classical flotation technique using electrolytically generated gas bubbles as separating agents. These two techniques can be considered as physical separations achieved by electrochemical means. The following two techniques are chemical in nature, involving either direct reaction at the electrode or with a reagent generated electrolytically. At the negative electrode, the cathode, reduction occurs, and in the present context the most relevant of these is the reduction of a metal ion to its elemental state. When this occurs with a metal such as copper, not only are the ions removed but they are extracted as the metal, that is, in the most desirable form. This remains one of the strongest points in favor of such electrochemical techniques. At the positive electrode, the anode, oxidation occurs either directly or by oxidizing agents generated at the anode. Oxidation can destroy unwanted organic compounds and can frequently render a toxic compound harmless.

II. ELECTRODIALYSIS AND THE CONCENTRATION OF EFFLUENT STREAMS

The concentration of effluent streams prior to treatment can clearly be desirable. Indeed, a surprisingly large number of "effluents" become commercially valuable once they are concentrated and purified. Otherwise, such solutions as dumping or lagooning all become cheaper when concentrated solutions are involved. Lastly, the removal of pollutants—as contrasted with their neutralization—enables waste waters to be recycled. This is becoming more important and financially desirable with every passing year. Such a separation of water from its contaminants may be achieved by a variety of methods, including filtration, centrifuging, and sedimentation, with or without flocculant addition. Distillation and freezing are two further methods which appear to be competitive, but only on a very large scale, and in the situation where waste heat is available. For smaller quantities of liquids two techniques currently appear most promising. These are electrodialysis and reverse osmosis (RO). In both cases contaminants are separated from water by its passage through a membrane. In one case the driving force for this separation is electric potential, while in the other it is hydraulic pressure. While it is not possible to predict how these two methods, so similar in many ways, stand or will stand in relation to one another, electrodialysis is already so important that it is difficult to conceive that it will not find a permanent place in effluent treatment. The principle is illustrated in Fig. 1, which shows a section across a cell. It will be seen that a cell stack contains pairs of cells, termed "concentrate cells" and "dilute cells." Under the potential gradient, ions of positive or negative charge migrate in the direction of the electric field, out of the diluate stream and into the concentrate stream. The diluate (clean) and concentrate (foul) streams are separated by "ion-exchange membranes," that is, membranes which permit only the passage of charged species (ions). In fact, they permit only the passage of ions of a given sign, and this gives rise to the nomenclature "cation-exchange membranes" (+*ve* ions) and "anion-exchange membranes" (−*ve* ions). Further details of engineering construction, costs, etc., may be found in Ref. 2. The most widespread application of electrodialysis (ED) plants is in the desalination of brackish waters. It will be appreciated that in some methods of desalination the water is removed from the contaminant (e.g., flash distillation, freezing), while in ED plants the contaminants are extracted from the water. The

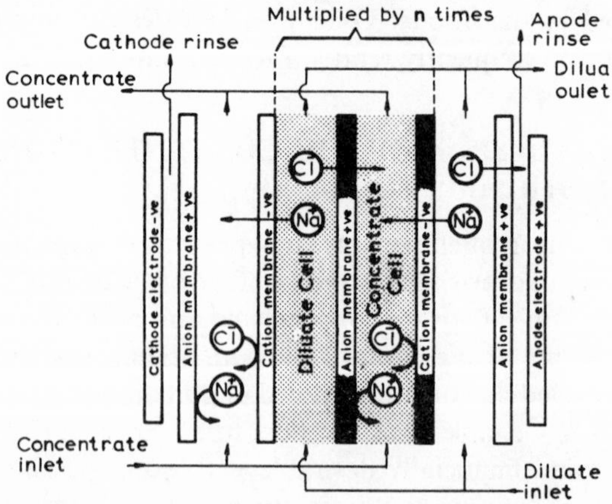

Figure 1. Schematic diagram of electrodialysis cell.

economic implications of this are such that for severe contaminations the former methods are preferable, at least on large-scale plants. Where a relatively minor amount of contaminant is to be removed, electrodialysis may be the preferred method. This is the case with slightly brackish waters, where the problem resolves itself to one of lowering the level of dissolved salts to around the 500-ppm mark, which is the approximate level for potability. This concept leads directly to the problem which will increasingly face advanced civilization in coming years. It is now appreciated that water, in between the rainfall which initially deposits it on the earth surface, and its drainage to the oceans, may be used more than once, in the sense that between Montana and Louisiana the same water may many times be used, fouled, repurified, and reused. Indeed, there is the concept of total recycling, in which a community has its stock of water which it recycles endlessly with only small additions to make good unavoidable losses. From a hygiene point of view, present techniques of filtration, etc., followed by chlorination–ozonization are perfectly adequate. But one problem which such conventional techniques do not tackle is the fact that at each "pass" the level of mineral salts, e.g., from detergents, leaching from the earth, etc., increases, and once above the 500-ppm level this is sufficient to render the water unattractive. The

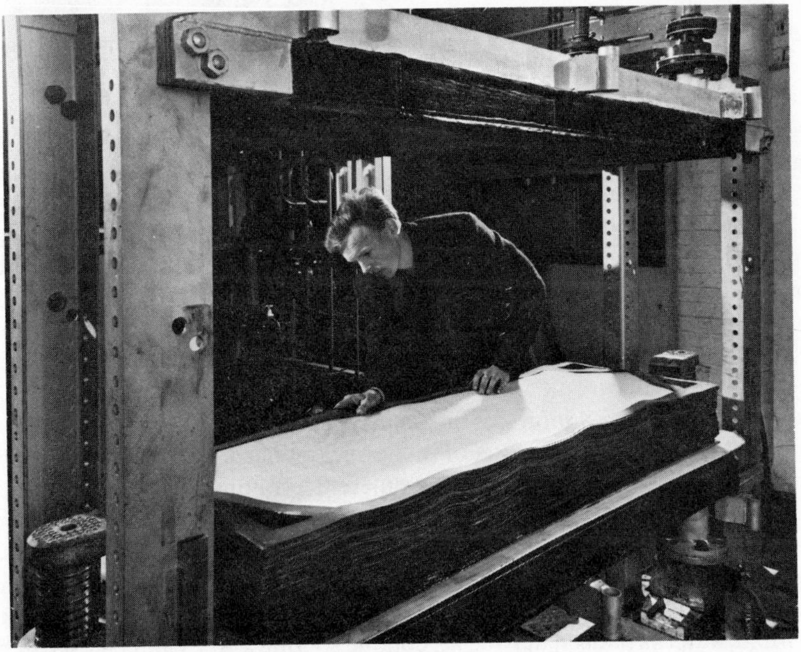

Figure 2. Assembly of electrodialysis cell (courtesy William Boby and Co., Ltd.).

Figure 3. The Benghazi electrodialysis plant—one of the largest in the world today (courtesy William Boby and Co., Ltd.).

incorporation of an ED plant in a conventional sewage works would solve this. Though the writer knows of no examples of this, it seems increasingly probable that this will become a large-scale application of ED.

Figure 2 illustrates the details of an actual ED cell, while Fig. 3 depicts one of the largest ED desalination plants in the world at Benghazi. The prospects of this process, together with RO, can only improve, as new membrane materials permit faster throughputs at less cost per unit area. Increased use of the process will permit longer manufacturing runs of the membranes, which in turn will become cheaper. An important point to take into consideration is that whereas a large desalination–purification plant operating under constant load may have one set of economic optima, a smaller plant, serving a remote community, may call for different economic factors. An actual example of this was recently described when the water requirements of holiday resort areas were considered. In such townships the water demand rises very steeply for a few months in the year, falling sharply back again in the off-season period. For such cases a high specific energy consumption per unit water produced may be acceptable so long as the capital cost can be kept low. Even for very saline waters, ED can meet such specifications.

III. ELECTROFLOTATION PROCESSES

Just as electrodialysis is a technique for "recasting" the problem of a dilute effluent, electroflotation can be invaluable as a means of separating dilute suspensions into slurries and clear liquid. The former can then be trucked away or even incinerated with little additional energy input; the liquid can be treated in the conventional manner. Such suspensions or colloidal oil–water mixtures can be treated by simply allowing them to stand in settling tanks, or by the addition of flocculating agents. But the former is capital intensive, for settling tanks are costly and also occupy valuable land space. Flocculating agents constitute an additional process cost, as well as requiring reliable dosing and metering equipment. It was found nearly a century ago that if electrolytically generated bubbles (which are very small and thus have a highly favorable surface-area-to-volume ratio) are allowed to rise up through a suspension or colloidal liquid, they speedily effect a separation. In Europe the leader in this particular field is St. Gobain, while in the U.K. Pollution Technical Services Ltd., Abingdon, Berks, have installed plants of this type, as has Simon-Hartley, Ltd. No such installations are available as yet in the U.S. A list of effluents which might typically be treated is shown in Table 1.

A schematic plant layout is shown in Fig. 4, while a photo of an actual installation, for clarification of paint-shop waste waters, can be seen in Fig. 5. The principles of operation are not altogether understood. In the main, the process is thought to be a mechanical one, the lifting of suspended particles by the minute bubbles, which, on account of their small size, are more efficient

Table 1. Typical Effluents Amenable to Electroflotation Treatment

Oil industry wastes
Engineering industry wastes (cutting oils, etc.)
Slaughter-house wastes
Food industry waste (vegetable–animal oils)
Dairy wastes (cheese wheys)
Cellulosic fibers (paper mills, board mills)
Glass fibers
Asbestosic wastes
Vegetable wastes (lucerne chaff, etc.)
Textile fiber wastes
Latex and rubber wastes
Polymeric wastes
Iron Oxide scale (rolling-mill waters)
Paper-mill white water ("loading matter")
Paint-shop waste water
Wool industry wash liquors

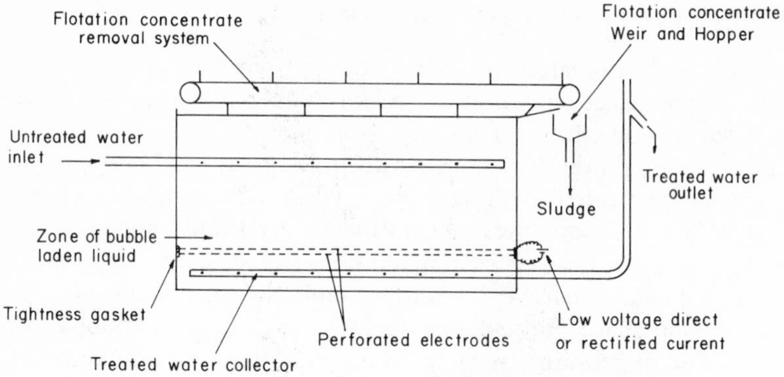

Figure 4. Schematic diagram of electroflotation cell (courtesy St. Gobain).

Figure 5. Electroflotation plant for removal of paint wastes from process waters, installed Bristol, England (courtesy William Press and Co., Ltd., Pollution Technical Services, Ltd.).

than larger bubbles, that is, the ratio of bubbles to suspended particles should be close to unity. Whether there is an additional effect relating to the known electrical charge on colloids, etc., which stabilizes them, is unknown, but at least in theory one might envisage such charge-stabilized systems being dispersed by a release of that charge at the electrodes. However, the idea of passing the suspension through electrodes or across them, in order to obtain charge dispersal, is incompatible with maximum gas-lift effect, which requires the bubbles to be generated well below the incoming suspension. A possible cell design utilizing both these effects would have primary and secondary sets of electrodes, the first serving to neutralize the colloids or suspensions and the second to generate bubbles to lift up the particulate matter. Alternating current of low frequency could serve for the primary electrodes in such a case.

1. Details of Cell Design and Performance

The cell, made of polythene or similar material, should be approximately 1 m high. A pair of electrodes is placed horizontally, and close together, near the bottom of the cell. Typically, they are 0.5–4 cm apart. This will depend partly on the conductivity of the waste water and partly on the nature of the suspension and the probable danger of the electrodes becoming short circuited. The clear water takeoff is situated in the quiescent region of the cell, close to the bottom. The solid or oily matter which rises to the top of the cell can then be removed by mechanical bladed conveyors or other devices. The cell voltage is 5–10 V; again this figure is very variable, and depends partly on the conductivity of the effluent and partly on the current density utilized—100 A/m^2 is a typical figure for the latter. Various cost figures have been quoted for operation, but 2.5 kWh/m^3 water is a typical figure, while the solids level in the treated water can usually be lowered to 30 mg/liter solid. The process is thus comparable with a settling tank or microstrainer rather than with a diatomaceous-earth filter bed. Tables 2 and 3 give actual examples.

In a further example (the purification of paint-bearing waters), a P.T.S. plant lowers solids from 22,000 to 200 ppm in the outlet water at a cost of

Table 2. Electroflotation Treatment of Steel-Rolling-Mill Wastes

Waste water spec.:	solids (mainly Fe_3O_4)	1500–350 mg/liter
	oils	600–300 mg/liter
	flowrate	75 m^3/h
Treated water spec.:	solids	30 mg/liter
	oils	40 mg/liter

Plant size: 25m^2 electrode area, 25m^3 cell volume. Pt–Ti electrodes operating at 100 A/m^2. 8.0 V. Power consumption 275 Wh/m^3.

Table 3. Paper-Mill Effluent Electroflotation Plant

Waste water spec.:	Insoluble matter (kaolin, fibrous)	1000 mg/liter
	flowrate	100 m^3/h
Treated water:	insoluble matter	30 mg/liter[a]

Plant size: as Table 2. 18/10 low carbon–steel electrodes, 80 A/m^2. Power consumption 200 Wh/m^3. Sludge water content 90–95 %.

[a]Some pretreatment with flocculating agents.

0.7 cents/1000 gal, which includes the cost of chemical dosing agents and electrode replacement costs.

2. Electrodes and Electrolytes

The electrodes may be platinized titanium (as manufactured by Marston Excelsior of Wolverhampton, U.K.) or carbon steel, stainless steel, or lead dioxide. A process for the deposition of the latter on titanium screens or sheets has advantages, such as ease of installation. Wright,[3] describes such electrodes and cites French patent 1,483,489 as covering the manufacture of PbO_2–Ti electrodes assigned to the Pacific Engineering Corp., Henderson, Nevada. Such electrodes, and also magnetite coatings, are probably the best for this process. A suggestion has also been made that aluminum electrodes could be usable. Their slow dissolution would form a method of supplying a flocculating agent. However, there are many obvious drawbacks to this concept. The power consumption is largely a function of the electrolyte conductivity; where this is very low it can be raised by addition of a conducting salt such as common salt. This forms the basis of a patent assigned to the Fairbanks Morse Co. (British patent 1,149,362), while the whole concept of electroflotation as an effluent treatment process was described in British patent 676,854 (to Metallgesellschaft). In this process, as in others such as electrodialysis and electrolytic sewage treatment (see below), problems of electrode fouling can arise. These can be handled by periodic reversal of electrode polarity. The cause is thought to be formation of regions of high pH which allow alkaline earth hydroxides to precipitate. Such polarity-reversal equipment can rule out the use of certain electrode materials with a given effluent stream. Thus, PbO_2 does not take kindly to polarity reversal, nor do many iron electrodes.

IV. CATHODIC PROCESSES—RECOVERY OF METALS

At an electrode of negative polarity (European sign convention) many metals can be plated out from dilute solutions of their ions. There are normally two incentives for this. The first is the accruing from metal savings; the second is the simplification of effluent disposal. Dilute solutions of copper (100 ppm or less) are still highly undesirable effluents, and have a damaging effect on the organisms which are used in normal biological effluent treatment plants. The electrochemist can remove such metal ions by electrodeposition, but he faces two problems. The first of these is the competitive reaction of hydrogen evolution. The second problem is that of rates of electrodeposition limited by mass-transport problems, resulting from the very low concentrations of the metal ion. The latter problem also occurs in a consideration of anodic processes for effluent treatment, and to some extent both can be treated here. In general, the less noble the metal the more difficult it will be to plate out, and either low current densities or less acidic solutions will have to be used. The theoretical situation regarding this competition between hydrogen gassing and metal plating can be easily assessed by an inspection of "Pourbaix diagrams," which show, for each metal, the potential at which hydrogen may start to evolve as a function of pH. Figure 6 shows such a diagram.

1. Recovery of Copper

Far and away the most common application of metal recovery by electrochemical methods is found in the copper-manufacturing industry. Freshly rolled, drawn, or fabricated copper is pickled in a sulfuric acid tank. This dissolves the scale, copper oxide, which converts the acid into sulfate of copper. In time, therefore, the pickle liquor loses "bite." Formerly, this was either discharged to waste or chemically worked up into copper sulfate.

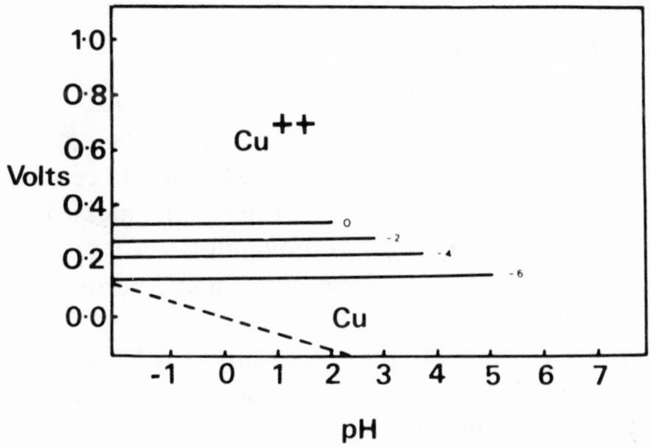

Figure 6. A simplified Pourbaix diagram for the copper–copper ion system. Solid lines, Cu/Cu^{2+} potential at Cu^{2+} concentrations shown mole/liter × 10^x; dashed line, hydrogen evolution potential.

Figure 7. A copper removal cell—Canning design. Note the grown sheets of recovered copper being removed (courtesy William Canning Co., Birmingham).

All modern plants are now equipped with electrolysis units which electrolyze the copper sulfate solution and convert it to metallic copper and sulfuric acid, with oxygen also evolved at the anode.

(*i*) *Conventional Cell Designs*

Two different types of such cells are shown in Figs. 7 and 8. Figure 9 shows a plant flowsheet. This is the normal practice—cycling the liquor through the recovery cell and back to the pickle tank. In certain installations the recovery electrodes are situated in a corner of the pickle tank itself. Yet a different type of copper recovery cell is shown in Fig. 10. The point of interest here is the rocking anode. This stirs the solution and permits higher working rates.*

Copper recovery plants are important not only because there are more of them in operation than any other type of electrochemical recovery cell, but also because we may use the situation as a model or a baseline for discussion of other processes of a similar nature.

*Ti starter cathodes are now being used in place of Cu or stainless. New Metals Group of IMI, Birmingham, U.K., report finer-grain deposits, easier parting of Cu, and less pitting of starter in Cl^- solutions.[26]

It is important to appreciate that the copper recovery system is a closed loop. The pickle liquor goes round and round, from pickle tank to recovery unit and back again. And the vital clue is that while copper sulfate must be converted back to acid and copper at each pass through the recovery cell, there is no need for the level of $CuSO_4$ to be reduced to near zero or anything like it. This means that the copper recovery takes place in a solution of relatively high copper concentration, with the effect of high efficiency of removal. The situation must at all costs be distinguished from those in which an effluent goes to waste and must be completely "cleaned up." The problem of treatment in a recycle system is thus much easier than in an open-ended system. Nevertheless, there are many such situations where recycling is entirely acceptable. Normal copper recovery plants are operated under the conditions shown in Table 4. For full details of the literature on the process the reader is referred to Ref. 1.

Figure 8. A copper removal cell—the APV–Kestner design (courtesy APV–Kestner Engineering Co.).

Table 4. Operating Conditions for Copper Recovery Cell

Anodes	PbO_2 (1–2% Ag)
Cathodes	stainless-steel or copper
Cell temp.	48–65°C
Current density	4–16 mA/cm^2
Cell voltage	2.5–3.5 V
Cu concn. (in)	7%
(out)	2%
Efficiency (C.E.)	70–90 m
Electrode gap	1–4 cm
Energy expenditure	900 AH–4.5 kWh/kg Cu

(ii) *Fluidized Bed Cells*

While Table 4 describes a normal copper removal plant, a number of projects exist aimed at designing cells which will remove copper down to some 10 ppm, at acceptable efficiencies. Perhaps the best known of these is the fluidized bed cell, currently being developed by Constructors John Brown, Ltd. The principle here is that the electrode, instead of being a flat sheet or plate, consists of a bed of beads, either of copper or metallized glass. These are restrained and held in place by means of a diaphragm, which separates the bed from the anode compartment (which may itself be a second bed electrode). The effluent liquor streams through the copper bead bed, and, because of the

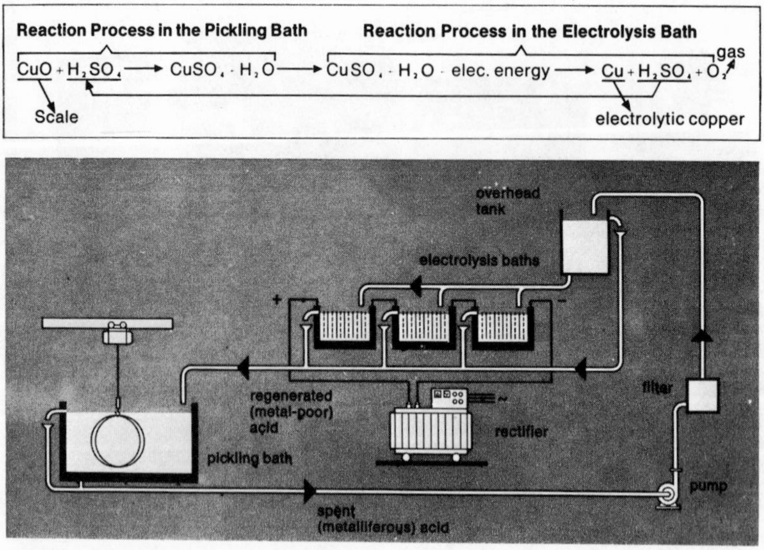

Figure 9. Plant flowsheet for the copper removal process (courtesy Demag G.m.b.H.).

Figure 10. A copper removal cell with rocking anodes for greater diffusion rates (courtesy Demag G.m.b.H.).

packing of the system and the turbulence resulting, copper ions are able to deposit at high efficiencies, even from a dilute solution. Thus, efficiencies of 10% can be obtained even at concentrations of 10 ppm. Figure 11 shows the economics of copper recovery from such dilute solutions with a bed electrode. There are, however, one or two problems. While copper costs around $1500/ton in bar form, to obtain it in bead form suitable for such fluidized beds one has to pay double this price. As the copper plates out the diameter of the beads grows, until they must be withdrawn. The copper is thus primarily reclaimed in a bead form. This may be melted down into ingot form, then rolled for sheet, etc. But, in this way, one has discarded the value latent in the shaping of the beads in the first place. It is thought that the probable answer

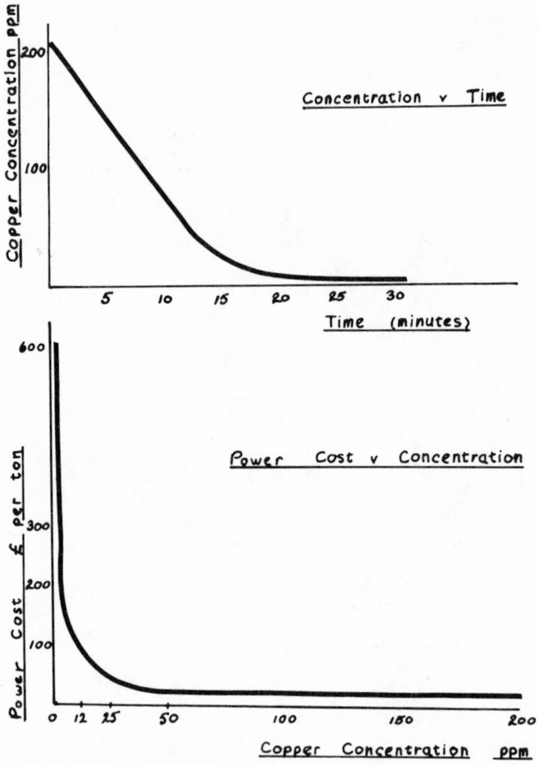

Figure 11. The economics of copper removal with fluidized bed cell (courtesy Constructors John Brown, Ltd.).

to the problem here will be to cycle the growing copper beads to a second cell—and if they are truly fluidized, this constitutes no problem—where they can be stripped down to their original diameter, the stripped off copper being plated out in sheet form suitable for direct reuse. We thus have a much more efficient process than the type of cell shown in Figs. 7, 8, and 10. But it also is a more sophisticated one, requiring better control and supervision. Clearly, there will be plants where such supervision is available. But in small, simple operations, such supervision may not be feasible.

(iii) Rotating Cathode System

Alternative cell designs may prove equally efficient and more convenient. Lancy Laboratories, of Zelienopile, Pennsylvania, has patented a rotating cathode system, which is highly efficient. Like the older established cells of Figs. 7 and 8, and unlike the fluidized bed cell, it reclaims copper in a directly useable form. That is to say, if the rotating cathode is made of stainless steel, the accruing copper can be peeled off and fed straight into a plating tank as an anode. (See footnote, page 108, Ref. 26.)

(iv) Force-Flow Cell

Another system which could be promising, is the force-flow cell. This scores by having no moving parts (unlike the Lancy cell), but the same effect results by forcing the effluent liquor through a narrow-gap cell by means of a pump. Early development work has taken place in the U.K., and results are shown in Fig. 12. This work is related to acidified copper sulfate recovery in an undivided cell of $\frac{1}{4}$-in. interelectrode gap. The results are capable of further optimization by regulation of gap size, and also of electrolyte flow rate. A given loss of efficiency also results from nonadherent metallic copper (which

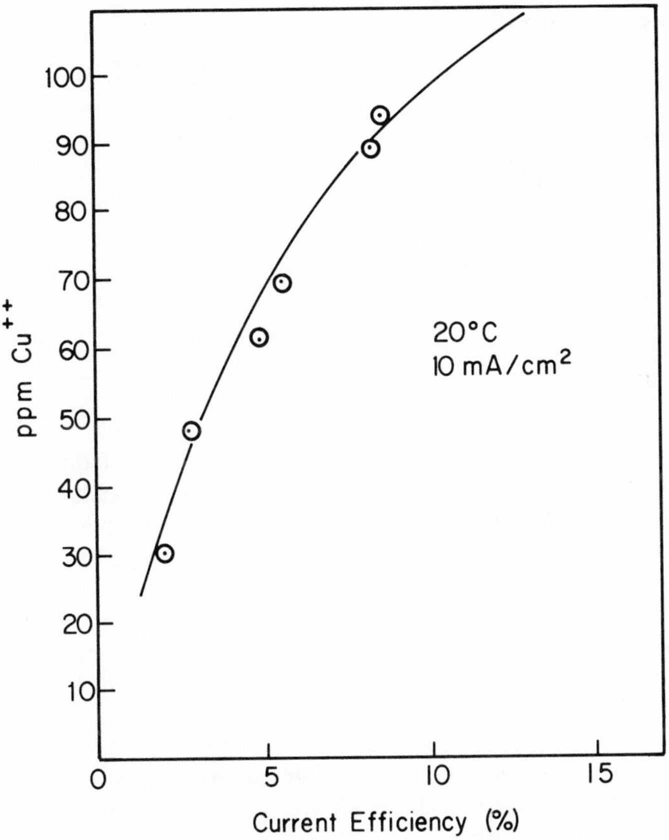

Figure 12. Efficiency of dilute copper removal with force-flow cell.

tends to form as a black powder). Part of this drifts to the anode and is there redissolved. However, this can be avoided by incorporation of a diaphragm in the cell, either of a cloth-type fabric or a "chemical" diaphragm such as that formed by dichromate additions in chlorate cell technology. Such diaphragms are discussed fully in Ref. 4. The force-flow cell is of special interest in that the anode is a massive one and therefore better fitted to perform anodic destruction reactions (e.g., cyanide destructions) than fluidized bed cells, where oxide films on the particles tend to increase the ohmic resistance of the system. Data has been published which indicates that such force-flow systems can destroy 100-ppm cyanide at 10% efficiency, approximately. This is a valuable development which should be pursued.

Other types of advanced cells are discussed in the section dealing with anodic processes, and further details may be found in the work of Jackson.[5]

2. Recovery of Other Metals

While the electrolytic recovery of copper is the most important process of its kind in use, other metals can be, and are, recovered electrolytically. For example, silver is reclaimed from photographic film processing laboratories, and gold is recovered from works where it is used. Nickel can be reclaimed by the "Lancy process," though it is more normal to use this process in a configuration where the end product is the hydroxide or salt rather than the metal itself. Again, a fuller summary of ideas which has been advanced in this field is found in Ref. 1. Because of the sheer size of the market, one should perhaps not conclude without reference to iron and, in particular, the recovery of iron from waste pickle liquors.

(*i*) *Iron*

There is a vast body of literature relating to the electrolytic recovery of iron from pickle liquors. This work goes back to the 1920's and is still proceeding today. Nevertheless, one can say that no major plant is presently in operation.

(*ii*) *Tin and Iron*

A process which was successful, in its time, was the electrolytic detinning of cans from municipal scrap and refuse. The detinned cans were then further dissolved to give iron powder. However (and thanks largely to the work of electrochemists), the amount of tin required to protect a can has been reduced so drastically that its recovery is no longer thought worthwhile. As for the iron powder, though much is still made electrolytically, nonelectrochemical routes are now proving more economical, except where very high-purity grades are required.

(iii) Reclaiming Metals from Junked Cars and Appliances

The question remains as to what is the best way for disposal of the vast quantities of ferrous and mixed-metal wastes which society annually produces. In particular, the problem of junked automobiles, domestic appliances, and cans comes to mind. The most efficient way of disposing of automobiles and washing machines, etc., depends on the location. In urban areas, where a large throughput of such vehicles is expected, the Proler machine appears to be the currently most favored solution. This shreds, by mechanical means, the entire automobile into nuggets of a 1-in. size, approximately. These are then subjected to a magnetic separation operation, for removal of the more valuable nonferrous materials. Out of urban areas the scrap dealer performs a similar function manually by removal of those parts of the vehicle which he knows contain such nonferrous metals. The efficiency of both methods is similar, for in the Proler process, however finely the matter is shredded, some ferrous and nonferrous material is inevitably inseparable. Having separated the metals, the copper, etc., is melted down and electrolytically refined. One might equally well eliminate the melting operation and feed the scrap directly into titanium anode baskets. There are pros and cons on either side of this argument. The ferrous metal is sent for remelting.

To what extent is there likely to be further development in the electrolytic processing of the ferrous scrap? It will be seen that since electrochemical methods involve processes in which each layer of iron has to be deposited on top of the next one, the work done (and the time it takes) are directly related to the thickness of the final object, that is to say, the cathode. In smelting and melting processes the thickness is irrelevant. We can thus see that the thinner the desired steel, in its reclaimed form, the better placed is an electrochemical process. In the U.K. work done by B.I.S.R.A. and the Electricity Council has examined high-rate electroforming of iron foil from scrap. A product of a few mil thickness is thus formed, perfectly free from pinholes and completely isotropic in its properties. By contrast, the same foil produced by rolling (and there are lower limits to the thicknesses which can be so produced) would be strained in the axis of rolling, and might well be pinholed. The electroformed rolled foil is commercially attractive, and the work is proceeding. However, it may never be viable to treat an automobile by simply immersing it "in toto" in a cell, followed by selective dissolution of the different metals. The logic behind this is that the distances from one extreme of the auto to the other are large (17 ft) so that even with extensive use of auxiliary electrodes the ohmic drops across such distances will smear the potential differences required for selective dissolution. Moreover, as any home mechanic knows the various lumps of metal which constitute an automobile are in very variable electric contact with one another, and while some adjacent parts are electrically conducting, there are many others which appear to be electrically isolated. Lastly, we must not overlook the growth of plastic usage in autos. All these facts argue strongly that the modern way to handle automobile scrap is by initial shredding. Reference 1 also discusses attempts made to enable tin cans to be reclaimed in the home, by electrochemical means.

(*iv*) *Chromium*

An additional area of technology, also insufficiently far advanced to consider at this time, is the electrochemical treatment of chromic acid wastes and their reduction to chromic salts. The investigation of all these processes presents an exciting challenge to the electrochemist who wishes to assist in the creation of a cleaner environment, for such wastes can no longer be rejected into rivers as in the halcyon past.

V. ANODIC DESTRUCTION PROCESSES

1. Fundamentals of the Process

While, as we have seen, metal ions may be removed from solution and the metals themselves reclaimed, at the cathode a number of toxic and undesirable species may be removed by application of anodic treatment to the effluent stream. There is, unfortunately, never any positive payoff in such processes, unless it be the savings of dumping fees and fines. However, there can be no anodes without cathodes, and the electrolytic treatment of effluents such as copper cyanide will still yield valuable copper at the cathode, even though only cyanide destruction occurs at the anode. In a consideration of anodic processes, the electrochemist has two basic alternatives in most cases. He can generate oxygen or add sodium chloride and generate free chlorine or hypochlorite by electrochemical means and use these oxidants to destroy or sterilize his noxious species. Alternatively, he can seek to treat and destroy his noxious species by direct electrochemical reaction at the anode.

(*i*) *Indirect Oxidation*

This solution is often preferable because, especially in dealing with dilute effluents, the electrogenerated chlorine can diffuse in three dimensions to react with the effluent instead of having to restrict oneself to a situation where a two-dimensional contact of effluent with planar electrode is called for. The disadvantage of the method, however, is that where water is recycled one has to accept a level of chloride-ion concentration of several percent.

This may or may not interfere with whatever processes the water is required for elsewhere in the circuit.

In the situation where chlorine is formed (or hypochlorite) by addition of sodium chloride (or sea water), the anode reaction is simply

$$2Cl^- = Cl_2 + 2e \tag{1}$$

or where this is done in a cell in which anode and cathode products are not separated, the product of the reaction is sodium hypochlorite, NaClO.

Thus, in such an undivided cell the anode reaction is Eq. (1) and the cathode reaction is

$$2e + 2H_2O = 2OH^- + H_2 \tag{2}$$

which, on mixing, gives

$$2OH^- + Cl_2 = Cl^- + ClO^- + H_2O \tag{3}$$

Both the "hypochlorite" and the free chlorine are sterilants, and also chemically reactive species in their own right. This last point is important for the following reason. Effluent streams often contain the pollutant in low concentration. Thus, if the method of treatment consisted of a reaction of this dilute species at the electrode, the reaction could well be mass-transport limited. (For a discussion of this phenomenon the reader is referred to one of the more readable textbooks on electrochemistry, such as Ref. 6.) If, however, we add to the effluent stream an inexpensive species such as common salt, this will react at the anode, following Eq. (1), and the reactive species so formed will itself diffuse away through the bulk of the solution and react there, without calling for diffusional migration of the dilute effluent species. What is more, the species NaClO, which is reasonably stable, will persist until it "meets" an oxidizable compound. Thus, the treatment will continue working not only in the cell, but also in the succeeding storage tanks, etc.

(ii) Direct Oxidation

The anodic oxidation of a species such as ethane, or many other organic species, has been extensively studied; one of the better surveys is that of Piersma and Gileadi.[7] There is, however, a common misconception which must be corrected here. This is the view that the electrooxidation of organic (and other) compounds is achieved by electrogenerated oxygen. Not only is this view mechanistically wrong—that is the least of the problems—but it is factually disastrous. As Piersma and Gileadi show, organic compounds are anodically oxidized in the potential range 0.4–0.9 V (r.h.e.). At this point, on Pt electrodes, a passivation sets in and increasing the potential brings *decreased* rates of electrooxidation. Further increase of voltage brings evolution of oxygen, and the cell current will increase strongly. But the little evidence that exists indicates that little if any oxidation of organic compounds occurs under these conditions. Thus, Kuhn and Sunderland,[8] in a screening survey, found no evidence of any electrochemical oxidation of a wide range of hydrocarbons, over a range of potentials, pH's, and electrode materials, in these super-oxygen-potential regions. *This explains the lack of success of several proposed processes*, and demands that the field be examined anew by studying the reaction of organic toxic species in the potential regions which are known to be favorable. Notwithstanding the lack of knowledge in this area, a number of processes exist, indeed flourish, for the treatment of effluent streams. These will now be discussed in detail.

2. Cyanide Destruction

These lethal compounds are widely used in the electrofinishing industry, electroplating and electroforming, and also in heat treatment plants for treatment of forgings and other components in the engineering industry. In the former context the effluent stream is usually dilute in cyanide, while in the latter it is far more concentrated. Two electrochemical techniques can cope well with either situation. In the first, which is more suited to strong cyanide solutions, the cyanide is electrooxidized at the anode. The reaction is, probably,

$$2CN^- = 2CN^{\cdot} + 2e \qquad (4)$$

followed by dimerization then alkaline hydrolysis:

$$2CN^{\cdot} = (CN)_2 \qquad (5)$$

$$(CN)_2 + 2NaOH = NaCN + NaCNO + H_2O \qquad (6)$$

The cyanate then rapidly reacts with further alkali in solution to give NH_4HCO_3, Na_2CO_3, and NH_4OH, as described by Naumann[9] or Lister.[10] It is interesting to note that, according to most authors, not only simple cyanides but also complex cyanides suffer decomposition at the anode. Though the latter are far less toxic, it is often overlooked that under certain conditions they may convert back to simple cyanides.

(i) *Practical Cyanide Destruction Plants*

The first such plant was described by Oyler[11] and Sperry and Caldwell[12] simultaneously. The former used copper electrodes in a simple tank and laid down the following conditions:

(1) solution very hot (200°F);
(2) vigorous air agitation;
(3) copper or steel electrodes;
(4) highest possible C.D.;
(5) high ratio of anode to cathode area.

These rules are reproduced here for purely historical reasons—cells today differ considerably. Sperry and Caldwell used strainless-steel anodes and brass cathodes. Connard and Beardsley recommend carbon anodes.[13]

A typical design would be a mild-steel tank, as used in electroplating, fitted with a series of stainless-steel anodes on hanger bars or attached in parallel with braided copper wire (see Fig. 13). Alternating with these are the cathodes, which are also in sheet form. The material for these depends largely on the nature of the actual cyanide. Thus, in many plants copper cyanide is destroyed, with simultaneous recovery of the copper, which is, as it were, a "bonus." In this case, copper cathodes are used, probably sheets $\frac{1}{32}$-in. thick, very much like the "starters" employed in copper refining.[26] Between anodes and cathodes or, better still, at either side of them are positioned the steam heating coils. Provision should be made for agitation. Air agitation is cheapest and best, though it increases the costs of the plant ventilation installation. In the Lancy design the cells are surrounded with polythene curtains which can be seen to have many advantages. In normal industrial practice, the cell is operated at around $4A/cm^2$ (that is, 38A/sq. ft.), and the process is operated batchwise. It will be appreciated that in such a batch process which starts with an aqueous solution of, say, 100,000-ppm CN^- and finishes with 0.4–0.1-ppm CN^-, the highest current which can be *efficiently* used will decrease as the species which is the reactant—that is, the cyanide—also decreases. In a large plant the optimum system would be a cascaded one, in which successively larger tanks are operated at successively lower C.D.'s. In most plants, however, the cell stands at constant C.D. for 7–20 days. The result is that as it is operated the majority of the electrical power goes to produce oxygen and hydrogen from water. Apart from the suggestion made above there are a number of ideas which would ameliorate this. Firstly is the use of a conventional tank-type cell but operated under a current regime in which the power was progressively reduced over the weeks. This could be done manually or electronically, and in either case the current reduction could be effected in accordance with a preset program or as a response to automatic determinations of CN^- concentration. Now that ion-selective CN^- electrodes exist, the latter would not be difficult. But the other possibility lies in newer cell designs. In reduction of CN^- levels to the 0.1-ppm mark, mass transport is everything, and the flowpaths through the

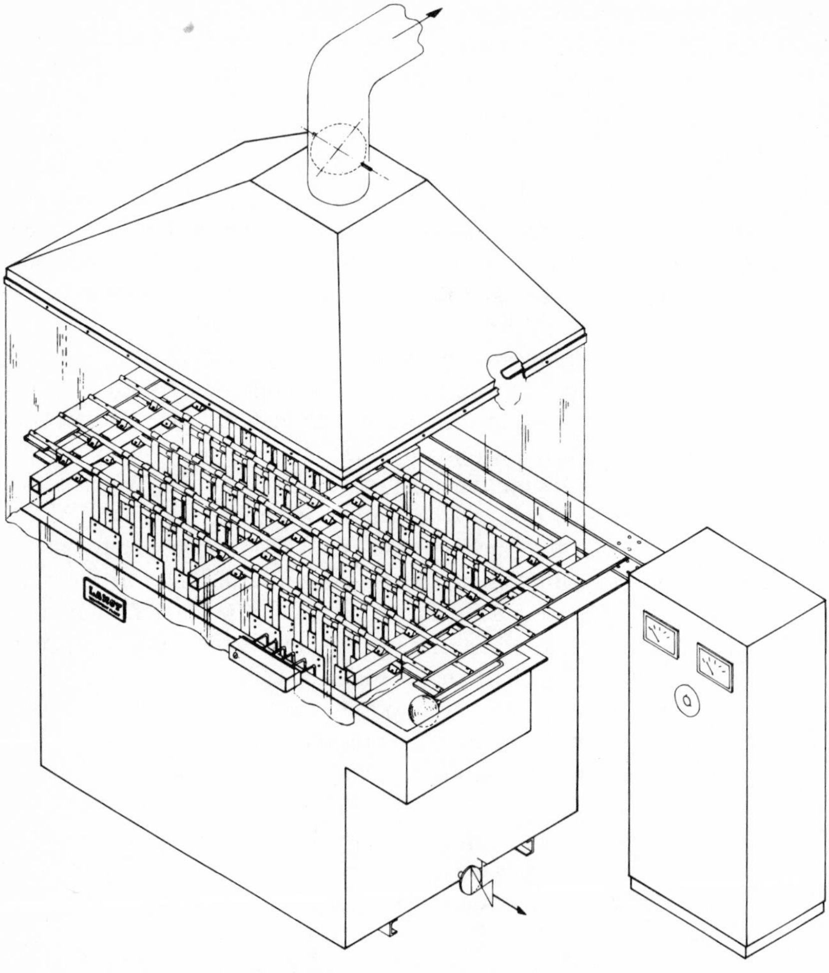

Figure 13. Cyanide destruction cell (courtesy Lancy Laboratories).

cell should be very carefully considered. In one approach the anodes and cathodes were brought much closer together. Not only does this reduce the cell voltage, but, also, when plates come within $\frac{1}{4}$-in. of one another a completely different hydrodynamic picture is set up. In this case there has to be a forced circulation of electrolyte through the cell; otherwise, wide variations of CN^- concentration occur within the cell. Also, problems due to gas locking can give rise to sudden and dramatic increases in cell voltage. It has been demonstrated that in such cells the efficiency of the reaction, under conditions of forced flow, can be very much superior to that in more conventional cells. This must be largely ascribed to the turbulent flow conditions which then are obtained. The second radically different approach to cell design lies in the use of "bed" electrodes, either "fluidized beds" or "static beds." These offer a very high area-to-volume ratio, and have demonstrated their superiority in electrodeposition of copper from dilute solutions. In this particular reaction it should be possible to use graphite beds, which, provided the potential does not rise too anodic, do not suffer undue wear. Platinum-coated glass beads are also readily available, and a number of other metals such as silver would also provide possible substrates for the process.

3. The Chloride–Cyanide Process

As shown before, the addition of chloride ions overcomes a number of problems. It raises the electrical conductance of the solution thus lowering cell voltage, and it provides an active species which can react in the bulk of the solution. This method has found favor in the Soviet bloc countries, especially when dealing with more dilute cyanide solutions. In a system where a feed of cyanide is continuously dosed with NaCl and passed into a cell and out again, there is no need to operate the process batchwise, for, providing the chlorine or hypochlorite is being generated at the same rate at which CN^- is passing into the system (in practice a very considerable excess of NaClO would be adopted), the reaction need not take place in the cell itself but in the receiving tanks which follow it. The technology of hypochlorite cells is well described in the literature (see, for example, Regner[15]), and these cells are very similar. Because the basic principles of this process (as indeed of the preceding one) are so simple, and largely free from patent protection, there is an understandable reluctance on the part of cell manufacturers to divulge actual details. Thus, the "CYNOX" process is described in Ref. 16, but the article leaves many questions unanswered. Three baths are placed close to one another. The first is the actual (cyanide) plating bath, the second is the detoxification bath, and the third and last is a water rinse bath which is not contaminated by operation. The wash in the second, detoxification bath, is continuously pumped round to a cell where the hypochlorite is generated. The level of active chlorine in the detoxification bath is held at 80–200 mg/liter, and the pH is controlled at 10–11:

$$2NaCN + 2NaOCl = 2NaCNO + 2NaCl \tag{7}$$

$$2NaCNO + 3NaOCl + H_2O = 2CO_2 + N_2 + 2NaOH + 3NaCl \tag{8}$$

$$2NaCN + 5NaOCl + H_2O = 2CO_2 + N_2 + 2NaOH + 5NaCl \tag{9}$$

It is seen that the chlorine-containing species are, in theory at least, continuously recycled. In practice, a considerable volume of slime is formed in the cell, and to prevent this the electrolyte is pumped through into a settling tank before being returned to the wash detoxification bath. Provision is made for separate slime pumps to periodically pump out the settling tank. The CYNOX plant can be supplied fully automated or manually controlled. The makers claim that the economics of operation are definitely superior to plants using bottled chlorine gas or calcium hypochlorite. They specify 7 g of active chlorine for destruction of 1 g of CN, and the cell produces 1 kg of active chlorine per 5.5 kWh. From these figures the cost of CN^- destruction is seen to be approximately 40 cents/kg, figuring power at 1 cent·kWh, which is a very conservative estimate. Similar figures are reached by Byrne, Turnley, and Williams,[17] who quote a destruction rate of 0.08–0.10 lb CN^- per kWh. The same authors make an interesting comment on the relative merits of use of free chlorine and *in situ* generated hypochlorite. The former process, as it takes place in a commercial chlorine cell, is some 80 % more efficient than the process as it occurs in a hypochlorite cell, as a result of inefficient side reactions in the latter case. For large applications, the free-chlorine process, using electrochemical generation of chlorine which is then piped to the cyanide destruction tank, may be justified.

The foregoing references represent what is known in the open literature as regards this process. Soviet workers such as Lu're and Genkin[18] have reported experiments in which various anode materials were used, and their results confirm the value of graphite as anode in this application.

(*i*) *Construction of Cells*

The basic principles for construction of chloride–cyanide destruction cells are similar to those outlined above. However, the cell body must now be constructed of rubber-lined mild steel, or one of the more modern plastic materials. Polypropylene, chlorinated P.V.C. such as "Trovidur" or "Trovidur H.T." (registered trademark of Dynamit Nobel, Inc.), or else glass-fiber reinforced polyester-resin tanks. All of these will withstand free chlorine at higher temperatures. Anodes should be of graphite, although the newer platinized titanium anodes or those in which ruthenium oxides are coated onto titanium bring voltage savings and longer cell life through elimination of anode wear of the graphite. The wear occurs through chemical oxidation of the graphite to CO and CO_2, mellitic acids, and mechanical erosion processes. The modern anodes are available from such firms as Marston Excelsior, Ltd., Wolverhampton, U.K. (subsidiary of Imperial Metal Industries, Ltd.). Steam heating coils may be made of mild steel as before, but these should now be cathodically protected by tying them into the main cathodes, though this procedure carries with it the risk of hydrogen embrittlement, and the normal techniques of cathodic protection should be followed here. The rate at which the cell is operated is a matter for economic optimization, and should be arrived at by on-site trials. Basically, the factors affecting this are as follows. For a given effluent flow rate, a theoretical Cl' demand can be arrived at. For the given flow rate containing a known and constant concentration of CN^-, flowing through a cell of a certain size, one may say that the rate of destruction *in the cell itself* is related to the current passed through the cell. However, as a result of overvoltage factors and ohmic losses, etc., the cost per unit of free chlorine generated in the cell increases as the cell is driven harder. The operator is thus confronted with a choice, where on the one hand if he drives his cell hard he pays more in running costs, but on the other hand he saves in capital outlay on storage and reaction tanks following the cell. At the other extremity, we have the efficient cell with low running costs but requirements for a considerable tankage to be associated with it. The optimum depends on many factors including the ingoing and outgoing concentrations of CN^-. The procedure for NaCl dosage is fairly standard, based on a saturated brine solution and fed through a variable stroke pump to maintain a constant Cl^- concentration. If the solution is too close to saturation, mechanical problems will occur as a result of crusting of solid salt. Mild-steel pumps should be avoided, though they can be used if it is necessary.

4. Electrochemical Sterilization of Domestic Wastes

For many years the idea of *sterilization* of domestic sewage by electrolysis has been examined and assessed. Two entirely separate approaches to the problem were adopted. In the first, the wastes were electrolyzed as they arrived in the plant, with only lime addition to control pH. In the second concept the electrolysis of NaCl generated hypochlorite which then sterilized the sewage. In its earlier versions, this idea was applied by mixing sewage and sodium chloride, then electrolyzing the mixture. It was later appreciated that an electrolysis of the NaCl alone, with subsequent discharge of the chlorine-rich liquid into a mixing tank where it reacted with the sewage, avoided problems associated with fouling of electrodes.

The first of these two methods is now of purely historical interest. The second idea, based on the use of NaCl, is clearly limited in its applicability to locations where inexpensive brine supplies are available, that is, in coastal areas or on inland brine fields. The economics of sewage treatment in coastal areas are frequently different from those pertaining to inland sites. Coastal land (especially in resort areas) is more valuable, and the amenity value of the area must be conserved by elimination of odors and effluents. Though many resorts still discharge raw effluent through outfall pipes into the sea, legislation will increasingly make this thoroughly deplorable practice more difficult.

Figure 14. Artists illustration for location of an EST (electrolytic sewage treatment) plant (courtesy Constructors John Brown, Ltd.).

Mantell[19] summarizes the situation with regard to early plants based on simple oxygen evolution as a sterilant. He describes the Landreth plants at Santa Monica, California, and Oklahoma City, Oklahoma. These were fitted with over 1000 steel electrodes between which paddles rotated, partly to agitate the solution and partly to remove the scum formed. As far as is known after the closure of these plants between the wars no further installation of any size was constructed, based on this principle, though several patents have since appeared which are based on this concept.[20]

Plants based on *in situ* generation of sodium hypochlorite have been more successful, although even where brine is available it has been found that bottled chlorine gas can compete. Credit for early work in this field must go to Mendia and Foyn, whose cell is described in U.S. patent 3,035,992. The work of Constructors John Brown (CJB), Ltd., in developing this is well known, and after extensive pilot-plant tests at Cosham, Portsmouth, a full-sized plant was installed on the island of Guernsey. The plant has recently been expanded, and this fact, together with several articles written by members of the local government engineering department, testify to the success of the method—in this particular location and set of circumstances at least.

Figure 14 shows how such a plant can be accomodated in the very heart of a densely populated resort without being noticeable, while Fig. 15 shows the overall flowsheet of the process. Doubts were expressed early on, about both the effectiveness of the process and its viability. Figure 16 shows the survival times of coliform bacilli as a function of contact time with the electrolyzed sea water, and the efficaciousness of the method is amply demonstrated. Figure 17 shows the cost of an installation, and readers will notice that the graph shows a feature so common to electrochemical processes, that data obtained in the laboratory or on small scale is linearly related to larger plants, without the complexities of scaling factors. Other descriptions of the plant are found in Ref. 21. An unexpected benefit in operation resulted from the

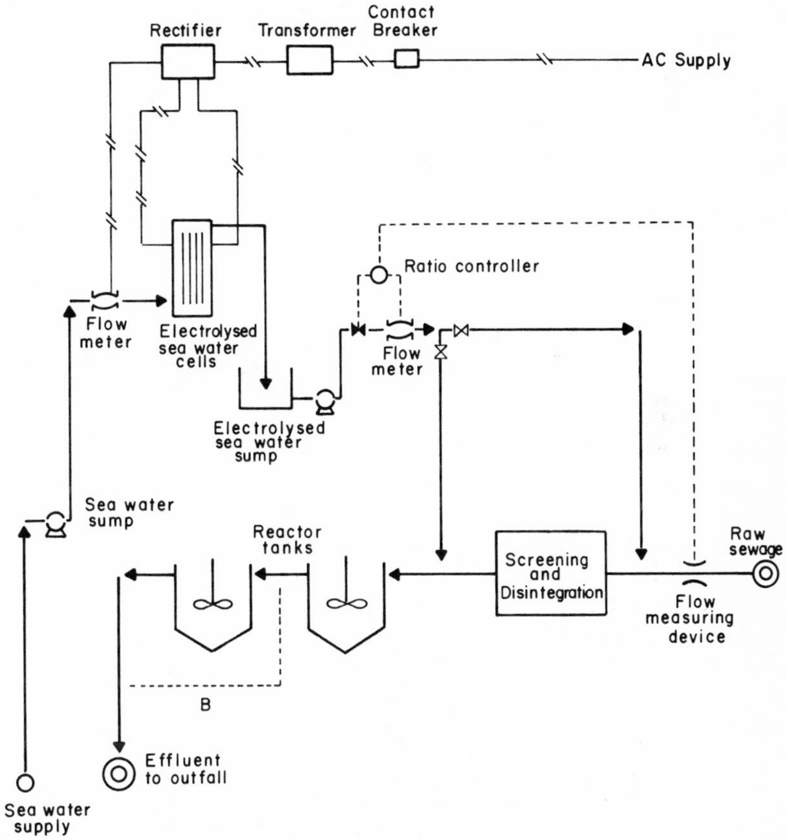

Figure 15. Flowsheet for the EST process (courtesy Constructors John Brown, Ltd.).

formation of magnesium hydroxides during electrolysis, which form useful flocculating agents. An unexpected problem was encountered—not with the Guernsey plant but with an installation made by another firm on the North Sea coast. Under the conditions obtaining there during the winter months servere Pt loss was found at the anodes when the temperature fell below 5°C. This has been explained in terms of the use of constant-current devices impressing ever higher overvoltages when performance, i.e., polarization, changed.

The cell design is straightforward and is diaphragmless with platinized titanium electrodes. A polarity-reversal program is sometimes used to reduce fouling of the electrodes by alkaline earth salts. Where reversal is not used, the Patterson Candy cell design employing bipolar electrodes is quite satisfactory. The evolution of hydrogen at the counterelectrode causes no apparent problems, in spite of a certain amount of hydrogen embrittlement of the titanium.

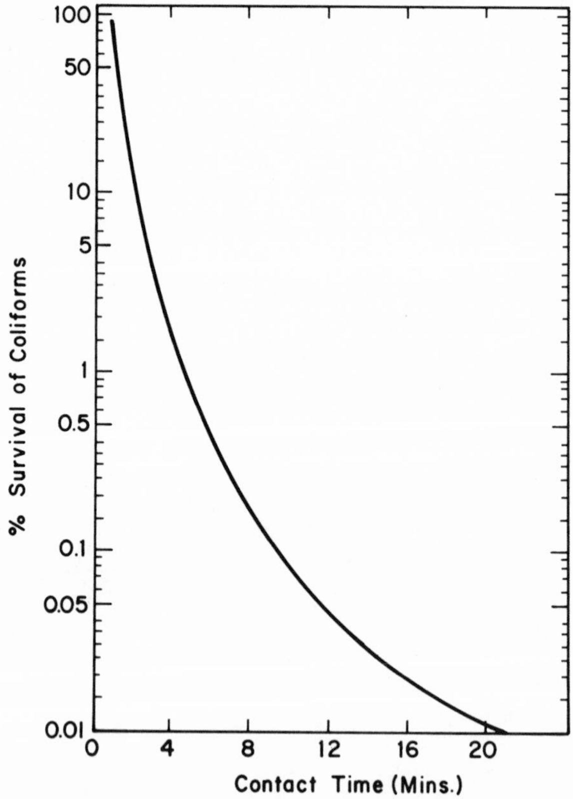

Figure 16. Survival time of coliform in electrolyzed sea water (courtesy Constructors John Brown, Ltd.).

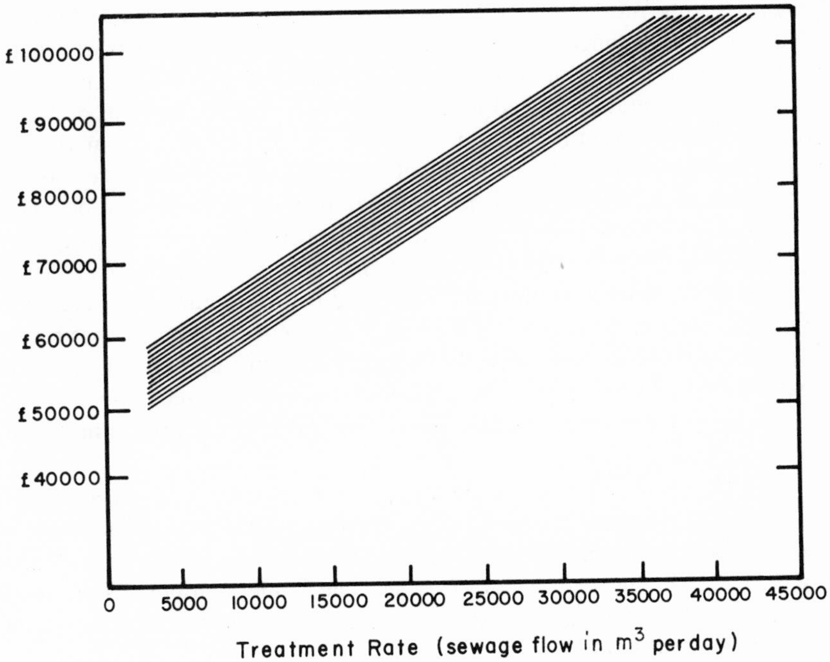

Figure 17. Process costs for EST plant (courtesy Constructors John Brown, Ltd.).

By platinizing both surfaces of a bipolar electrode, current reversal may be used. Other hypochlorite cells use mild-steel anodes with monopolar construction, and one foreseeable development is the use of mild-steel–titanium laminates in a bipolar design. Such composites are now technically and economically possible.

5. The Electrochemical Destruction of Other Organic Compounds

In addition to the destruction of cyanides, anodic oxidation offers a solution to problems of removal of many other species. While the reader will have to refer to the review of Piersma and Gileadi[6] or to such other works as Fichter[22] for the overall possibilities of the method, we shall discuss here some applications which have actually been studied. Surfleet[23] at the Electricity Council, Capenhurst, Cheshire, studied the destruction of acetate wastes from the perfumery and similar industries. His technique was based on the use of the Kolbe reaction. As he conceived it, the reaction

$$2RCOO^- = R{-}R + 2CO_2 + 2e \tag{10}$$

took place, converting the dissolved acetates to gaseous products. The hydrocarbon was flared off at a stack, though in a larger plant there would have been some more useful application of it, perhaps as an electrical energy producer. At high concentrations of acetate, the current efficiencies were high, approximately 80%. But as the process proceeded, the steady-state concentration of acetate dropped and oxygen evolution became the preferred process, with resultant poor efficiencies. One might consider the method as a first stage in a multistage reduction of c.o.d. process, but the figures quoted by Surfleet show that, as described by him, the process would not be viable in its own right. The inefficiencies at low concentrations are once again a manifestation of mass-transport problems. Whether a better cell design could improve matters is doubtful. Fluidized bed electrodes do not seem to operate so well in the regions of high anodic potential, where the resistance of the oxide film on the individual spheres becomes appreciable, nor are they tolerant of a situation where gaseous reaction products are evolved, for the latter force apart the spheres of the bed and increase ohmic drop across the bed.

The anodic destruction of phenols has been studied, both by Surfleet[24] and by Soviet workers, whom he cites. The main line of attack has been the use of free-chlorine species, as with cyanide destruction. These form chlorophenols and can also cause ring cleavage. The former compounds are more objectionable, tastewise, than their unchlorinated analogs, and no actual use of this process is known.

Other reports describe the destruction of cresols, cyanates, and thiocyanates with a technology almost identical to the cyanide treatment with chloride ions. A technology based on two- or three-compartment cells using porous or ion-selective membranes has been built up by the Gas Industry in the U.K. and is described in various patents. Again, further details of this are found in Ref. 1.

VI. CONCLUSIONS

This chapter has endeavored to present a picture of existing processes which serve to preserve a cleaner environment. It would appear that two major developments outside the control of the pure scientist will certainly occur. The first is the development of better engineered cells in which electrochemical reactions may be accomplished. We have considered the fluidized bed cell and the force-flow cell in the context of copper recovery. But their application to other reactions is equally valid. One interesting development recently disclosed, but not fully evaluated, is the porous bed cell of RCI (Resources Control, Inc.) now working in collaboration with Stauffer Chemicals. This is illustrated in Fig. 18 with a conventional cell in contrast. The relationship between the RCI cell and the fluidized bed cell is an interesting one. Because the latter uses a bed of highly conductive material, the ohmic

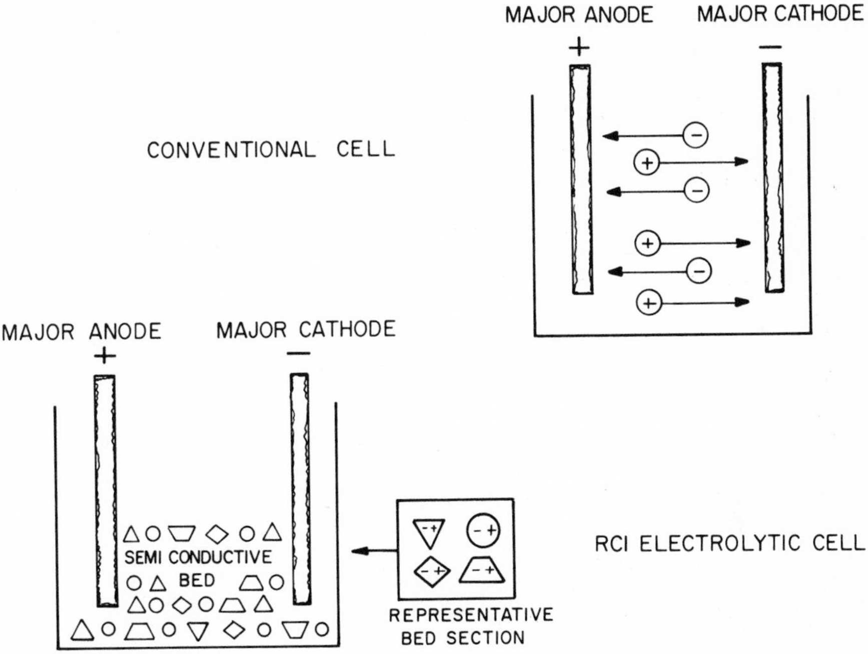

Figure 18. Schematic representation of RCI cell (courtesy Resources Control, Inc.).

drop across the bed is held to a minimum, and although the metal–solution potential of the particles in various regions of the bed (closer to and further away from the feeder electrode) is far from equipotential, the aim is to try to achieve this as nearly as possible. In the RCI bed, by contrast, the particles themselves are poorly conducting (described as "semiconducting") and the metal–solution potential varies, not just from one particle to the next, but quite probably at various points on a given particle. In this case it is possible to envisage both anodic and cathodic reactions taking place on the same particle, and the bed can be regarded as a set of cells in electrical series connection. The effectiveness of this arrangement is not known, though the cost figures quoted by the manufacturers are such as to suggest that this method could eclipse all other forms of cyanide destructive treatment. To some extent this is a cell in which an inefficient process at low voltage has been replaced by an efficient one at high voltage. It is as if a series—a cascade—of cells operating at very low current densities were connected in series. Further results from this cell must be awaited with the greatest of interest.

Beyond the control of the electrochemist lie such factors as the diminishing stock of fossil fuels known to man, and such questions are dealt with elsewhere in this volume. But it should be remembered that the exercise of costing a process, or two competitive processes, produces a result which is only valid at a given point in time and in a given situation. The indications are that electrochemical methods of treating aqueous effluents will become increasingly competitive in the years ahead.

REFERENCES

[1] A. T. Kuhn, *Modern Aspects of Electrochemistry, No. 7*, Eds., J. O'M. Bockris and B. E. Conway, Plenum Press, New York, in press.
[2] G. S. Solt, in *Industrial Electrochemical Processes*, Elsevier, Amsterdam, London, New York, 1971.
[3] P. M. Wright, *ibid.*
[4] C. Jackson, *ibid.*
[5] C. Jackson, *ibid.*
[6] J. O'M. Bockris and A. K. N. Reddy, *Modern Electrochemistry*, Plenum Press, New York, 1970.
[7] B. J. Piersma and E. Gileadi, *Modern Aspects of Electrochemistry, No. 4*, Ed., J. O'M. Bockris, Plenum Press, New York, 1966, p. 47.
[8] A. T. Kuhn and G. Sunderland, unpublished observations.
[9] R. Naumann, *Z. Elektrochem.* **16** (1910) 191.
[10] M. W. Lister, *Can. J. Chem.* **33** (1955) 426.
[11] R. W. Oyler, *Plating*, April 1949, 343.
[12] L. B. Sperry and M. R. Caldwell, *Plating*, April 1949, 341.
[13] J. M. Connard and G. P. Beardsley, *Metal Finishing*, May 1961, 54.
[14] *New Scientist*, June 26, 1969, 704.
[15] A. Regner, *Industrial Electrochemistry*, Artia Press, Prague, 1959.
[16] W. Kurz and W. Weber, *Galvanotechnik u. Oberflachenschutz*, **3** (1962) 92.
[17] J. T. Byrne, W. S. Turnley, and A. K. Williams, *J. Electrochem. Soc.* **105** (1958) 607.
[18] Yu. Yu. Lu're and V. E. Genkin, *Zh. Priklad. Khim.* **33** (1960) 384.
[19] C. L. Mantell, *Electroorganic Chemical Processing*, Noyes Development Corp., Parkridge, N.J., 1968.
[20] French Pat. 1,321,895 and addition to same 82,434.
[21] *Chem. Engineer* **73** (1966) 98; B. G. Frampton, *J. Inst. Municipal Engineers* **86** (1969) 92; also, *Surveyor and Municipal Engineer*, July 23, 1966.
[22] F. Fichter, *Organische Elektrochemie*, Steinkopf Verlag, 1942, reprinted University of Salford Bookshop, Salford, Lancs., 1970.
[23] B. Surfleet, Electricity Council (U.K.), Report ECRC R/165.
[24] B. Surfleet, Electricity Council (U.K.), Report ECRC R/204.
[25] British Pats. 888,654, and 901, 204.
[26] J. P. A. Wortley, *Proc. Ann. Mtg. A.I.M.E.*, March 2, 1971.

Chapter 5

THE ELECTROFILTRATION OF PARTICULATES FROM GASES

Edmund C. Potter

Division of Mineral Chemistry
Commonwealth Scientific and Industrial Research Organization
Sydney, Australia

I. INTRODUCTION

It is part of our more profound recognition of the world about us that matter and energy cannot be satisfyingly described without reference to electric charges and their movements. Such a statement is justified by the dominant roles of electricity and of electronics in our everyday lives, adding up to a wealth of man-made experiences that take precedence over natural manifestations of electrical phenomena. In this class stands electrostatic attraction, which is exploited on the large scale for the efficient removal of particulate nonconductors from suspension in gases. The *electrostatic precipitation* of dusts, mists, smoke, and fume is an established process for cleaning gases permitted to enter the atmosphere, and *electrostatic precipitators* (or *electrofilters*) play an important part in stopping excessive air pollution from combustion equipment and by various industrial processes including cement manufacture and ore smelting. This chapter describes the principles of electrostatic precipitation and examines the manner of their application in practice.

II. THE BROAD PRINCIPLE OF ELECTROSTATIC PRECIPITATION

The gas to be cleaned traverses a space where the suspended solid or liquid particles become electrically charged in an applied field of several kilovolts per centimeter. The charged particles then migrate towards a vertical earthed electrode where they are discharged, forming either mobile beads and films of liquid or deposits of powder. When the deposits have strength enough to maintain coherence if detached, they may be dislodged from the earthed electrode into collecting vessels for convenient bulk disposal.

Given an adequate high-voltage source it is not difficult to construct an effective bench-scale electrostatic precipitator. The simplest geometry is cylindrical, and a suitable arrangement is shown in Fig. 1. The metal tube *A* (10–25 cm diam) contains a wire *B* stretched down its center and well insulated from it. The gas to be cleaned enters at port *P* and passes up the tube beneath which is a collecting vessel *C*. If the cleaned gas is to be discharged immediately to the atmosphere, the top of the precipitator can be

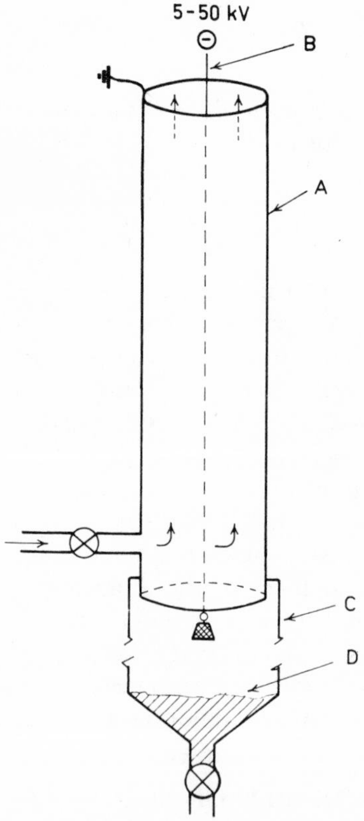

Figure 1. Simple arrangement for a cylindrical electrostatic precipitator.

open; otherwise, there is an exit port and the top is closed. The tube is earthed, and the wire is brought to a high potential. By about 5 kV a corona forms around the wire and negative and positive entities form in its immediate vicinity. With the wire negative* the positive ions cannot stray far down the voltage gradient; but the electrons stream away from the corona region, ionizing some gas molecules and charging all the particles suspended in the gas. The higher the applied field strength between wire and tube, the more rapid is the charging of the suspended particles and the faster is their outward drift to the earthed tube wall. The limit of field strength is reached at around 5 kV/cm when the insulation of the gas breaks down and a spark bridges the gap between wire and tube causing the field to collapse in the neighborhood. At field strengths up to this limit, however, the migrating particles eventually lose their charge to the tube wall and cling there until their own combined weight compels a slip or fall down the tube to accumulate as dust or fluid *D* in the collecting vessel.

In the case of solid particles there is some subtlety in this dislodgement step of the process. If the deposits loosen prematurely they will be too slender and weak to resist fragmentation, and some particles will inevitably be reentrained. If, however, the deposits are reluctant to detach they eventually obstruct the tube, and, more importantly, the electrical discharge of particles must take place at and through a layer of powder. Should the electrical conductivity of this powder be insufficient the discharge process is retarded, and the deposition of new particles is hindered and eventually stopped. In addition, an appreciable part of the potential gradient between wire and tube appears uselessly across the powder deposit, sufficient at the extreme to cause sparking across interstitial gas spaces. To avoid these various events, which can ruin the precipitation process, it is usual to induce dislodgement of accumulated dust by judiciously rapping or otherwise vibrating the collecting and discharge electrodes or, where possible, by irrigating the precipitator with water sprays.

The process of electrostatic precipitation of particles from a flowing gas stream is progressive along the length of the collecting surface. The first areas of surface to encounter the dust-laden gas collect particles at a greater rate than subsequent areas, and in an efficient precipitator the last areas will have few particles to collect. Since the precipitator handles discrete numbers of particles, it is formally possible to collect every one of them even though the supporting gas is flowing. In practice, the flow is turbulent, and full particle collection is never achieved. It thus becomes technologically important to examine the principal factors controlling precipitator behavior so that efficient and economic large-scale gas-cleaning plants can be designed.

*A positive potential can also be used, but the corona is less stable and sparkover is more easily encountered.

III. THE DEUTSCH EQUATION

Figure 2 represents a section of an earthed cylindrical duct through which dusty gas is passing. Dust is progressively deposited on the inside of the cylinder under the influence of the corona discharge wire on the longitudinal axis. Consider the two parallel planes an infinitesimal distance dl apart. In unit time the number dN of particles collected between the planes at distances l and $l + dl$ from the duct entrance is given by

$$dN = -\pi r^2 v \frac{dn_l}{dl} dl \tag{1}$$

where r is the radius of the cylinder, v is the overall forward velocity of the dusty gas, and n_l is the number of particles per unit volume of gas passing the plane at l. All dN of these collected particles were situated within the small distance w of the inside surface of the cylinder and were contained in the corresponding cylindrical shell of gas that passed plane l in unit time, where w is the mean radial component of velocity of the particles towards the cylinder wall (the particle migration velocity). Assuming that the particles are thoroughly mixed across any plane such as l, and taking the probability of particles sticking at the cylinder wall to be unity, it follows that

$$dN = 2\pi r n_l dl w \tag{2}$$

From (1) and (2)

$$\frac{dn_l}{n_l} = -\frac{2w}{rv} dl \tag{3}$$

Integrating (3),

$$ln \frac{n_l}{n_0} = -\frac{2wl}{rv} \tag{4}$$

where l is the length of precipitator duct under consideration, and n_0 is the numerical particle concentration at the inlet of the duct. Noting that the fractional slip Σ of particles through the duct is n_l/n_0, and that the fractional efficiency ε of particle collection is $(n_0 - n_l)/n_0$, we have

$$\ln \Sigma = \ln(1 - \varepsilon) = -2wl/rv \tag{5a}$$

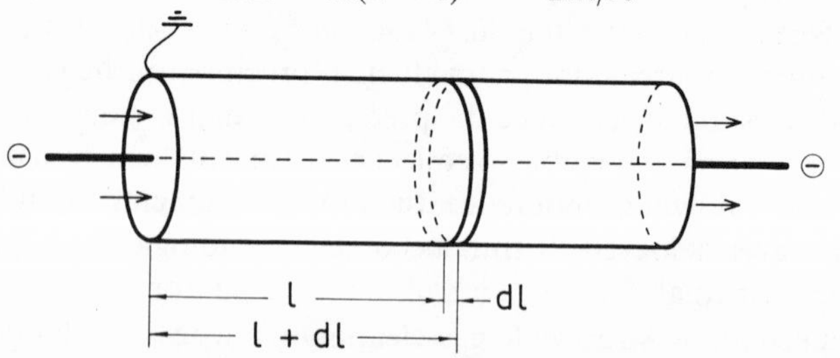

Figure 2. Model for charged particles suspended in a carrier gas and precipitating in an earthed cylindrical duct.

Equation (5a) was deduced by Deutsch[1] in 1922, and indicates that the efficiency of the turbulent-flow precipitator is logarithmically related to the duct length and to the reciprocals of duct radius and gas velocity. Thus, if a particular duct gives a 90% efficiency of dust collection, then 99% efficiency should be attained by doubling the duct length or by halving the gas velocity in the original duct, and this result should be independent of the inlet dust concentration in the gas. The same should hold for a rectangular, i.e., plate-type, duct (with the corona wire stretched centrally distance d from parallel collector plates), for which the appropriate form of Eq. (5a) is

$$\ln \Sigma = \ln (1 - \varepsilon) = -wl/dv \qquad (5b)$$

Both Eqs. (5a) and (5b) are easily shown to unify to the common equation

$$\ln \Sigma = \ln (1 - \varepsilon) = -Aw/V \qquad (6)$$

where A is the collecting area of the precipitator, and V is the volumetric rate of gas flow through it.

Equation (6) is the principal relation used by designers of electrostatic precipitators. Since the equation excludes 100% dust removal with the gas flowing, some compromise must be used to obtain a practical outcome. To this end the efficiency ε is usually fixed to satisfy statutory or acceptable pollution limits, and the attainable migration velocity w is selected on the basis of accumulated experience. The parameter A/V is then calculated and virtually decides the size of the precipitator. The electrostatic part of Eq. (6) is contained in the parameter w, which it is instructive to examine before considering the agreement between observed and expected precipitator behavior.

IV. THE PARTICLE MIGRATION VELOCITY

In a working precipitator, as in Fig. 1, the applied voltage provides the field that both charges the particles and moves them towards the collecting surface. However, it is possible (though uncommon) to separate the charging from the migration, relying upon the particles to retain their charge while they precipitate in a duct without applied field. The charging process is described later, but is reasonably complete within 0.1 sec of particles entering the charging field, and there is little error in assuming all suspended particles are charged to saturation in a typical case where the supporting gas takes several seconds to traverse the duct with applied field.

The saturation charge q_s acquired by a spherical particle is determined by its size and by the magnitude of the surrounding field, which is distorted by the presence of the particle, depending on its dielectric constant. Thus, following White,[2]

$$q_s = a^2 E_c\{1 + [2(D - 1)/(D + 2)]\} \qquad (7)$$

where a is the radius of the particle, D is its dielectric constant, and E_c is the charging field. For well-conducting particles or for droplets of water the factor in the large bracket in (7) is close to 3, its upper limiting value; whereas for insulators such as vitreous materials the factor is expected to lie between $1\frac{1}{2}$ and $2\frac{1}{2}$ (corresponding to D values from 2 to 10). The force F_1 on a particle carrying the saturation charge in a precipitating field E_p is

$$F_1 = q_s E_p \tag{8}$$

Since both the time taken by dust particles to accelerate to their migration velocity and the influence of gravity are negligible, the force F_1 is balanced by the viscous drag F_2 on the particles in the gas (viscosity η). For spherical particles above about 1 μm in diameter Stokes' law may be usefully applied, that is,

$$F_2 = 6\pi a \eta w \tag{9}$$

where $\pi \simeq \frac{22}{7}$, and it is assumed that the turbulent gas motion carries the particles with it except for the component w of velocity towards the collecting surface. Substituting (7) into (8) and equating to (9), it follows that

$$w = \frac{aE_c E_p}{6\pi\eta}\{1 + [2(D - 1)/(D + 2)]\} \tag{10}$$

Equations (6) and (10) may be combined to give a complete expression for the efficiency of an electrostatic precipitator, but it is not possible to use this expression for accurate absolute estimates of efficiency for practical systems because these depart from the ideal model on which the equations are based. These departures from ideality arise from unavoidable nonuniformities of dust distribution, gas flow, electric field, and particle geometry. Various improved equations take some of these factors into account,[3–7] but their practical benefit is difficult to gauge in view of the imprecision of control and measurement encountered with a full-scale precipitator plant. However, it is convenient to have a simplified expression, and for this purpose we make the approximation $E_c = E_p = E$ and put $D = 4.3$ (a reasonable value for many insulating particles of ceramic or cementiferous character). It then follows from Eqs. (6) and (10) that

$$\log \Sigma = \log(1 - \varepsilon) \simeq -aE^2A/4\eta V \tag{11}$$

Equation (11) shows the importance to efficient precipitation of elevating the values of the parameters a, E, and A/V, although usually only the last two are controllable in practice. The most important parameter is clearly the applied voltage, which through E^2 is operative as its square and for this reason large-scale electrostatic precipitators are worked at the highest steady voltage short of continuous electrical breakdown of the gas. Additionally, increased precipitation efficiency can be attained by raising A/V, but if the collecting area A has already been fixed, improvement in this way is only achieved by lowering the gas velocity, i.e., sacrificing throughput of gas. The effect of particle size is significant in that larger particles are easier to precipitate, so that using Eqs. (10) or (11), a reasonable efficiency of 90% for spherical particles 20 μm across becomes only about 20% for the same substance as particles 2 μm across. An adjustment of the operating conditions so that the efficiency for the 20-μm particles is raised (for example) to 99% causes an improvement for 2-μm particles from 20 to 63% efficiency—a considerably better performance.

V. AGREEMENT BETWEEN THEORY AND OBSERVATION

As will emerge later there are various mechanical and electrical reasons to expect deviations in practice from the ideal relations deduced above, so that full conformity with the Deutsch equation (6) or with its elaboration using Eq. (10) is the unlikely event. Indeed, many performance data from particular precipitators do not conform completely to the Deutsch equation, but in spite of this it is the common technological practice to use the equation for the calculation of an *effective* value of w in the interests of designing equipment more reliably to specification. By a further simplification this effective migration velocity (usually 3–15 cm/sec) is sometimes offered alone as a sufficient measure of precipitator effectiveness, even though Eq. (10) shows that migration velocity is strongly dependent on field strength and consequently on applied voltage.

1. The Effect of Applied Voltage

A complication in comparing theory with observation is that the theory applies only to the *electrostatic* precipitation phenomena, whereas practical observations always include a substantial contribution due to *mechanical* settling of dust in recesses and passages where the field does not penetrate. In practice, this mechanical efficiency is a valuable bonus amounting to as much as 80% of the dust burden entering a precipitator, but it is not always evaluated separately and reported efficiencies are hardly ever corrected for it. For our present purpose the penalty for failing to make this correction is numerically small at high efficiency (the usual technical situation). For example, if the observed efficiency is 99.5% and the corresponding mechanical efficiency is 50%, then the corrected (i.e., electrostatic) efficiency is (99.5–50)/(100–50), which is 99.0%. The difference is large, however, at observed efficiencies not greatly in excess of the mechanical efficiency, this being a consequence of the semilogarithmic form of the efficiency equation, as seen in the following argument.

Let the electrostatic component ε_e of the observed efficiency be expressed by an equation of the form (11), so that $\log(1 - \varepsilon_e) = mE^2$, where m is constant under steady conditions. If the mechanical component of the observed efficiency is denoted by ε_m, then we may put $\log(1 - \varepsilon) = \log[(1 - \varepsilon_e)(1 - \varepsilon_m)] = mE^2 + \log(1 - \varepsilon_m)$. It follows that the effect of the mechanical component of efficiency on a plot of $\log(1 - \varepsilon)$ against E^2 is to shift the line upwards by $\log(1 - \varepsilon_m)$, leaving the slope unchanged. The mechanical efficiency may be directly found from observations on an electrically dead precipitator, but from the above reasoning it may also be estimated by extrapolation to zero voltage.

Figure 3 shows the relation between $\log(1 - \varepsilon)$ and the square of the applied voltage for a particular plate-type precipitator collecting pulverized-coal ash at two different gas velocities. The straight line expected from Eq. (11) is reasonably well supported, and the lower velocity is associated with

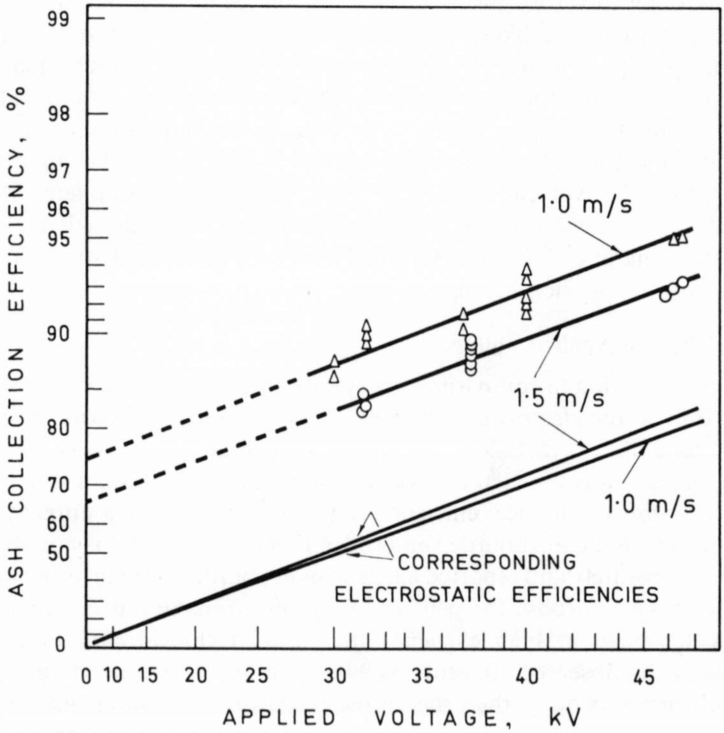

Figure 3. Relation between particle collection efficiency [scale of $\log(1 - \varepsilon)$] and applied voltage (scale of voltage squared) at carrier gas velocities of 1.0 and 1.5 m/sec, including data corrected for mechanical collection of dust. (Calculated from observations made on a plate-type research precipitator by staff of Electricité de France.)

the higher efficiencies of precipitation, also as expected. The intercepts of the straight lines on the vertical axis give the mechanical efficiencies, approximately 60% for the higher gas velocity and about 75% for the lower gas velocity. Clearly, the higher velocity keeps more dust particles aloft in the supporting gas (air at 20°C on this occasion). On shifting the two lines downwards so that they commence at the origin, the two lower lines are obtained for which the scale of the vertical axis now denotes the electrostatic component of efficiency directly. These two corrected lines are nearly coincident for the two velocities, a finding not in accord with Eq. (11), which indicates that the slope of the line for the higher velocity should be $\frac{2}{3}$ that of the line for the lower velocity. The probable reason for the disagreement emerges below.

2. The Effect of Particle Size

A second complication in comparing observation with theory as expressed by (11) arises from the fact that dust particles, even if they are the same shape, are not normally the same size unless special techniques for production or sorting are used. Usually, therefore, a particle-size distribution typical of industrial powders is encountered, and the electrostatic precipitator is required to operate with particles that may cover a size range of 50–1, say 0.5–25 μm. In such a mixture of particles each size has its own migration velocity in the electric field and the resultant precipitation efficiency is a summation over all sizes and size distributions of particles. Equation (11) shows that the smaller particles have greater difficulty in collecting, so that a precipitator working with a flowing mixture of spherical particles is selective in the sense that the first collecting areas receive a deposit enriched in the large particles and the final areas collect a deposit enriched in fines, if the precipitator is long enough. Referring back to Fig. 3, it can now be appreciated that since the larger particles preferentially settled by mechanical fallout, the precipitator was electrostatically handling a greater proportion of fines at the lower gas velocity, with the result that the expected difference between the efficiencies for the two gas velocities was obscured. The expected effect may be found enhanced, however, if the gas path through a precipitator is so contorted that the *higher* gas velocities encourage fallout of larger particles by a centrifugal action when the gas changes direction.

Figure 4 shows the change of particle-size distribution occurring within a particular cylindrical electrostatic precipitator. The tinted area upper left shows that the residual dust passing out of the precipitator contained much more material below a 6-μm diameter than the inlet dust. The tinted area lower right shows that particles above 12 μm in diameter were nearly all trapped. The precipitator in question was constructed in three identical stages in series, and particle-size distributions at the various stages were measured by a gravimetric centrifugal technique using samples of dust extracted isokinetically at appropriate points. In addition, the dust-collection efficiencies were measured for each stage in the usual gravimetric way, and corrected for mechanical entrainment. In order to deduce the electrostatic performance for any particle size, the entire particle-size range was split into a number of narrow bands (usually 1 μm wide), and the measured gravimetric distributions were reallocated graphically to these bands. The results, in conjunction with the corresponding stage efficiencies, were used to calculate the electrostatic efficiencies for each narrow band of particle size in each of the three stages. On plotting these efficiencies logarithmically against the central value of each size band, a test could be made of Eq. (11).

The result is shown in Fig. 5, in which the three lines describe the respective performance of one-third, two-thirds, and the whole of the precipitator. In each case the expected linear relation between $\log(1 - \varepsilon)$ and a is found,

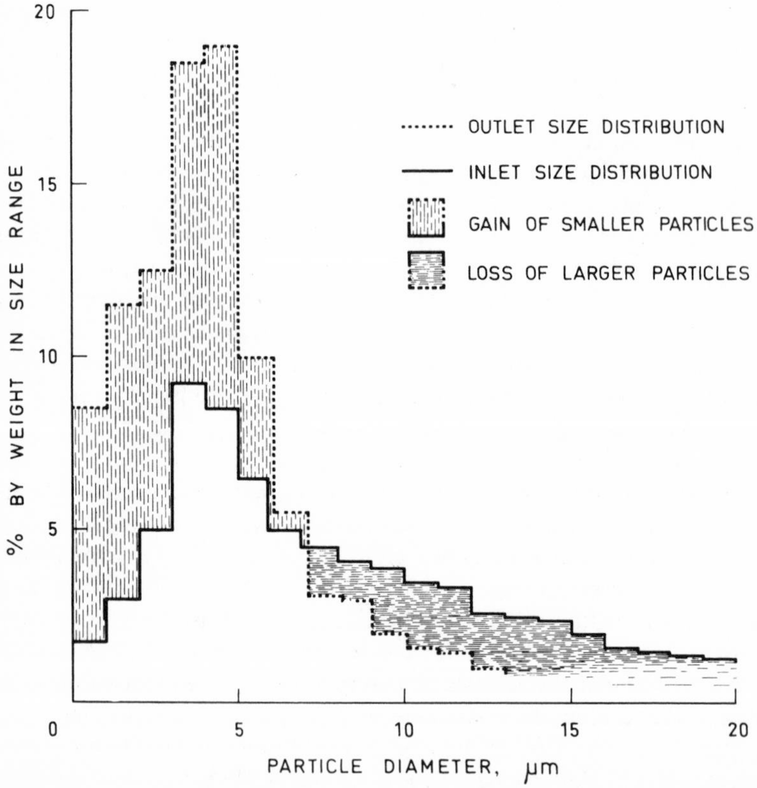

Figure 4. Change of particle-size distribution within a cylindrical electrostatic precipitator.

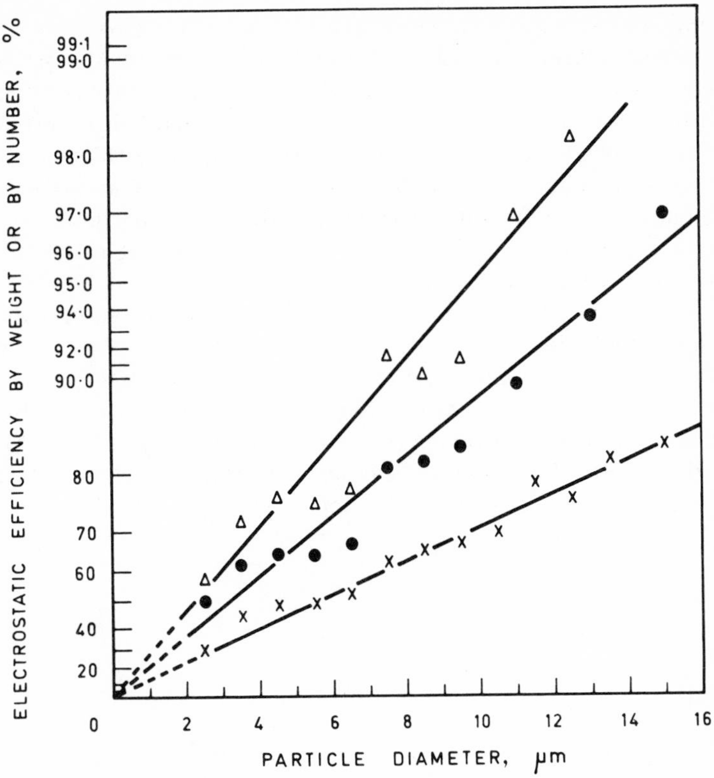

Figure 5. Efficiency of electrostatic precipitation for spherical particles of different sizes collected in admixture in a cylindrical precipitator consisting of three identical stages in series:
X—X, first stage, efficiency by weight over all particles 61 %;
●—●, first and second stages, efficiency by weight over all particles 74 %;
Δ—Δ, all three stages, efficiency by weight over all particles 82 %.

and the precipitation efficiency tends to zero as the particle size vanishes. If Eq. (11) is followed the slopes of the three lines should be in the ratio 1:2:3. The actual ratios obtained from Fig. 5 are 1:1.8:2.5, showing that the precipitator exhibited a small progressive loss of expected performance throughout its length.* This behavior is shown directly for the same precipitator in Fig. 6, where the electrostatic efficiency as measured (i.e., over all sizes of particle) is plotted logarithmically against the length of the precipitator for a selection of applied voltages and gas velocities. The slight curvature of the lines shows that the performance near the inlet is not sustained, though the subsequent loss is not great as is seen on considering the

*For a precipitator of uniform geometry, the length is directly proportional to the area of the collecting surface.

uppermost line in Fig. 6. This line was obtained at the maximum applied voltage short of flashover, and the full length of the precipitator yielded an efficiency of 99.65% as opposed to the 99.9% that would have been reached if the better performance of the inlet areas had been maintained throughout the precipitator. On the basis of Fig. 6 alone it might be thought that the loss of performance could be wholly explained by the fact that the later areas of precipitator were left with smaller (i.e., more difficult) particles to collect. Figure 5 shows that even for particles of uniform size there remained a loss of performance through the precipitator.

It is a tedious and delicate procedure to obtain information of the type given in Fig. 5. The wider the band (i.e., size range) selected for the calculations, the less detail can be seen of the variation of efficiency with particle size; whereas too narrow a band divides the numerical data into imprecise fragments. If the data pertain to high collection efficiency (say $>95\%$ averaging all particle sizes), the numerical closeness of associated numerators and denominators in the calculations demands exceptional accuracy if absurdi-

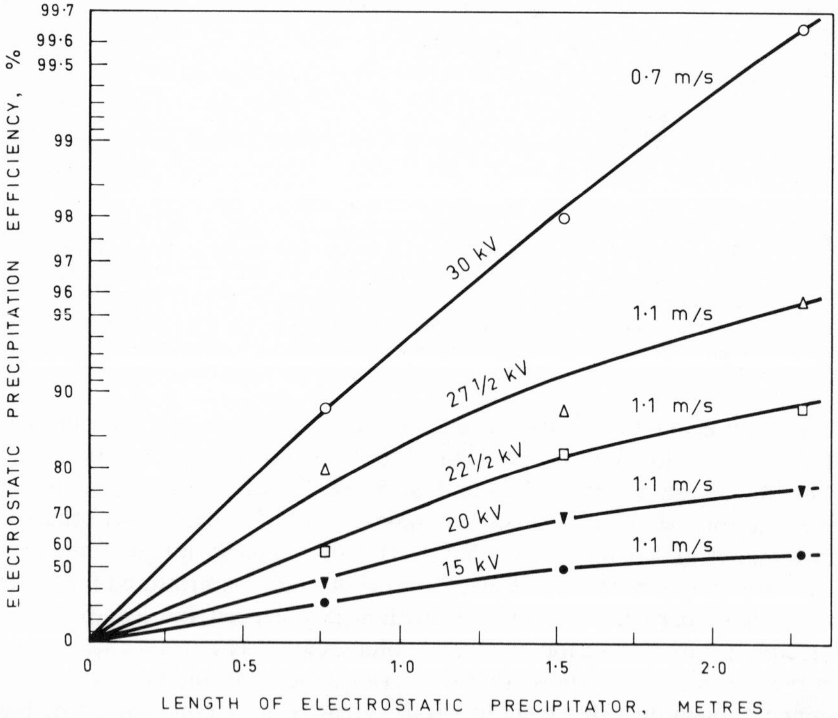

Figure 6. Efficiency of electrostatic precipitation as a function of cylinder length at various gas velocities and applied voltages. Cylinder diameter 15 cm, temperature 120°C. For uppermost line, inlet dust burden 4.1 grains/ft^3, and A/V in Eq. (6) 440 ft^2/1000 ft^3·min.

ties such as $\varepsilon > 1$ are not to intrude. Even so the expanding nature of the logarithmic scale can produce exaggerated scattering of results above 99% efficiency, and the size region below about 2 μm is particularly difficult to cover.

It should also be recognized that Eqs. (5a), (5b), (6), and (11) were deduced for *numerical* efficiency, whereas it is the *weight* efficiency that is measured in practice. However, the two types of efficiency are identical for particles of uniform size and the same density, and this is also theoretically true in the presence of particles of other sizes. Thus, in Fig. 5 the ordinate corresponds to efficiencies either by weight or by number since the separate particle sizes can be selected on the abscissa and all particles of the same size are known to have the same density. When, however, the efficiency for all particles at once is considered, the summing process gives a different result whether weights or numbers are used. The caption to Fig. 5 gives the overall weight efficiencies for each of the three lines, as directly deduced from the original gravimetric measurements. Advantageously, these figures are little affected by the presence of the smallest particles, which make a minor contribution to the weight of a sample of the mixed particles. On the other hand, the smallest particles are the most prolific and dominate the overall numerical efficiency appropriate to each of the three lines. Usually the method used for particle sizing cannot resolve the distribution of the smallest sizes and the numerical efficiencies are thus inaccessible, being probably about half the corresponding weight efficiencies. In practice, no difficulties arise from this inability to discern numerical efficiency, but gravimetric data from particle mixtures have limited use in theoretical studies.

The trend of the lines in Fig. 5 towards the origin is unmistakeable, but no reliable measurements exist in the dotted sections close to the origin. The expectation is that submicron particles are but poorly collected in electrostatic precipitators, and it is of practical interest to examine this proposition. Presumably, the moving particles inside a precipitator, being charged alike, tend to travel in parallel paths even though their trajectories twist and whirl with the turbulence of the supporting gas. However, the differing drift velocities in the electric field must continually cause larger particles to catch up smaller ones [see Eq. (10)]. Nevertheless, actual surface-to-surface collisions between charged particles are likely to be so rare that aggregation of particles, including those with moist or sticky surfaces, is inappreciable. On the other hand, uncharged particles that have not yet entered the electric field can collide closely and form attachments secured by van der Waals or surface-tension forces. Despite continual buffeting a proportion of these attachments can be expected to survive long enough for the component particles to be charged as a single unit and as such to remain stable for the duration of their presence suspended in the precipitator.

The probable effect of prior particle aggregation on collection efficiency may be deduced using the data on which Fig. 5 is based. In this case if every particle had effectively doubled its diameter by attachment to others, then the efficiency of 82% observed over all three stages of the precipitator would have been raised to 91%. Alternatively, the same improvement would have been obtained by eliminating all particles less than 4 μm in diameter, either by forming at least 15-μm agglomerates or by attachment to particles of that size or larger. Since the centrifugal method of determining the particle-size distribution of collected dust is almost certain to break up particle aggregates cohering by van der Waals or surface-tension forces, the finding of a relation between particle diameter and $\log(1 - \varepsilon)$ that deviates from a course towards the origin and even exhibits a minimum[8,9] is evidence of particle agglomeration. The electron microscope may also be used to detect agglomeration, as in the case of iron oxide fume, which forms long threads out of its submicron particles[10] and is therefore readily collected at high efficiency (ca. 99.9%) in electrostatic precipitators.[11]

VI. ELECTRICAL ASPECTS OF PRECIPITATION

The test described above for conformity of the electrical part of Eq. (11) with observation assumed that the field strength E was equal to the mean gradient of applied voltage and was identifiable with both the charging and the precipitating fields. In this writer's experience the linearity of $\log(1 - \varepsilon)$ plotted against the square of applied voltage (see Fig. 3) represents common precipitator behavior, discerned as it is in practice against a scatter of results that obscures any more exact relationship between these variables. It must be admitted, however, that the field strength is not uniform in the space between corona wire and collector, and does not vary linearly across the gap. It is, therefore, improbable that a spatially nonuniform field would influence alike the totally different processes of particle charging and particle migration at all levels of other relevant variables such as particle size and turbulent gas velocity. The theoretical difficulties of an adequate analysis of these problems are formidable, and the doubtful practical rewards of success seem to create little incentive for the attempt. It is, nevertheless, useful to obtain a semiquantitative appreciation of the electrical aspects of precipitation.

1. The Particle Charging Process

The negative corona in electronegative gases* produces a multitude of electrons and negative gaseous ions streaming outwards towards an earthed electrode or plate. The electrons create more gaseous ions as they collide with gas molecules during their journey towards the plate, and those

*For example, air or oxygen, but not pure nitrogen or pure hydrogen.

that survive are easily deflected by suspended particles that have already acquired some negative charge. It is considered, therefore, that free electrons play a negligible role in particle charging and that the particles acquire their surface charges solely by collisions with the far more massive gaseous ions (e.g., O_2^-, SO_2^-), it being left open whether only the charge or the mass as well is captured by the particles. The observed current consists of differing contributions due to the arrival of electrons, gaseous ions, and charged particles at the earthed collecting electrode.

Particles destined for electrostatic precipitation acquire their full charge in proportion to their surface areas, the commonest range being approximately 10^2–10^5 electronic charges per particle.[2] These saturation charges are approached quickly but at a diminishing rate, since particles already some way towards saturation respond only to that dwindling proportion of collisions vigorous enough to overcome the field already generated. Having become charged, the larger particles (ca. $>1\ \mu m$) sweep a path down the field gradient towards the earthed plate, remaining undeflected by ionic and molecular bombardment at all times. The smaller particles, however, being slight enough to undergo Brownian motion through molecular collision, pursue an erratic course during ionic bombardment, and the influence of the applied field is less conspicuous.

The task of analyzing the above dynamic situation is mathematically complex, but there are two simplified treatments. The first, usually accorded the description "ion bombardment," was set out by Pauthenier and Moreau-Hanot[12] for spherical particles and treats the problem electrostatically with the particle motion dominated by the field. The result may be expressed as

$$q_t/q_s = t/(t + \tau) \tag{12}$$

in which q_t is the charge acquired after time t, q_s is the saturation charge given by Eq. (7), and τ is a time constant equal to $1/\pi\kappa\rho$, where κ is the ionic mobility and ρ is the ionic space-charge density. This treatment takes no account of random thermal motions experienced by the charging particles and thus cannot be expected to be accurate for small particles (say $<1\ \mu m$). On the other hand, the second treatment, originally attempted by Deutsch[1] and usually termed "ion-diffusion charging," assumes that the ionic bombardment is wholly calculable from the kinetic theory of gases, with the ionic concentrations around the spherical particles being diminished as their accumulating charges increasingly repel approaching ions. Since the calculation neglects the migration of the particles in the applied field it must be inaccurate, especially for the larger particles. The result as deduced by White[13] has the form

$$q_t = k_1 \log(1 + k_2 t) \tag{13}$$

where k_1 and k_2 are constants at fixed values of temperature, gas pressure, particle size, and ionic space-charge density.

The agreement of Eqs. (12) and (13) with observation has been broadly established by various workers,[13–15] and the fact that the equations respectively suit the larger and the submicron particles seems to have been taken as a confirmation of two distinct mechanisms of particle charging, namely, ion bombardment and ion-diffusion charging. However, the distinction is hard to justify, particularly since both supposedly separate mechanisms have ion bombardment in common as the single means for charging particles.

For the present purpose it suffices to consider the common size range for particles (1–50 μm), for which Eq. (12) is the appropriate theoretical charging equation. This shows that the proportion of full charge attained in a given time by a particle is independent of its size, although the magnitude of the saturation charge varies greatly, being directly proportional to the square of the particle radius [see Eq. (7)]. In order to calculate $\tau = 1/\pi\kappa\rho$, we recall that the ionic mobility depends on the nature of the gas ions, and a value for air at atmospheric pressure of 2 cm/sec per V/cm (600 esu) is taken here. The value of ρ is taken as 6.28×10^{-2} esu, using the data for a cylindrical precipitator, radius 12.5 cm, obtained in Section VI.2 below. Hence, $\tau = 0.0084$ sec, and from Eq. (12) the time taken for the particles to attain 95% of their saturation charge is $19\tau = 0.16$ sec, which is short compared with the time particles take to traverse the precipitator (usually several seconds). It is, therefore, customary to assume that particles are fully charged when they are collected in precipitators and that no significant lack of collection efficiency can be ascribed to inadequate particle charging.

2. Distribution of Field

In the immediate vicinity of the discharge electrode a high field strength marks the corona region. Up to the applied voltage where the corona starts the current is virtually zero, and the field strength (as calculated from electrostatics for wire-in-cylinder geometry) is inversely proportional to the distance from the wire [see Eq. (14) with the first term neglected]. Above the corona starting voltage the ions between the concentric electrodes create a space charge to alter the field distribution. Allowance for this effect leads to the form of equation developed by Townsend[16] and by Pauthenier and Moreau-Hanot[17]:

$$E_r = [2I/\kappa + (E_0 r_0/r)^2]^{0.5} \tag{14}$$

where I is the current per unit length of discharge wire, E_0 is the corona starting field, r_0 is the radius of the corona region, and E_r is the field at distance r from the center of the cylinder. The second term may be neglected at sufficient values of r, when (14) simplifies to

$$E_r = (2I/\kappa)^{0.5} \tag{15}$$

Equation (15), which has experimental backing,[18] indicates that the field is uniform beyond a sufficient distance from the central wire, especially at the more elevated corona currents. This supports the assumption (see the beginning of Section VI) made to test Eq. (11).

To illustrate the use of Eq. (15) the cylindrical precipitator is assumed to be operating with air at room temperature and pressure, for which κ is taken as 600 esu. The field at the cylinder wall is taken to be 3 kV/cm, so that $E_r = 10$ esu. Hence, $I = 3 \times 10^4$ esu/cm of discharge wire, or 10 μA/cm. If the radius of the cylinder is 12.5 cm, the current density i at the wall is $I/(2\pi \times 12.5) = 3.77 \times 10^2$ esu or 0.126 μA/cm^2. However, $i = \rho\kappa E_r$, whence ρ is 6.28×10^{-2} esu/cm^3. Since the electronic charge is 4.8×10^{-10} esu, the number concentration of singly charged ions at the cylinder wall is $6.28 \times 10^{-2}/4.8 \times 10^{-10}$, that is 1.3×10^8 ions/cm^3. This ionic concentration lies in the typical range (10^7–10^9 ions/cm^3) for an effective electrostatic precipitator and may be compared with an extreme numerical concentration for particles of 10^6/cm^3 at the precipitator inlet, assuming a normal particulate burden of 10 μg/cm^3 to consist entirely of 2-μm-diam spheres of sp. gr. 2.4 (approximating to aluminosilicate coal ash). The ions therefore greatly outnumber the particles even at the precipitator inlet. Each 2-μm particle, however, carries about 500 electronic charges when fully charged [from Eq. (7)], so that at the precipitator inlet the space charge due to the particles is 5×10^8 electronic charges/cm^3—approximately the same as for that due to the gas ions. However, the currents corresponding to movements of particles and of gas ions are heavily in favor of the latter, because these are migrating at a much faster rate. For example, in the present case the ions are moving at $\kappa E = 6000$ cm/sec, whereas the migration velocity for 2-μm-diam spheres is calculated from Eq. (10) to be only 1.74 cm/sec, assuming $\eta = 1.85 \times 10^{-4}$ P, $D = 4$, and $E_c = E_p = 10$ esu/cm as above. This does not mean that the particles always have a negligible influence on electrical conditions in the precipitator, as is seen below.

3. Electrical Effects of Suspended Particles

Strictly speaking, Eq. (14) applies to dust-free conditions, and needs further elaboration to take into account the space charge due to the particles. The precipitator operates to remove these particles, so that, at a given distance from the cylinder wall, the alteration of the field strength by the particulate space charge is most at the inlet to the precipitator and quickly decreases from that point as the particle concentration undergoes approximately

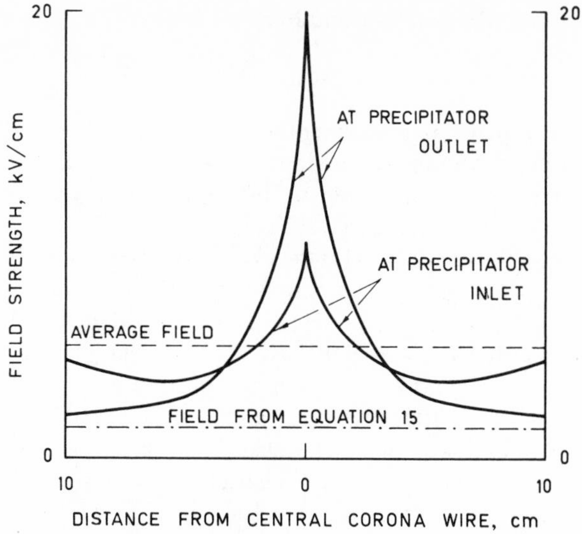

Figure 7. Illustrative examples of deduced field strength in a cylindrical electrostatic precipitator. (Numerical axes for guidance only.)

exponential reduction with length. A complicated expression for the variation of field strength with particle-area concentration and with distance from the cylinder wall can be derived,[19] and representative calculations have been made by Lowe and Lucas.[20] Their results are presented in illustrative form in Fig. 7 and include two horizontal lines showing, respectively, the average field (applied voltage/cylinder radius) and the field estimated from the simplified equation (15). Figure 7 illustrates the steep fall in field strength close to the corona wire whether dust is present or not, and shows that Eq. (15) underestimates the field and that averaging the field overestimates it except near the corona wire. It is seen from the two curves that over the outer two-thirds of the cylinder the field does not vary widely throughout the length of the precipitator. The presence of the dust clearly has a suppressive effect on the field in the corona region. As a result suspended dust decreases the corona current and increases the corona starting voltage. A condition may indeed be reached where the dust so limits ion concentration as to interfere with its own charging. This would show as a poorer performance in the early parts of the precipitator, that is, the opposite of the trend shown in Fig. 6.

In spite of the largely successful efforts by theoretical physicists to quantify field and current relations in corona fields, the position seems to be that limited application can be made of this work to the complicated dynamics of the electrostatic precipitator, so that refinements of the basic equations (6) and (10) are difficult to exploit usefully in practice. This view seems all the more justified when the electrical and other effects of the *collected* dust are considered.

4. Electrical Effects of Collected Dust

Using Lowe's typical data[21] for coal-ash production and precipitator size in a large modern power station, and assuming a mean bulk density for the ash of 0.9 g/cm^3, it is readily calculated that the *average* rate of thickening of ash deposits throughout such a working precipitator is close to 7 μm/min with a mean inlet dust burden of 15 g/m^3 (6.5 grains/ft^3).

Since in theory the collection rate of ash falls exponentially with distance from the precipitator inlet, the thickening rate of deposits in a high-efficiency precipitator is considerably above the average near the inlet and well below the average at the outlet. An estimate of the thickening rate at a given spot in the precipitator can be made from the type of information shown in Fig. 6, using the known rate of supply of dust to the precipitator. For this calculation the precipitator is notionally divided into a number of narrow sections, and the dust collected in each section (as estimated from the interpolated cumulative efficiencies at beginning and end of the section) provides the thickening rate of the dust deposit at the midpoint of that section. In this way Fig. 8 has been constructed from the uppermost line in Fig. 6, and it is seen that the dust deposit thickens at the inlet at 9 μm/min and that this rate has diminished to the overall average (1.2 μm/min)* at a distance from the inlet that is only $\frac{1}{5}$ of the precipitator length. If the precipitator is highly efficient as in the present case (99.65% by weight), then the rate of deposit thickening at the outlet must be near the vanishing point, for example, <0.01 μm/min as here. The mechanical implications of these findings are discussed later, but there may also be electrical consequences.

Dust sufficient to obscure the collecting surface of a precipitator is a physical barrier that can become an electrical one under certain circumstances. The current across the gas space of a precipitator has been seen above to be almost wholly ionic, and originates independently of the presence of dust on the collecting electrode. It is expected, however, that the ions will have such a high probability of striking dust particles piled many layers deep on the collecting electrode that the conduction across the deposit will proceed by surface paths over touching particles, supplemented by any conduction through the substance of the particles themselves. The current density over the collecting area is typically 10^{-8} A/cm^2, and to maintain this the applied voltage gradient redistributes appropriately across any collected dust layer. The voltage gradient across the dust may attain a value of 5 kV/cm,

*Since the average thickening rate is directly proportional to the inlet dust burden, this figure is compatible with the full-scale value given above (7 μm/min).

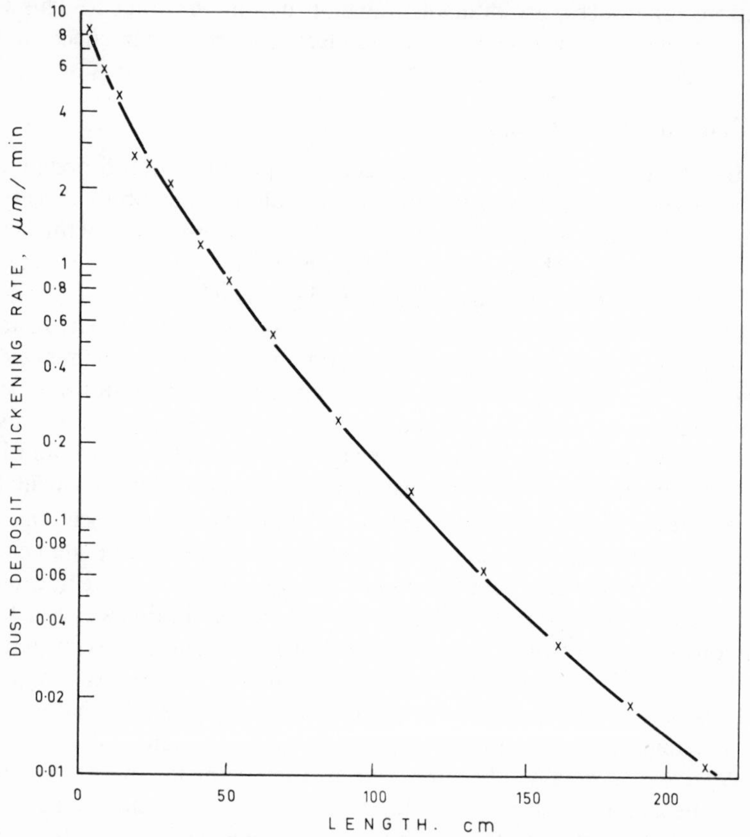

Figure 8. Deduced thickening rate of a dust deposit as a function of length of a cylindrical electrostatic precipitator. Conditions as for uppermost line in Fig. 6.

for example, before electrical breakdown of its interstitial gas brings a different conduction regime into play. If, therefore, the dust as collected has a resistivity as high as $5 \times 10^3/10^{-8} = 5 \times 10^{11}\ \Omega\text{·cm}$, the typical current density is only just being maintained. A greater resistivity will begin to cause electrical congestion, so that some ions and charged particles may have to be rejected at the affected area and their charges shed elsewhere in the precipitator where the dust deposit is more tenuous and less obstructive.

There is a widespread belief among precipitation technologists that there is an order of deposit resistivity above which precipitation is seriously impaired. The above argument suggests that this resistivity limit may be near $5 \times 10^{11}\ \Omega\text{·cm}$, but a figure of $10^{10}\ \Omega\text{·cm}$ has a popular acceptance.[22] It is difficult to find concordant practical evidence for adverse effects of excessive resistivity on precipitation behavior, and indeed the idea has been challenged by some authors.[23] Opinions also vary regarding the electrical events caused by highly resistant dust on collector surfaces. Busby and Darby[24] suggest that the voltage gradient is adjusted across such a dust deposit at the expense of the field in the gas space, and the precipitation is hindered through a corresponding fall in the particle migration velocity [see Eqs. (6) and (10)]. They realize, however, that the dust layer is not uniform and suffers abrupt changes in profile when dislodged, so that the average effect on the field in the gas space may be insignificant, especially if the electrical abnormalities are only occasional and local. Another view arises from the consideration that the dust deposit, being built of particles that originally arrived fully charged at its surface, will itself be virtually fully charged if it is at the limit of its ability to conduct the precipitator current. At this stage[25] not only is the cohesive bonding in the deposit being tested by the electrostatic repulsion between its component particles, but approaching particles lacking the requisite kinetic energy or a favorable trajectory are repulsed rather than accepted by the existing deposit. This is the equivalent of a reduction in the sticking probability of particles reaching the deposit surface.

A confirmed effect of excessive deposit resistivity is the phenomenon of "back ionization" or "back corona."[26] Here the field strength has risen to a value that is sufficient to cause electrical breakdown of the interstitial gas in the deposit. The gas thus becomes ionized and is the preferred conduction path, being associated with a second and rival discharge in the precipitator. New negative ions inside the dust deposit negotiate the field to earth as usual, but the corresponding positive ions are ejected and are capable of discharging negative particles and ions taking part in the normal precipitator function. It is a characteristic of corona discharges that they originate preferentially from raised points and asperities where the field is more intense, and this explains the popularity of barbs, serrations, and sharp edges for discharge wires in large-scale precipitators. Hence, in the case of back corona the discharge prefers protruding parts of the deposit originating at fracture edges and collector-plate roughness, and the result is a scattered cratering of the deposit from which streams of positive ions emerge.

The onset of back corona is alleged to be one of the most serious of precipitator disorders, but Lowe and his coworkers[20,23] claim that efficiency need be little affected, attributing such near immunity to the confinement of the harmful positive ions to a limited number of tubular streams. In contrast, Lowe, Dalmon, and Hignett[23] (see also Busby and Darby[24]) attach much importance to dust that collects on the *discharge* wire—an unexpected effect at first sight. However, Hignett[27] has considered the effect of turbulence on the motions of charged particles of different sizes, and finds that particles are increasingly dominated below about 10 μm by the gas turbulence and can be swept into the corona region close to the discharge wire. Some of these finer particles acquire a positive charge and move the short distance to deposit on the negative discharge wire. Eventually a coating of fine particles obstructs the corona in the same way as would a porous insulating coating such as varnish. In agreement with these arguments, the deposition of fine particles on discharge wires is recognized as so certain to occur in full-scale practice that steps are taken to dislodge the deposits before the corona is seriously suppressed.

(i) Conditioning to Counteract High Resistivity of Ash

Probably the largest application of electrostatic precipitation is to remove the fine ash from the flue gases in power stations based on the combustion of powdered coal. This so-called fly ash, which is essentially a glassy aluminosilicate containing a wide variety of minor elements,[28] has usually originated from the molten state at 1200–1600°C as spheres from 0.5 to 50 μm in diameter. The ash contains small quantities of water-soluble mineral salts and reaches the precipitator suspended in a gas, principally nitrogen, containing by volume approximately 14% carbon dioxide, 4% oxygen, 7% water vapor, up to 0.15% nitrogen oxides (mostly nitric oxide), up to 0.3% sulfur dioxide, and up to 0.003% sulfur trioxide. The fly ash is hygroscopic, and is normally precipitated at 105–150°C, i.e., above the dew point of sulfuric acid in the flue gas. As a powder the ash has a measurable electrical resistivity, which depends not only on its composition but also on temperature, particle-size distribution, compaction, and humidity. Reflecting belief in the importance of the matter,[29] determined efforts have been made to measure dust resistivities, and values from 10^6–10^{14} Ω·cm have been recorded. Much ingenuity has gone into the development of equipment[30,31] so that resistivity measurements may be as representative as possible of conditions on the collecting surface itself. These devices have included the use of miniature auxiliary precipitators operating in conjunction with the full-scale plant, but so far no direct observation of the resistivity of fly ash freshly deposited on the full scale has provided the ultimate test of reliability of these measurements. Moreover, any measurements have to be interpreted against the numerical vagueness of a dividing line between harmful and harmless levels of resistivity.

A common type of resistivity variation seen with fly ash (and cement dust) is shown semiquantitatively in Fig. 9. If the measurement begins at *A* in air at room temperature, the ash has already adsorbed (and probably also absorbed) some moisture, and the surfaces of the particles are covered by a microscopic layer of water rendered electrolytically conducting by solution of mineral salts from the ash. When the temperature is raised, the conductivity of this surface electrolyte improves and the ash resistivity accordingly decreases severalfold (along *AB*). Approaching 100°C the electrolyte is

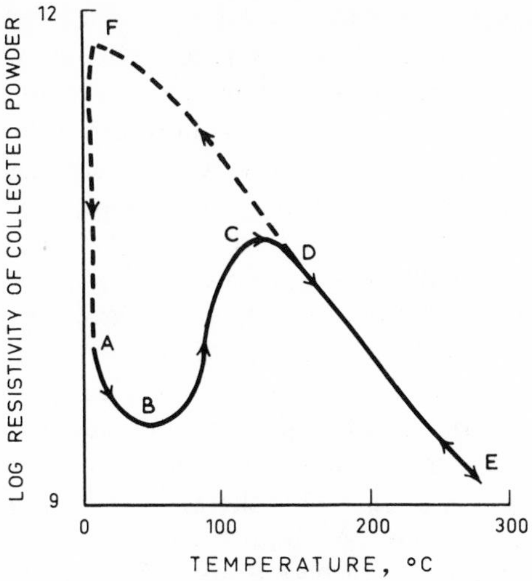

Figure 9. Semiquantitative illustration of variations with temperature of the electrical resistivity of a coal fly ash in bulk in moist atmosphere.

drying out and the dust resistivity begins to increase (along *BC*). At somewhat above 100°C the drying out is complete and the behavior of the dry ash is seen for the first time, being a considerable decrease in resistivity with increasing temperature (along *DE*). On cooling the dry ash, moisture is not immediately picked up and the resistivity curve runs steeply upwards (along *EDF*), returning eventually down to the original starting point (at *A*) as the ash stands in air at room temperature. In the range of common operating temperatures just above 100°C the difference between the two curves in Fig. 9 is most marked, being up to a ratio of 100:1 at a selected temperature. In practice, the fly ash is subjected from the moment of its formation at about 1500°C to cooling in moist flue gas, but it is not clear whether the time of passage of hot ash from flame to precipitator (5-10 sec) is sufficient for an electrolytic surface film to form on the particles in all cases.

In this connection there has been considerable emphasis on the role of the sulfur dioxide content of the flue gas in ensuring adequate surface resistivity for the fly ash. Sulfur dioxide undergoes some catalytic oxidation to sulfur trioxide in flue gas, and thus it is reasonable that a highly conducting film of strong aqueous sulfuric acid can attach to the ash surface even at 200°C. Indeed, the content of sulfate (0.1–1.5% by weight[24]) can be readily leached by cold water from fly ash, although the resulting solution is not particularly acid by the time the ash is recovered. The sulfate originates from the sulfur present in the original coal, but since nearly all of this sulfur is oxidized to sulfur dioxide and as such is expelled (regrettably) with the flue gas, there is at best a poor correlation between the concentrations of sulfate in ash and of sulfur in coal. The desire to have a neat correlation between ash resistivity and precipitation behavior through the influence of aqueous sulfate films has been so strong, however, that a view is prevalent associating poor ash precipitation with relatively low sulfur content of the parent coal. Good comparative evidence for this idea is elusive, and the only apparently sound information[32] is described by its authors as "a very rough guide." In the present author's experience low-sulfur coals (even down to 0.5 %) can produce fly ash that precipitates faultlessly in spite of a resistivity measured at the level of 10^{14} Ω·cm at operating temperature (120°C). Most U.S. coals are high in sulfur (3–5 %), but reports of high precipitation efficiency (>99.0 %) seem uncommon in U.S. practice, and thus the question of correlation between the two variables remains open.

In spite of these doubts, however, there is good evidence that the deliberate addition of trace amounts of sulfur trioxide to ash-laden flue gases can greatly improve precipitation where this is difficult. White[29] seems first to have demonstrated the potential of sulfur trioxide for lowering ash resistivity and easing precipitation, but Busby and Darby[24] brought such ash conditioning to practical application. In two independent full-scale plants these workers found that the continuous addition of 16 ppm by weight of sulfur trioxide to the flue gas raised the precipitation efficiency of ash from an unacceptable 95 % to at least 99.3 %, and a useful improvement to 98 % was secured by the use of only 4 ppm. Other successful experiences with sulfur trioxide conditioning have been reported,[33,34] but there have been failures and other conditioning agents have been sought among substances that could generate electrolytically conducting films on ash particles. Some success has been reported with water itself,[24] ammonia,[35,36] and indirect sources of sulfur trioxide such as sulfuric acid, sulfamic acid, ammonium bisulfate, and ammonium sulfate.[21] More full-scale testing is required before the best measures become clear, so that at present conditioning agents seem to be retained as a last resort, to be used when a wide range of operational adjustments have been tried and rejected. It is usually considered that a conditioner, to be acceptable, must be effective at a concentration in the flue gas of no more than 25 ppm, because fly ash ought not to be excessively contaminated whether it is to be used as a constructional material or dumped. Economic considerations are also important, for sulfur trioxide becomes costly to use (i.e., to throw away) if continuous additions above about 25 ppm by volume are required. At this level of addition a 2000-MW power station would demand sulfur trioxide at approximately 20 tons/day (equivalent to 25 tons/day of 98 % sulfuric acid). Considerable though this demand is, it represents less than 1 % of the equivalent of sulfuric acid emitted by the power station burning coal at a typical U.S. sulfur level of 3 %.

VII. STRUCTURE AND DISLODGMENT OF COLLECTED DUST

Experience shows that dust layers cohere well and cling strongly to vertical collecting surfaces in precipitators. Deposits 1 cm thick can readily build up before the powder fractures under it own weight, causing a wedge-shaped portion to slip and fall through the oncoming dusty gas into the hopper beneath. If Fig. 8 is representative of the thickening rates of deposits, and assuming that an *average* thickness of 1 mm over a 1-cm length into the precipitator is the most slender wedge that can reasonably be expected to cohere under free fall, it follows that the time taken to form this wedge is least at the entrance to the precipitator and is equal to 1000/9 min, that is, about 2 h. In the middle of the precipitator relevant to Fig. 8 the same wedge takes over 2 days to form, a period that lengthens to 1 month in the final 1 cm of the precipitator. These times denote the greatest reasonable frequencies of dust dislodgment over a small selected area, and illustrate the remarkably slow deposition rates in the second half of the precipitator. Indeed, if the dust were to accumulate to a depth of 1 cm without falling from the final collecting area of the precipitator, there would be almost a year of continuous operation between successive falls of dust.

In reality dust profiles and dislodgment must be much more disordered than the above argument suggests, if only for the reason that delicately poised deposits will respond erratically to the random attentions of turbulence, vibration, and adjacent falls of dust. Starting from a dust-free collecting surface, the first natural fall of dust can be expected to take place close to the inlet of the precipitator, and the second fall is likely to occur over the next adjacent collecting area. In the case of a cylindrical precipitator with upward gas flow, this second fall of dust may scavenge some or most of the new dust that has collected over the adjacent lower area since the first fall. However, in the case of the more common plate-type precipitator (with vertical collector plates and horizontal gas flow) the second fall of dust acts like the first in avalanching over lower areas that are the same distance in from the entrance to the precipitator. Depending upon the vigor and places of origin of successive avalanches, an ever-changing irregularity of dust profile develops throughout the precipitator, and the original "logarithmic" outline of the deposit is gradually lost. In addition, since dust deposits are extremely fragile, especially at exposed edges, the shock of fracture combines with the gas turbulence to cause some fragmentation. The particles thereby reentrained are soon recharged and precipitated again, but the thickening rate of deposits is correspondingly raised in the succeeding area of the precipitator. Reentrainment of collected particles is a direct waste of successful precipitation but seems seldom to be the reason for vital loss of collection efficiency, because the later collecting areas of a precipitator have capacity to cope with puffs of dust from gas turbulence.

The slip and fracture of dust sheets clinging to vertical surfaces are an aspect of particle mechanics that should have attracted much more basic examination, because a correct collection step is pointless in long-term operation of an electrostatic precipitator unless it is accompanied by the clear fall of judiciously dislodged dust layers. One way of ensuring this is to irrigate the precipitator with sprays of water, either intermittently, or continuously if circumstances permit. Irrigation keeps the collecting surfaces clear and stops reentrainment of dust but cannot be used at elevated temperatures or where water is incompatible with the supporting gas or with the dust and its disposal. When irrigation is impractical, the problem is to stop dust deposits growing to such thickness that the precipitator field is impaired, either by excessive electrical resistance or by physical obstruction. With nearly all dusts unaided and regular dislodgment cannot be guaranteed, and the practice has necessarily grown up of forcing the dust to dislodge by some form of adjustable vibrating or rapping mechanism. It must be admitted that this procedure is the most inelegant sector of precipitation technology, although it has attracted a large amount of empirical investigation.

An assembly of dry dust particles makes a very weak structure, particularly if they are rounded and do not interlock. The strength resides in the van der Waals attractions at "touching" contacts between the particles, and the closer and more numerous these contacts the stronger is the powder. Hence, the strength is increased by compaction and by adding some smaller particles.* The deposit of fly ash in a power-station precipitator is an approximation to an assembly of spheres of various sizes.[37] There is no opportunity to secure the degree of compression of such a deposit that would usefully strengthen it, so that it is the contribution of the smaller particles in the size distribution that mainly secures the strength of the deposit.

The tensile strength of fly ash at 100–150°C[38] lies in the approximate range 0.1–1 g/cm^2. These values are so small that a pendant column of such powder would break under its own weight at an unsupported length of about 0.1–1 cm, depending on tensile strength.† Clinging in practice to the rusty surfaces of collector plates up to 40 ft high, a narrow column of fly ash finds support through its own neighboring bulk and by its shear strength at or close to the collector-plate surface. Neglecting the support from the neighboring bulk in the case of powder clinging at the leading edge of a plate, it appears as a first approximation that the powder shears at a thickness of $2S/\Delta$, where S is the shear strength of powder at bulk density Δ. Since the shear strength has the same range of values as the tensile strength, and taking

*The use of compression to make tablets and the "stickiness" of fine, dry powders like flour illustrate these points.

†A column of powder slowly pushed out of a vertical tube would appear to drop off the open end.

Δ as 0.9 g/cm^3, the powder shears at a thickness between 0.22 and 2.2 cm. When the powder shears with the plate free of vibration, the opportunity arises for the fly-ash sheet to remain intact as it slides and gathers speed, deriving support from friction against the plate. As the sheet slides off the bottom edge of the plate, the pendant portions may have time before the hopper is reached to fracture into segments 0.1–1 cm long, depending on tensile strength. If, however, the powder fractures because the plate has been vibrated or rapped normal to its vertical plane, a likely separation of the deposit from the plate at this time removes the advantage of frictional support and the dislodged sheet is liable to segmentation (or near disintegration) inside the precipitator proper. On the other hand, vibration or rapping in the plane of the vertical plate will, if not too violent, preserve the shearing mode of dislodgment and give less opportunity for disintegration of the powder and reentrainment of particles. For this reason rapping in the plane of the plates (i.e., shear rapping) is preferred[21] in large-scale practice.

The variation in dust thickening rate shown in Fig. 8 is so great that the forced dislodgment of deposits by rapping cannot be uniformly practiced throughout the precipitator. An individual collector plate might extend for one-third of the length of the precipitator, and Fig. 8 shows that the dust thickening rate may vary by a factor of 10 from the near to the remote end of this plate. As explained earlier the dust profile inevitably becomes irregular over the plate surface, and numerous sections of the dust become ready for dislodgment at different times. Because of this the plate has to be rapped more often than the frequency calculated from Fig. 8 as appropriate to any selected column of deposit. The usual procedure is to seek the optimum rapping rate for each bank of plates by trial when the precipitator is commissioned—a problem that generates an unlimited scope for optimism when efficiency is disappointing and tempts the precipitator manufacturer to reticence regarding his technique for programming rapping mechanisms. Lowe[21] quotes a rapping interval of 10 min for the inlet zone of the full-scale precipitator and 1 h near the outlet, with the discharge wires being rapped lightly every 30 sec. Diverse variants of this rapping program exist, and the impression is readily obtained that precipitators have a fortunate tolerance to rapping imperfections.

It is not only the frequency of rapping that assists successful dust dislodgment but also the force of the blow struck on the collector surface. The impact of a hammer on the top supports of a collector plate cannot be expected to produce a uniform shock over the whole area of the plate, but it can be argued that the hammer blow should be preadjusted so that the impact is effective over only those areas where the deposit is ready to be dislodged. This means optimizing not only the frequency but also the force of the hammer blow—a procedure that is occasionally attempted when previous experience has not given the desired result. Popularly the efficiency of a rapping blow is judged against the peak acceleration (in gravity units) it produces in the plate carrying the deposit to be dislodged. Values from 20–50 g are common in large-scale practice, but much higher values around 1000 g have been measured.

One laboratory study allowing interesting interpretation of the published results has been made by Sproull,[39] who weighed the dust dislodged and retained following a controlled rapping blow against a flat plate collecting particles of various kinds. The observations for a fly ash at 150°C, assuming a bulk density of 0.96 g/cm^3, yield the results given in Fig. 10. It is seen that the dislodgment requires a greater peak acceleration when the precipitating field is kept on during rapping, an effect also reported by Lowe.[21] This is also a common experience with large-scale precipitators undergoing intermittent use, where the rapping is usefully continued for some time after the field has been discontinued. Figure 10 shows, however, that the part of the fly-ash deposit most affected in this way by the field is immediately adjacent

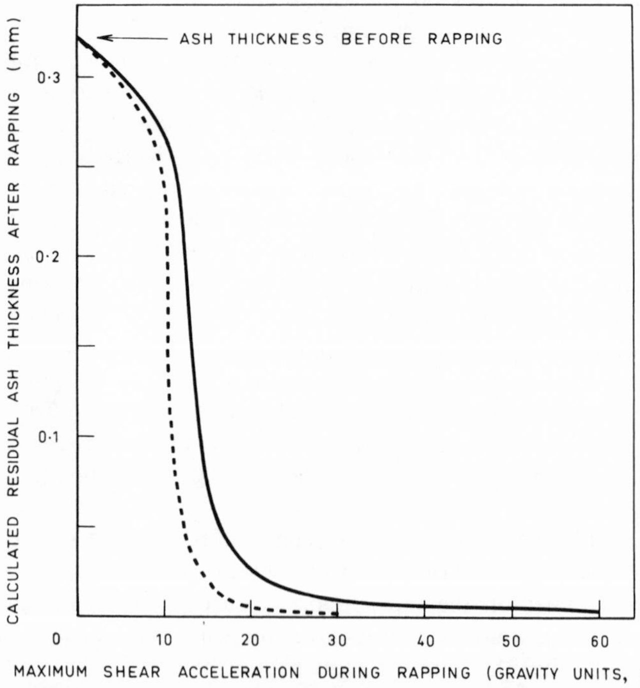

Figure 10. Effect of shear-rapping intensity on dislodgment of powder layers precipitated electrostatically on an earthed plate. (Calculated from Sproull's data.[39])

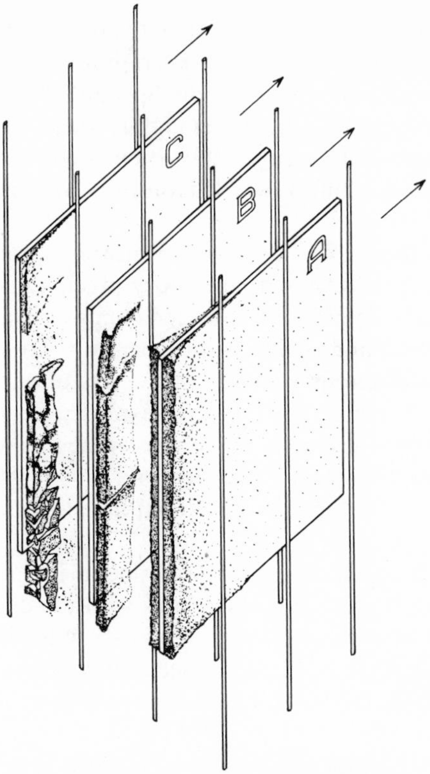

Figure 11. Plate-type electrostatic precipitator showing modes of collection and dislodgment of dust: A, dust ready to dislodge; B, tidy dust dislodgment; C, untidy dust dislodgment.

to the collector plate, and is tenaciously held (>20 g to dislodge) even when the field is off. This part of the deposit represents little more than the first layer of particles. It is also deduced from the top left of Fig. 11 that the last particles to deposit before rapping are relatively easy to dislodge whether the field is on or off, and this suggests that the "outside" of the deposit consists of a "fluff" of particles. Beneath this is the main bulk of the dust deposit corresponding to the steep sections of the two curves in Fig. 10. The steepness suggests that the strength of this consolidated bulk of the deposit is reasonably uniform, although by precise selection of the peak acceleration it is seen to be possible to shear the deposit at any predetermined depth.

It is also evident that little additional dust remains to be dislodged above 20–25 g, so that the rapping blow may usefully be moderated to this peak value, gaining thereby a longer life for the metal components, which are subject to fatigue failure through the continual hammering. On the other hand, no worthwhile dust is dislodged at up to about 12 g with the field on, so that there is a penalty in having too weak a rapping blow. Another feature of the three-layer dust deposit deduced from Sproull's observations is the difficulty of ensuring that a blow strong enough to shear the deposit at a plane close to the collector plate is so quickly attenuated as not to reentrain the exposed particles (the "fluff"), which require a far weaker blow for their dislodgment.

This physical picture of a dust deposit in three layers has independent support.[40] Fly ash collected in a laboratory precipitator is seen under the optical microscope to consist of chains and clumps of particles precariously attached to a consolidated agglomeration beneath. To understand the very strong attachment of the first one or two layers of particles, it suffices to recall that collector plates have a surface roughness (whether bright or rusted) comparable in linear scale to the sizes of the smallest fly-ash particles. These will wedge into notches, scratches, and pits in the plate surface and form a particulate underskin resisting all but the most violent of mechanical shocks.

Representations of modes of dust dislodgment are seen in Fig. 11, which depicts three vertical collector plates with the discharge wires interspaced as in large-scale practice. The arrows show the horizontal direction of the gas through the array. Plate A is seen carrying the dust deposit, thick and ready to fall from the leading areas but becoming sparse further along the passage between each adjacent pair of plates (cf. Fig. 8). Plate B shows the leading wedge of dust slipping tidily down the plate. The deposit has suffered some fracture during dislodgment, but the cohesion of the pieces is good enough to ensure clean fall into the hopper beneath (not shown). Plate C is undergoing an untidy dislodgment of dust. Not only has some of the dust wedge remained on the plate, but the falling part has suffered multiple fracturing causing clouds of dust to reentrain.

VIII. THE PRESENT TECHNICAL OUTLOOK

There can be little argument among user, devotee, or unaligned critic but that the electrostatic precipitator, or electrofilter, is an effective defence against most forms of particulate air pollution. This type of plant is routinely designed and operated for more than 99 % removal efficiency, which, at a gas throughput that may reach 1 ton/sec, is a creditably close approach to perfection for a large-scale industrial process. As may be judged from the principles described above (see also, Figs. 1, 2, and 11), the electrostatic precipitator is uncomplicated in construction and operates with a minimum of mechanical elaboration, being built largely of common materials such as simple steel shapes and concrete. Various elegant modifications to improve

Table 1. Comparative Collection Efficiencies for Fly-Ash from Four Representative Bituminous Coals from Eastern Australia

Origin of coal	Ash yield on combustion, %	Total sulfur content, %	Collection efficiency, %[a]	
			$A/V = 200$[b]	$A/V = 400$[b]
New South Wales	15.6	0.4	96.2	99.7
New South Wales	27.1	0.4	89.0	97.6
Queensland	13.5	2.5	89.1	97.5
Queensland	15.3	0.2	86.0	95.8

[a]Temperature 120°C: mean field 2.4 kV/cm.
[b]Units of A/V (specific collecting area): ft^2 per 1000 ft^3 per min gas throughput.

performance have been worked out, like moveable or ingeniously shaped electrodes and peaking high-voltage waveforms, but the advantages of these devices (even if significant) may be outweighed by the additional costs and by a fear of more breakdowns and increased maintenance. Thus, the traditional designs of precipitator persist and continue to be scaled up successfully to meet the calls for more throughput and higher efficiency as industrial activity grows and antipollution regulations intensify.

In rare cases there are difficulties with the electrostatic precipitation of certain coal ashes, in particular, some originating from southern Scotland and eastern Australia.[24] Table 1 illustrates the strictly comparable situation with four Australian coals and shows that their total sulfur contents have no apparent bearing on the precipitation performance of the ash. In the worst cases the electrostatic function of the precipitator is hardly noticeable and remains so after additive has been used to improve the electrical resistivity of the collected ash. The difficulties are connected with some specific effect of individual coal ashes, since the poor precipitation does not appear until some hours after starting operations with a freshly cleaned precipitator. Present research is aimed at pinpointing the exact causes so that more practical remedies than frequent plant cleaning can be applied.

Speaking generally, no basically new application of electrostatics to effluent gas cleaning seems likely, unless it be associated with a dramatic reduction in the size and cost of precipitators of a given capacity and performance. In the area of large industrial coal-fired furnaces, there are few coals with so little ash and few countries so insensitive to smoke that flue-gas cleaning can be ignored. Thus, these large plants are not complete unless they have gas cleaning, and it is an unconsoling exercise to learn what sum of money need not have been spent if the obligation to remove smoke and ash had not existed. The question of cost, however, has considerable relevance when a choice of the method of gas cleaning has to be made or when plants have to instal gas cleaning after being constructed without it.

1. The Cost of Electrostatic Precipitation

The more complete the gas cleaning required or imposed, the more expensive is the precipitator. As a rough guide it may be assumed that the cost is directly proportional to the size of the precipitator. If the Deutsch equation (6) sufficiently expresses the precipitator performance, then, for a given throughput and type of dust at fixed applied voltage (i.e., at constant V and w), the ratio of costs for fractional efficiencies of ε_1 and ε_2 is given by $\log(1 - \varepsilon_1)/\log(1 - \varepsilon_2)$. Hence, for example, the cost of 99% efficiency in a given situation is estimated to be double that for 90% efficiency. Also, by increasing the cost by one-quarter a mediocre efficiency of 98% can be raised to 99.3%, which is a normal requirement for modern British precipitators in coal-fired power stations. Lowe[21] estimates for British conditions that a precipitator with an efficiency of 99.3% costs $3\frac{1}{4}$% of the price of the power station it serves, corresponding to an equivalent $20,000 U.S. for each million cubic feet per hour of gas throughput. O'Connor and Citarella[41]

Table 2. Comparative Attributes of Various Methods of Industrial Dust Arrestment at a Gas Treatment Level of 4×10^6 ft^3/h[a]

Method	Arrestment efficiency for the same silica dust, %	Comparative volumetric size of plant	Comparative pressure-drop through plant	Comparative power cost to run plant	Comparative total costs (capital + running) per annum or per unit gas volume
Cyclone (medium efficiency)	65	1	4	2	1
Cyclone (high efficiency)	84	2	6	2	2
Scrubber (jet impingement)	88	2	9	4	3
Electrofilter	99	8	1	1	4
Scrubber (annular throat, low energy)	99.5	3	14	6	4
Fabric filter (low velocity)	99.8	7	2	2	4
Scrubber (venturi, high energy)	99.9	3	35	15	8

[a]Adapted from Stairmand's data.[42]

have analyzed costs for U.S. conditions, and their figures suggest that the precipitator cost for 99.3 % efficiency in 1967 was $15,000 U.S. for each million cubic feet per hour of gas throughput. They also calculated that the obligation to clean flue gases before emission to the atmosphere increased the individual domestic power bill no more than 1 %—a modest surcharge for freedom from dust nuisance.

Following a cooperative survey of western European practices for industrial gas cleaning, Stairmand[42] listed the costs of various methods of dust arrestment. As might be expected comparisons become more complicated the more performance detail is introduced. For example, the various methods have different abilities to cope with particles of differing sizes and with elevated temperatures, and the pressure drops incurred with the various methods range widely. A selection of the data assembled by Stairmand is given in Table 2, converted to a comparative form for easy broad assessment of the various methods. In this way it is seen that cyclones are the least bulky and cheapest devices, but have an inferior efficiency. Electrostatic precipitators (electrofilters) are among the largest devices, but present the least obstruction to gas flow. Even so, taking their size into account, they are *relatively* cheaper than cyclones, and are much more efficient. Scrubbers and fabric filters attain the highest efficiencies and can be much less bulky than electrofilters, but they cost at least as much and some types cause a considerable pressure drop reflected in the increased power costs to drive the gas through. The electrofilter has the least power cost, the actual electrical function requiring only about 4 kW for each million cubic feet per hour of gas cleaned.

ACKNOWLEDGMENTS

The writer wishes to thank Dr. R. A. Durie for scrutinizing his manuscript and making valuable comments. He is also grateful for facilities given by the Commonwealth Scientific and Industrial Research Organization, and is indebted to his colleagues P. R. C. Goard, C. A. J. Paulson, and S. G. Szirmai for frequent technical discussions.

REFERENCES

1. W. Deutsch, *Ann. Phys.* **68** (1922) 335.
2. H. J. White, *Industrial Electrostatic Precipitation*, Pergamon Press, London, 1963, p. 135.
3. J. C. Williams and R. Jackson, *Proc. Symp. on Interaction between Fluids and Particles*, June 1962, Instn. Chem. Engrs, London, pp. 282–288.
4. C. Allander and S. Matts, *Staub* **52** (1957) 738.
5. M. Robinson, *Atmospheric Environment* **1** (1967) 193.
6. J. Petroll, *Staub-Reinhalt. Luft* **29** (1969) 139, 364.
7. P. Cooperman, Paper 66-124 to *59th Annual General Meeting of Air Pollution Control Association*, San Francisco, Calif., 1966; *J. Air Pollution Control Assoc.*, **20** (1970) 828.

[8]K. Darby, *Colloquium on Electrostatic Precipitators*, London, February 19, 1965, Inst. Elect. Engrs. Preprints, pp. 21–26.
[9]K. R. Parker, *Indian J. Power and River Valley Dev.*, Symp. on Steam Power Stations, 1965, pp. 58–72.
[10]K. H. Sargent, *Recent Advances with Oxygen in Iron and Steel Making*, Eds., Chater and Harrison, Butterworth, London 1964, Chap. 7; E. Bulba, *Solutions* **1** (1969) 6; H. L. Green and W. R. Lane, *Particulate Clouds: Dusts, Smokes and Mists*, 2nd ed., Spon, London, 1964, p. 149.
[11]K. Darby, private communication, July 1968.
[12]M. Pauthenier and M. Moreau-Hanot, *J. Phys. Radium* **3** (1932) 590.
[13]H. J. White, *Trans. AIEE* **70** (1951) 1186.
[14]G. W. Hewitt, *Trans AIEE* **76** (1957) 300.
[15]V. G. Drozin and V. K. La Mer, *J. Colloid Sci.* **14** (1959) 74.
[16]J. S. Townsend, *Phil. Mag.* **28** (1914) 83.
[17]M. Pauthenier and M. Moreau-Hanot, *J. Phys. Radium* **6** (1935) 257.
[18]J. S. Lagarias, *AIEE* **78** (1959) 427.
[19]H. E. Rose and A. J. Wood, *Introduction to Electrostatic Precipitation in Theory and Practice*, 2nd ed., Constable and Co., London, 1966, p. 61.
[20]H. J. Lowe and D. H. Lucas, *Brit. J. Appl. Phys.* Suppl. No. 2 (1953) S40.
[21]H. J. Lowe, *Phil. Trans. Roy. Soc. London* **265A** (1969) 301.
[22]O. Güpner, *Zement-Kalk-Gips* **57** (1967) 199; see, also, Ref. 19, p. 127 and Ref. 2, pp. 315, 327.
[23]H. J. Lowe, J. Dalmon, and E. T. Hignett, *Colloquium on Electrostatic Precipitators*, London, February 19, 1965, Inst. Elect. Engrs. Preprints, pp. 5–11.
[24]H. G. Trevor Busby and K. Darby, *J. Inst. Fuel* **36** (1963) 184.
[25]Ref. 19, p. 124.
[26]Ref. 2, p. 319 *et seq.*
[27]E. T. Hignett, *Colloquium on Electrostatic Precipitators*, London, February 19, 1965, Inst. Elect. Engrs. Preprint, pp. 12–18.
[28]A. C. Smith, *J. appl. Chem.* **8** (1958) 636; H. S. Simons and J. W. Jeffery, *J. appl. Chem.* **10** (1960) 328; J. D. Watt and D. J. Thorne, *J. appl. Chem.* **15** (1965) 585.
[29]H. J. White, *Air Repair* **3** (1953) 79; E. V. Harlow, U.S. Pat. 2 746 563, May 22, 1956.
[30]H. G. Eishold, *Staub-Reinhalt. Luft* **26** (1966) 11.
[31]Anon., *J. Air Pollution Control Assoc.* **15** (1965) 256.
[32]K. Darby and D. O. Heinrich, *Staub-Reinhalt. Luft* **26** (1966) 12.
[33]K. Schrader, *Mitt. VGB* **48** (1968) 430.
[34]J. Coutaller and C. Richard, *Pollution Atmosphérique* No. 33 (1967) 9.
[35]K. S. Watson and K. J. Blecher, *Proc. Clean Air Conf.*, Sydney, 1965, Paper No. 10, New South Wales University Press, Sydney.
[36]J. T. Reese and J. Greco, *J. Air Pollution Control Assoc.* **18** (1968) 523.
[37]J. F. Corcoran, *Fuel* **49** (1970) 331.
[38]E. C. Potter and S. G. Szirmai, unpublished work.
[39]W. T. Sproull, *J. Air Pollution Control Assoc.* **15** (1965) 50.
[40]E. C. Potter and P. R. C. Goard, unpublished work.
[41]J. R. O'Connor and J. F. Citarella, *J. Air Pollution Control Assoc.* **20** (1970) 283.
[42]C. J. Stairmand, *The Chem. Engr.* No. 221 (1968) CE257-61.

Chapter 6

ELECTROCHEMICAL METHODS OF POLLUTION ANALYSIS

B. D. Epstein

Gulf General Atomic Company
P.O. Box 608, San Diego, California

I. INTRODUCTION

The material in this chapter is intended to serve as a guide to the electrochemical methods that have been or, in some cases, can be used to determine aspects of air and water quality. It would really be insufficient to talk about the analysis of pollutants by themselves inasmuch as dissolved oxygen, for instance, is not a water pollutant even though the presence of insufficient dissolved oxygen in a natural water is often indicative of an unhealthy or "polluted" condition.

The work has been divided into two major areas—air monitoring and water monitoring. Although examples have been included of electrochemical techniques used to analyze for certain environmental pollutants on an individual basis under laboratory conditions, it should be stressed that the need at the present time is for real-time monitoring equipment. To give meaningful results, the equipment should require a minimum of maintenance, have a high reliability, and be essentially free from interferences (both thermal and chemical).

The biggest shortcoming of "redox" electrochemical techniques, voltammetry, amperometry, and coulometry, is that they often require extensive sample pretreatment in the form of reagent addition and deoxygenation. Sometimes this drawback can be eliminated by continual reuse or regeneration of reagents or by self-contained generation of hydrogen to be used for deoxygenation. In the case of selective ion electrodes, specificity is not always what it might be, and there are additional effects from complexation and ionic strength that obscure the analysis.

Another important point to be considered is that while an electrochemical system might function admirably in the laboratory under "tap-water" conditions it still is a long way from having the system operate in sewage or natural water. Suspended matter, grease, slime buildup, and bacterial action can all serve to rapidly degrade the quality of a one-line analysis.

We have not discussed, for the most part, parallel analyses using non-electrochemical techniques. It is not intended to slight the other techniques or to say that they are inferior to electrochemical ones. However, the scope of work must be limited to the keynote of this volume, *Electrochemical Aspects of a Cleaner Environment*. In essence, this has been written for the individual who has a more-than-passing interest in electrochemistry and who is interested in environmental problems and the role that electroanalytical chemistry is playing to help solve them.

A word of clarification regarding the units used for measurement of analytes in the field of air and water: in the field of air monitoring, concentration is generally based on a volume–volume ratio. Thus, ppm for air would mean 1 volume of material measured contained in 10^6 volumes of air. The other often-used measure for particulate, or solid related material, is milligrams/cubic meter (mg/m^3). In water analysis, on the other hand, ppm refers to mg/liter, in essence assuming that the density of the sample liquid is unity.

By way of a point of reference, some of the air-quality standards set by the Los Angeles County Air Pollution Control District (LACAPCD) and the California State Air Resources Board (CSARB) are listed in Table 1 for a condition of a first-state smog alert, or "adverse day."

Table 2, which is an exerpt from Ref. 1, lists the Public Health Service drinking-water standards (1962). In addition, the following maximum allowable limits have been set:

Substance	Limit, ppm
Arsenic	0.05
Cadmium	0.01
Chromium (VI)	0.05
Cyanide	0.2
Lead	0.05

The reader is directed also to annual reviews containing information on electrochemical techniques as applied to pollution-analysis problems (Refs. 2 and 3). A comprehensive review of modern electrochemical methods may be found in a number of sources (see, for example, Refs. 4 and 5). These

Table 1. Air-Quality Standards

Pollutant	LACAPCD[a]	CSARB[b]
Ozone, ppm	0.50	0.10 for 1 h
Carbon monoxide, ppm	50.0	10.0 for 10 h
Oxides of nitrogen, ppm	3.0	0.25 for 1 h (NO_2)
Sulfur dioxide, ppm	–	0.50 for 1 h

[a]Based on peak reading.
[b]Based on average reading over a given time period.

Table 2. Public Health Service Drinking-Water Standards (1962)

Substance	Concentration, mg/liter
Alkyl benzene sulfonate	0.5
Arsenic	0.01
Chloride	250
Copper	1
Carbon chloroform extract	0.2
Cyanide	0.01
Iron	0.3
Manganese	0.05
Nitrate (NO_3)[a]	45
Phenols	0.001
Sulfate (SO_4)	250
Total dissolved solids	500
Zinc	5

[a]In areas in which the nitrate content of water is known to be in excess of the listed concentration, the public should be warned of the potential dangers of using the water for infant feeding.

works are primarily concerned with theory and methodology rather than specific applications that are directly relevant to environmental analysis.

II. AIR POLLUTION MONITORING

1. Ozone

Ozone as an atmospheric pollutant has become extremely important. Ozone arises from photochemical reactions in the upper atmosphere and from various industrial sources. It represents a real health hazard to the residents of the Los Angeles basin, where levels of ozone are carefully monitored. Outdoor schoolyard activities are curtailed when the ozone content of the air reaches 0.35 ppm.

Because ozone is such a powerful oxidizing agent ($O_3 + 2H^+ + 2e^- = O_2 + H_2O$; $E^0 = 2.07$ V vs NHE),[6] it is quite reactive with many natural products. In fact, ozone has been used successfully as a disinfectant in the treatment of sewage. However, because this action is of limited duration in solution and because it is costly, it has had only limited use as a satisfactory disinfectant in the U.S., until just recently. Increasing production and usage of ozone in this area might prove to be another source of air pollution.

Most electrochemical methods for monitoring ozone in the atmosphere use its oxidative power to liberate I_2 from a buffered solution of I^-. One of the earliest instruments developed around this principle was described by Bowen and Regener.[7] In this instrument the air sample is passed at a controlled rate into a flow cell containing KI and a known concentration of thiosulfate. The solution passes by an anode which is set to generate iodine. Dual polarized platinum electrodes serve as the sensing system which controls the rate of production of iodine at the generator anode. A small constant iodine concentration is maintained at the amperometric sensor by producing a slightly greater amount of iodine than is necessary to completely react with the thiosulfate. Since ozone liberates iodine from the reaction mixture, less current is required at the generator anode to maintain the iodine level. The decrease in current is then proportional to the ozone concentration. To take into account other oxidants that might be present in the air sample, the inlet air stream is divided into two parts and passed into essentially identical absorbing and titrating cells. However, one of the air streams is heated to a temperature sufficient to decompose ozone. Thus, one cell gives the value for total oxidants and the other for oxidants other than ozone.

Britaev[8] has reported on a conductivity monitor to measure ozone by using the change in conductance of a potassium iodide solution. This is accomplished through the reaction of O_3 with I^- in water to liberate a much more mobile OH^- ion. The same principle is presumably employed in a patent granted to Mast Development Co. (Davenport, Iowa),[9] which uses different specific reagents to monitor trace amounts of ozone, H_2S, or SO_2. Variations in gas composition will often contain contaminants that can produce or consume ions which will interfere with conductivity measurements.

Mast Development Co. markets a coulometric ozone monitor that is based on the ozone–iodide reaction. The early work on this instrument, done by Brewer and Milford,[10] resulted in a patent assigned to Mast Development.[11] Mast and Saunders[12] describe the further development of this instrument. A schematic diagram of the electrochemical cell is shown in Fig. 1. The electrolyte solution is 2% KI, ~5% KBr, buffered to pH 7 with phosphate buffer. A thin film of this solution is continuously flowed down the cathode support to keep the coiled cathode and anode wires wet. The flow

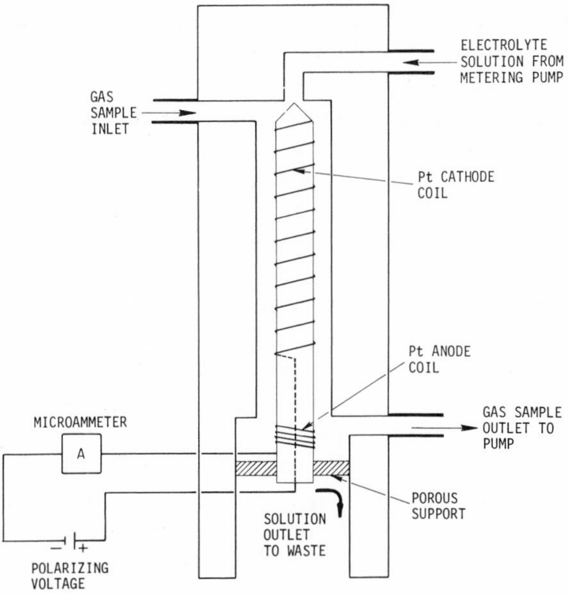

Figure 1. Mast ozone monitor.

rate is maintained constant at 1.25 ml/h using a diaphragm metering pump. The cathode region is long enough so that the air sample contacts the electrolyte for a time sufficient for most of the ozone to react. The air flow is controlled at 140 cm^3/min. The spent solution flows down past the anode carrying the anodically produced iodine to the waste receptacle. The iodine produced in the cathode region by the reaction $H_2O + O_3 + 2I^- \rightarrow O_2 + I_2 + 2OH^-$ is reduced at the platinum cathode to regenerate I^-. A potential difference of 0.25 V is imposed between the anode and cathode so that the resultant anode reaction is the oxidation of I^- to I_2. Because of the flow geometry, this iodine is swept away before it can diffuse back to the cathode. In this particular configuration, 0.05 ppm of ozone will give rise to a current of 1 μA. Response time for 1.0-ppm full-scale operation is ~30 sec.

Potter and Duckworth[13] have reported on field tests of the Mast ozone instrument. They state that the NO_2 efficiency is 10%, which means that 1 ppm NO_2 would simulate the presence of 0.1 ppm ozone. One major problem they encountered was that the growth of fungus and traces of yeast and algae in the cathode compartment reduced the sensitivity and blocked the solution flow. The use of a low concentration of fungicide in the electrolyte solution helped. By modifying the electrolyte reservoir they achieved an average unattended operating time of 7 days, with a maximum of 11 days.

Agreement with their colorimetric procedure was quite good. Other workers[14] have found that aging of the Mast sensor in continuous field use could result in a difference from colorimetric procedures as great as 60%.

Brewer and Milford[10] describe another ozone instrument using a sacrificial anode such as silver or mercury. The anode reaction is depolarized by the formation of AgI or Hg_2I_2. Once again the electrolyte is neutral buffered KI. The air sample is bubbled at a rate of 200 to 500 cc/min through a cup-like cell containing a few milliliters of solution. A platinum gauze cathode is used to reduce the iodine that is liberated by the gaseous oxidants. A polarizing voltage of 0.42 V is used when silver is the anode, 0.32 V for mercury. Evaporation of water from the cell was one of the major problems, and response time was not as good as the thin-electrolyte-film cell. Buswell and Keen[15] employed a similar bubbling cell with a sample flow rate of 250 cc/min. They were able to determine ozone to within 0.02 ppm over a range of 0–0.30 ppm using a silver anode.

A galvanic ozone detector is described by Hersch and Deuringer.[16] The electrolyte is made up of 3*M* NaBr, ~1m*M* NaI, 0.1*M* NaH_2PO_4, and 0.1*M* Na_2HPO_4. Ozone liberates bromine from the solution via a reaction which is postulated as $O_3 + 2Br^- \rightarrow O_2 + O^{2-} + Br_2$. The bromine is then reduced at a platinum mesh cathode: $2e^- + Br_2 \rightarrow 2Br^-$. The anode in this case is a paste of activated charcoal (Darco G 60*) made up with electrolyte. The reported anode reaction is given by $—C + O^{2-} \rightarrow —C—O + 2e^-$. The formation of a long-lived O^{2-} species is unlikely. The mechanism might involve active oxygenated species, as will be described later. However, using a charcoal of this type, with an ash content of ~9–13% and probably a high surface hydroquinone-like oxygen content, it seems likely that the carbon anode is quite electroactive and might well undergo a reaction of the sort $C—OH \rightarrow C{=}O + H^+ + e^-$. The authors say that the carbon does have a finite life and that it can be reactivated by cathodic treatment. The electrochemical oxidation reduction of carbon surfaces has been discussed previously on the basis of a quinone–hydroquinone-like redox couple.[17]

In any case, the galvanic ozone monitor does give rise to a current that can be measured by a galvanometer connected across the two electrodes. Nitric oxide does not interfere, and NO_2 has only a 3% equivalent signal to ozone. Interfering reductants such as SO_2 are removed by scrubbing the sample gas with acidified $KMnO_4$. The background level of gaseous oxidant formed via reactions of the sample with the oxidant scrubber is quite small. Olefins are removed by scrubbing the sample gas with mercuric perchlorate and perchloric acid.

A diagram of the Hersch–Deuringer cell[16] is given in Fig. 2. Sample gas is bubbled into the side arm, which serves as a "gas lift pump" to circulate

*Atlas Chemical Industries, Inc. (Wilmington, Delaware).

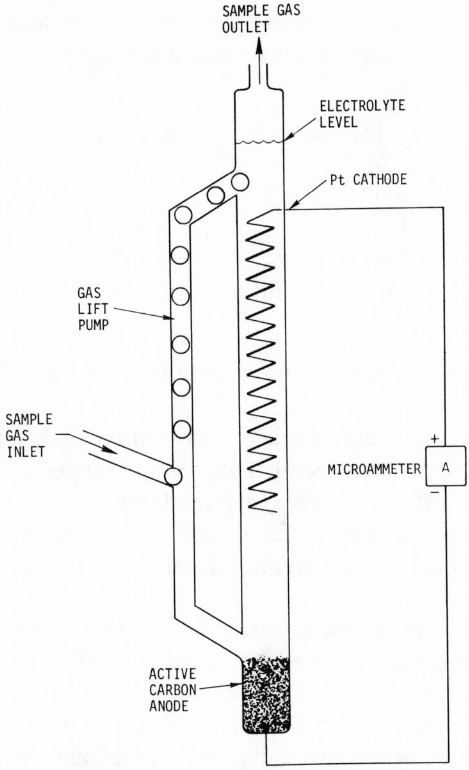

Figure 2. Hersch–Deuringer galvanic ozone detector.

the electrolyte past the cathode and anode. For low-level ozone content (up to ~2 ppm) the carbon anode is used, but for concentrations up to 100 ppm a silver anode is employed. It is also suggested that gas dilution or selective decomposition of some of the ozone might be used to accommodate higher concentrations. Response time to 90% is ~40 sec. The electrolyte is recirculated so that no external reagent addition is needed, nor is any source of polarization voltage required since the cell operates spontaneously.

A patent for this cell[18] was awarded to Beckman Instruments (Fullerton, California). However, Beckman is currently marketing a total oxidants analyzer (Model 908) which uses the galvanic gas lift cell with activated carbon anode but has a neutral buffered KI electrolyte. Apparently, the electrolyte was chosen to enable the measurement of NO_2 and other oxidants as well as ozone. The instrument is equipped with a Cr_2O_3–sand scrubber to remove interfering reductants such as SO_2, H_2S, mercaptans, and sulfides. A carbon column enables a periodic zero correction by adsorbing most of the active oxidants in the sample gas stream. The instrument has a 0.2-ppm full-scale maximum sensitivity with a 4-ppb minimum detectable limit at this range. Given below is a list of maximum signal errors for interference at a concentration of 1 ppm.

Interference	Error, ppm
SO_2	0.1
H_2S	0.1
Mercaptans	0.1
Organic sulfides	0.1
Ammonia	0.01

An automatic water addition system maintains a constant electrolyte concentration.

Wartburg, Brewer, and Lodge[19] have discussed some interferences using a coulometric ozone detector. They found that peroxy acids and simple hydroperoxides interfere by giving high readings. Although this is true, very often it is necessary to get values for total oxidants, especially in areas where photochemical smog contains detrimental levels of peroxyorganic compounds.

An amperometric titration procedure for O_3 by Duffy and Pelton[20] employs an alkaline buffered (Na_2HPO_4) solution of KI containing a known amount of $S_2O_3^{2-}$. A controlled amount of sample gas is bubbled into the solution and the remaining $S_2O_3^{2-}$ is back titrated with coulometrically generated I_2 to an amperometric end point. At the higher pH the interference by NO_2 and SO_2 is reported to be reduced. Saltzman[21] states that the advantage of the alkaline-buffered method over the neutral-buffered method is the increased delay that is permissible between sampling and completion of the analysis. However, he mentions that SO_2 and H_2S are very serious interferences.

Hendricks and Larsen[22] have stated that a scrubbing train of granular chromic acid and potassium permanganate results in a loss of 15% of ozone at an 0.8-ppm concentration level. They also feel that all iodine methods lack sufficient specificity for industrial hygiene applications. Saltzman and Wartburg[23] have employed a glass-fiber filter impregnated with Cr_2O_3 and H_2SO_4 which gives good results for removal of SO_2. Silica gel saturated with 0.4*M* $Na_2Cr_2O_7$ and 0.72*M* H_2SO_4 could be used to remove NO_2.

Schulze[24] has described an amperometric unit to simultaneously determine ozone and SO_2. The basis for this dual-mode instrument is the use of scrubbers to selectively remove ozone and SO_2. In the analysis of ozone, SO_2 is removed by scrubbing the sample gas with quartz chips that have been soaked in a solution of equal weights of Cr_2O_3, H_2O, and 85% H_3PO_4. To determine SO_2, a scrubber of granular $FeSO_4 \cdot 7H_2O$ is used to remove O_3. The iodine liberated by ozone is determined amperometrically by dual polarized platinum wires using a polarization voltage of 25 mV. To determine SO_2, I_2 is coulometrically generated at a constant rate and the depletion of I_2 by reaction with SO_2 is measured amperometrically. Regeneration of the spent electrolyte containing excess I_2 is accomplished by reaction with nylon fibers in the electrolyte storage tank. Full-scale response to 0.10 ppm of O_3 or SO_2 is ~20 min. In actual practice the instrument contains two electrolytic cells and gas contactors, and the sample stream is split in two prior to selective scrubbing to provide simultaneous determination of SO_2 and O_3. Schulze claims that the "neutral unbuffered" NaI procedure is insensitive to NO_2.

Boyd, Willis, and Cyr[25] have examined the stoichiometry of the ozone–iodine method spectrophotometrically. They found that at pH 7, 1.5 moles of I_3^- were produced for each mole of ozone consumed. They cite earlier work[26,27] that would confirm this observation if a stoichiometry of 1:1 at pH 9 were assumed. The following are reactions of O_3 with I^- in alkaline solution that would yield a stoichiometric ratio of 1:1:

$$O_3 + I^- \rightarrow O_3^- + \tfrac{1}{2}I_2$$

$$O_3^- + I^- + H_2O \rightarrow 2OH^- + O_2 + \tfrac{1}{2}I_2$$

However, at lower pH the ozonide ion is postulated to protonate $O_3^- + H^+ \leftrightarrows HO_3$. Boyd and coworkers feel that the pK of HO_3 must be ~7–8. They propose the following set of reactions which would show a high yield of I_2 and a pH dependence:

$$HO_3 + I^- \rightarrow HO_3^- + \tfrac{1}{2}I_2$$

$$HO_3^- + I^- + 2H^+ \rightarrow HO_2^{\cdot} + H_2O + \tfrac{1}{2}I_2$$

$$HO_2^{\cdot} + I^- + H^+ \rightarrow H_2O_2 + \tfrac{1}{2}I_2$$

$$H_2O_2 + I^- \rightarrow OH^{\cdot} + OH^- + \tfrac{1}{2}I_2$$

$$OH^{\cdot} + I^- \rightarrow OH^- + \tfrac{1}{2}I_2$$

Many of these species are extremely unstable in solution and have very short lifetimes. In fact, the lifetime of the protonated form of superoxide, HO_2, has a bimolecular decay rate of 10^{13} liter·mole^{-1}·sec^{-1}. However, the ultimate formation of peroxide does seem likely and might account for the enhanced iodine formation.

On the whole, it appears that electrochemical devices for ozone and total oxidants have great utility. They are portable and easy to maintain, and with regenerating and recirculating electrolyte they should be suitable for long-term field monitoring.

2. Oxides of Nitrogen

The emission of oxides of nitrogen into the atmosphere is of great concern in air pollution control. The large sources of NO and NO_2 are heat engines, automobiles, power plants, and other high-temperature burner systems. The conversion of NO to NO_2 in the atmosphere is facilitated by hydrocarbons. These pollutants are active physiologically and contribute to respiratory system and eye irritation. They are important in atmospheric photochemical reactions and are responsible for the Los Angeles-type smog.

Several electrochemical approaches for NO_2, NO, and NO_x (total of NO_2, NO, and other oxides such as N_2O_4, N_2O_5, etc.) have been applied. These approaches have been used in the monitoring of nitrogen oxides at their source, automobile and stack gases, as well as of the pollutant dispersed in the atmosphere.

Sawyer and coworkers[28] investigated the polarographic behavior of NO_2 using a membrane-covered platinum electrode and aqueous electrolyte solution. A half-wave oxidation potential of $\sim +0.2$ V vs SCE and a sizable diffusion current were obtained for NO_2 diffusing from the gas phase. However, owing to the corrosive nature of the gas, the polyethylene membrane was attacked. Although the authors suggest the use of Teflon, no data for permeability are given for NO_2 in the gas phase. Water saturated with NO_2 showed an $E_{1/2}$ of $+0.66$ V vs SCE with a diffusion current of 1.6 μA for a 0.027-mm-thick Teflon membrane. Polyethylene (0.05 mm thick) yielded a current of about one-half the value for Teflon.

Another type of polarographic NO, NO_2 detector has been described by Chand and Cunningham,[29] who studied the oxidation of NO and NO_2 in several nonaqueous solvents by cyclic voltammetry. They found that 1,2 propanediol cyclic carbonate with an electrolyte of KPF_6 was most suitable for the oxidation of the oxides of nitrogen. They report that at a gold wire anode, at a potential of $+0.70$ V vs NHE, current densities of 232 mA/cm^2 for NO_2 oxidation and 17.2 mA/cm^2 for NO oxidation are obtained. Essentially no interference from SO_2 in this solvent system is obtained. A practical instrument built around the selective, nonaqueous, polarographic approach has not yet been fully developed.

The commerically available electrochemical transducer employs a sealed polarographic cell containing electrodes and an aqueous electrolyte. The gas to be measured diffuses across a membrane where it reacts at the sensing electrode. Since the cell is sealed, loss of solvent and contamination are minimized. Sample gas must be cooled to at least 110°F for operation. The NO and NO_2 ranges available are from 0–2 to 0–2000 ppm.* There is also a choice of transducer cells and analyzers, which are claimed to analyze

*Dynasciences, Inc. (Chatsworth, California); Environmetrics (Marina del Rey, California).

for either NO_2 or NO and NO_2 together. NO_2 is determined by reduction, while NO_x is determined by electrooxidation. The lowest range offered for ambient air measurement is 0–0.2 ppm full scale.

The same basic principle used in the Mast ozone detector has been applied to the electrochemical analysis of NO_2.[30] The device is described in Section II.1. The electrolyte is an acidified 10% KBr and ~3% ethylene glycol aqueous solution. The NO_2 liberates Br_2 from the electrolyte which is, in turn, reduced at a platinum coil cathode. A polarizing voltage of 0.56 V is used between the platinum anode and cathode. By varying the exact composition and pH of the KBr solution, a number of usable ranges for NO_2 estimation are achieved. For stack-gas monitoring, the acidified KBr seems particularly applicable where a range of 0–5000 ppm is attainable. Although SO_2 does appear to be a significant interference, it would seem that selective scrubbing might eliminate that problem.

Nitric oxide can be rapidly converted to NO_2 by reaction with ozone.[31] This approach has been used for the colorimetric estimation of NO and NO_2.[32] DiMartini[33] has used this photochemical conversion technique to determine NO and NO_2 electrochemically. Using a nitrate selective ion electrode, he determined the buildup of NO_3^- from the dissolution of NO_2, N_2O_4, and N_2O_5. A small mercury lamp served as the ozone source in the sample gas stream. The following reactions were proposed:

$$NO + O_3 \longrightarrow NO_2 + O_2$$

$$2NO_2 + O_3 \longrightarrow N_2O_5 + O_2$$

$$N_2O_5 + H_2O \longrightarrow 2HNO_3$$

$$3N_2O_4 + 2H_2O \longrightarrow 4HNO_3 + 2NO$$

With a high sample-gas flow rate and good gas–liquid contact, a concentration of oxides of nitrogen as low as 0.062 ppm could be measured.

It appears that the photochemical conversion method might be useful for monitoring of NO at low levels when applied to other electrochemical techniques, such as coulometry or polarography. The excess ozone generated would be removed thermally or catalytically so that only NO and NO_2 might be detected.

Beckman Instruments (Fullerton, California) has made a preliminary description of a trace-concentration NO_2 analyzer (Model 910) which uses its carbon–platinum galvanic cell. This cell, described in the Section II.1, basically employs the galvanic current generated from the liberation of I_2 by NO_2. Although no technical details have been provided, a claim is made for a scrubber that removes O_3, SO_2, H_2S, and mercaptans.

3. Analysis of SO_2 and Oxidizable Sulfur Contaminants

Sulfur dioxide is a chief pollutant in areas that contain heavy industry. The combustion of sulfur-containing fuels and the refinery and smelter operations contributed to the atmospheric pollution by SO_2. Hydrogen sulfide and organic sulfur compounds are a problem around large sewage treatment facilities, especially where some of the biologic action has become anaerobic. Ore roasting and industrial processes also contribute to the problem.

One of the earliest methods[34] is based on the change in conductivity of an acidified solution of H_2O_2 when SO_2 is bubbled into it. Obviously, any pollutant that dissolves in the test solution to liberate or consume ions will interfere. However, the method does have applicability in a wide variety of situations of monitoring ambient air where interferences are seldom present to an appreciable extent compared to SO_2. Booras and Zimmer[35] have compared the conductivity method with the standard West–Gaeke colorimetric method in the Chicago area. Conductivity gave reasonably good results in following the fluctuations and trends in SO_2 concentration but tended to give slightly greater readings (by about 20%) than the colorimetric procedure. The American Society for Testing Materials[36] (ASTM) has described conductivity methods for batch and continuous monitoring of SO_2. The electrolyte is made slightly ($\sim 0.00002N$) acidic with sulfuric acid and contains 3mM H_2O_2. In the batch procedure a known flow rate of gas is bubbled through a fixed-volume absorbing solution for a specified time. The conductivity is measured before and after gas introduction. Calibration may be performed absolutely from the calculated amount of H_2SO_4 liberated, or it may be done empirically by use of known gas mixtures. For continuous monitoring, the sample gas is contacted with the absorbing solution, which is flowing in a thin film at constant rate along the walls of a glass tube. The liquid containing absorbed SO_2 then passes through a small-conductivity flow cell. The conductance of the absorbing solution is measured before and after contact with the sample gas to give a differential measure of the increase in conductivity. The sensitivity for the SO_2 conductivity method ranges between 0.01 and 10 ppm. Continuous-conductance analyzers are commercially available.* Recent work[37] indicates that the conductivity method is unreliable below 1 ppm and should be used with caution.

Sulfur dioxide can be rapidly oxidized in solution by iodine or bromine. Basically, a four-electrode coulometric titration procedure has been used to monitor SO_2 levels. In the bromine oxidation technique, bromine is produced at a platinum anode in an acidified bromide electrolyte. A shielded platinum cathode is used to prevent uptake of Br_2 by electroreduction. A pair of indicator electrodes, platinum–calomel or other suitable reference, are used to

*Leeds & Northrup Co. (Sumneytown Pike, North Wales, Pennsylvania).

sense the Br_2 level in the electrolyte. Sample gas is bubbled through the electrolyte, and SO_2 and other oxidizable gases consume Br_2. The reduction in halogen concentration is detected by the sensor electrodes. The voltage output of the sensor pair is tied into a feedback loop which controls the current for the generator electrodes. Thus, as the sensor electrode voltage drops, the generator current is increased so as to reestablish the initial, preset sensor voltage. The electrolysis current is proportional to the rate at which pollutants are consuming Br_2 and is therefore, at constant gas flow, proportional to their concentration. This method has been described by the ASTM and is the basis for several continuous SO_2 monitors.[38–40] Equivalent response has been reported by the ASTM for the following gases: 10 ppm SO_2, 4 ppm H_2S, 8 ppm butyl mercaptan, and 12 ppm methyl disulfide. Oxidizing gases such as O_3 or NO_2 will interfere since they can liberate halogen in solution. The lower limit of sensitivity as described by the ASTM is 0.1 ppm SO_2.

Schulze[24] has described an instrument to monitor SO_2 and ozone. The basis unit is described in Section II.1. For SO_2 estimation he uses a pair of platinum generating electrodes to produce I_2 from a "neutral unbuffered" KI solution at a constant rate. Sample gas is sprayed into the electrolyte to react with the iodine. The solution which has been contacted by the gas flows past a pair of polarized platinum electrodes. Amperometric sensing of the loss of I_2, compared to gas scrubbed of SO_2, yields the value of SO_2 concentration in the sample. Because of the cell configuration, this device is somewhat slow in response to large variations in sulfur dioxide concentrations. A scrubber of $FeSO_4 \cdot 7H_2O$ crystals is used to remove oxidizing gases.

Beckman Instruments (Model 906A)[41] offers an iodimetric SO_2 analyzer with a function that differs somewhat from that of the ordinary coulometric analyzers. Iodine is generated at a platinum anode by a constant electrolysis current. The sample cell employs a gas lift principle in which the sample gas is contacted with the I_2 liberated at the anode. The SO_2 reacts with the I_2 to lower its concentration. The remainder of the unreacted iodine is passed by a platinum cathode where it is reduced to I^-. The anode–cathode circuit is operated at constant current so that the amount of unreacted I_2 is insufficient to carry the full current at the platinum cathode. An auxiliary carbon cathode, connected in parallel to the platinum cathode, is employed to make up the current equivalent to the reacted I_2. The manufacturer states that this reaction is C (oxidized state) + $ne^- \rightarrow$ C (reduced). Thus, the current flowing through the carbon electrode is proportional to the I_2 oxidation of SO_2. The minimum detectable limit is stated as 0.02 ppm. A selective scrubber is said to remove interference.

Membrane polarographic SO_2 detectors are the newest addition to the field of electroanalytical pollution devices. This type of detector employs a noble-metal anode at which the pollutant gas can be selectively oxidized. The cathode is an unpolarizable electrode which also serves as a reference electrode. A controlled potential is imposed between the anode and cathode so that the anode potential remains at a value that is selective for SO_2 oxidation. The electrodes and electrolyte solution are contained in a sealed cell. One wall of this cell is made from a thin, semipermeable membrane, which allows transport of gas such as SO_2 but minimizes contamination of the electrolyte. The anode is placed near the membrane to limit the diffusion time for SO_2 to be electrooxidized. Chand and Cunningham[42] have observed the oxidation of SO_2 in formamide–KPF_6 solution using cyclic voltammetry. They observed a peak potential of +0.90 V (vs NHE) at a gold electrode. At this potential the oxidation rates of NO and NO_2 are low and they do not interfere. Sealed polarographic SO_2 detectors have a usable range of from 0–0.2 ppm full scale to several thousand ppm full scale. Membrane polarographic SO_2 sensors are available from several commerical sources.*

The bromine coulometric oxidation approach has also been used to monitor Kraft mill gases containing SO_2, H_2S, mercaptans, organosulfides, and disulfides. Adams and Koppe[43] have used a bromine cell to analyze for these gases as they emerge from a gas chromatographic column. Thus, the pollutants can be separated chromatographically and their concentrations quantified coulometrically.

4. Carbon Monoxide

Although no commercial electrochemical CO detector has been described at present, several workers have investigated its voltammetry in aqueous solution. Ovenden[44] has studied the oxidation of CO on platinum, palladium, rhodium, and iridium vibrating electrodes in buffered solutions at pH 7 and 10. Palladium seemed to have the best response for CO oxidation in pH 10 borate buffer yielding an approximate linear relation between anodic current and a gas composition of $\sim 10\%$ CO $\rightarrow 100\%$ CO. Roberts and Sawyer[45] have observed similar results at a gold electrode in pH 12 solutions. They report the necessity for substantial anodic–cathodic pretreatment of the electrode to give reproducible and sensitive response to CO.

5. Fluoride and Fluorine

The determination of airborne fluorides from aluminum refineries and other chemical process industries is an area where electrochemical monitoring may prove useful. The advent of the fluoride selective ion electrode certainly would seem a good starting point. Although fluorine gas is very

*Dynasciences, Inc. (Chatsworth, California); Envirometrics, Inc. (Marina del Rey, California); Ericssen Instruments (Ossining, New York).

toxic and highly reactive, there does not appear to be a widespread need to monitor it. However, an example of an electrochemical approach that might be used is given below.

Elfers and Decker[46] have used a fluoride selective ion electrode for the determination of fluoride in ambient air and stack-gas samples. For ambient air sampling the air is contacted with a cellulose acetate membrane filter impregnated with sodium formate. After 4 h contact time, the trapped fluoride is solubilized with a sodium citrate solution. Fluoride activity is then measured with the selective ion electrode. In stack-gas sampling, gaseous fluorides react with a heated glass probe to give soluble fluosilicic acid. The remaining particulates are trapped on a filter. These portions are analyzed for F^- by direct potentiometry with the fluoride electrode. The Al^{3+} and Fe^{3+} are possible interferants since they would complex F^- and lower its activity. The lower limit would appear to be 0.25 ppb in ambient air.

The same principle of halogen replacement employed in ozone detection has been used in a fluorine analyzer.[47] The air sample is bubbled through an electrolyte of $\sim 0.2M$ LiCl, unbuffered, where the F_2 generates chlorine. The chlorine is reduced at a platinum cathode using an impressed potential and silver anode. The current flowing for chlorine reduction and silver oxidation gives a measure of the gaseous fluorine content. Ozone has no effect. The current output is linear to ~ 1000 ppm of F_2. No mention is made of the fouling of the system by CO_2.

III. WATER POLLUTION MONITORING

1. Three Most Common Electrometric Water Measurements: Conductivity, pH, and Oxidation–Reduction Potential

(i) Conductivity

The measurement of electrolytic conductivity is a general indicator of the free-ion content of a water sample. If the type of ions in a given monitoring situation remains constant, then conductance measurements can serve as a direct measure of ionic concentration. On the other hand, when the ionic composition of the sample water varies considerably, then the conductivity can only serve as a guideline as to the general quality of the water. In many drinking-water applications, the conductivity is related to the total dissolved solids (TDS) level. The ions generally associated with this parameter are Na^+, Ca^{2+}, Mg^{2+}, Cl^-, SO_4^{2-}, and HCO_3^-.

Ordinarily for individual sample measurement a typical two-electrode conductivity cell is used. The electrodes are generally platinized platinum and are spaced to give a cell constant that permits easy measurement of conductance over the range that is expected. Typical conductances found in monitoring applications may run from a few hundred micromhos per centimeter to tens of thousands of micromhos per centimeter. This range requires the use of more than one cell constant.

The difficulty with the use of the two-electrode system is that the electrodes can eventually become coated with slime. The measurement then reflects the conductivity of the surface film as well as the bulk solution.

To avoid the error caused by film buildup, the use of a four-electrode conductivity probe provides some advantages. In this approach, a pair of driver electrodes is used to pass a small ac current through the sample solution. Enough voltage compliance in the driver circuit must be available to keep the amplitude of the current constant. The two sensor electrodes are placed geometrically between the driver electrodes, with sufficient separation to permit reading the effective solution iR drop between them. The two sensor electrodes thus have an oscillating potential impressed between them which is proportional $i_{ac} \times R_{solution}$. The alternating voltage that is sensed can be used directly as a measure of solution resistance or it can be used as a servo to control the amplitude of the driver current, thus making the voltage output of the driver amplifier the measured function. If material should build up on the sensor electrodes, the effect would be small since no current actually flows through the resistive film. Such a device is marketed by Union Carbide Instruments (White Plains, New York).

An interesting discussion of the application of conductivity measurements to waste streams has been given by Corrigan and coworkers.[48]

Temperature compensation must be employed in any real monitoring situation to allow for meaningful analysis.

(ii) pH

The determination of pH has been discussed in an excellent book by Bates.[49] The pH measurement is a common and important one. Fish and other forms of marine life can live only in narrow pH limits. Corrosion rates are strongly affected by low pH values. Industrial wastes often affect the local pH near the point of discharge; consequently, it is necessary to know the effluent pH and the pH variation caused in the receiving water.

Although the measurement of pH using glass electrodes is ordinarily straightforward, continuous monitoring in natural and waste waters poses some problems. Temperature compensation is a requirement. Fouling of the salt-bridge tip of the reference electrode can also cause difficulties. However, the most serious difficulty lies in the formation and growth of slime films on the surface of the glass electrode. Without sufficiently frequent cleaning, the glass electrode rapidly becomes a slime electrode and only reflects changes in the pH of the slime. Unfortunately, this type of degradation is not obvious without a recalibration of the electrode.

(iii) Oxidation–Reduction Potential

The measurement of oxidation–reduction potential (ORP) of water systems is intended to give a measure of the dissolved oxidants or reductants. It is facilitated by the use of a platinum indicator electrode whose potential is measured with respect to a stable reference electrode, such as the saturated calomel electrode. Natural water systems are rarely poised with respect to one reversible redox couple; consequently, the potential measured is more often a mixed potential. The effect of the surface state of the platinum electrode can also influence the measured potential. The absolute potential measurement is therefore of little value. However, in any given water system, the variation of potential may serve as a guide to the relative oxidizing or reducing power of that system. Horne[50] shows that, under anoxic conditions in sea water, as the redox potential becomes more negative, the DO content drops and the biological reduction of SO_4^{2-} to S^{2-} becomes important. In this case, at a potential of -139 mV the DO has become essentially zero. It has been said that if the DO electrode should become fouled and fail, the ORP might be relied on to show trends in DO.[51]

In industrial waste-water situations the ORP may provide a reasonable indication of the effectiveness of treatment for removal of redox materials such as Fe(II) and Cr(VI).

Once again, the user of an ORP electrode system must be cautioned about the buildup of slime films on the electrodes.

2. Ion Selective Electrodes

The development of practical ion selective electrodes has made a great impact on the water-quality monitoring field. The scope of this field is sufficiently large that we must direct the reader elsewhere for a comprehensive review. Durst has edited an excellent book on the subject,[52] which covers both the theoretical and practical aspects of ion selective electrodes. Rechnitz,[53] Riseman,[54] and Andelman[55] have published reviews which cover important features of the application of these electrodes.

There are a few ion selective electrode systems which, because they open some new areas, justify a brief discussion here. Nagelberg, Braddock, and Barbero[54] have developed a divalent phosphate electrode using a liquid ion exchanger. Although the application was for physiological pH's, 7.0–7.5, it is conceivable that in a monitoring situation the sample water pH could be altered to lie within this range. The response for divalent phosphate was linear from 10^{-4} to $10^{-2} M$ in the presence of $10^{-2} M$ Cl^- and could go as low as $10^{-5} M$ HPO_4^{2-} in the absence of Cl^-. Guilbault and Brignac[57] have tried to perfect the Pungor-type phosphate–silicone rubber membrane electrode but have not achieved reasonable selectivity.

Guilbault and Montalvo[58] have developed an enzyme electrode system for the determination of urea. Guilbault and Hrabankova[59] have also described an electrode for the measurement of amino acids. These systems use an enzyme, urease in the case of urea detection, *L*-amino acid oxidase (AAO) in the case of amino acids, immobilized in a polymer gel. The gel is held in close contact with the sensing surface of an ammonium ion selective electrode. Urease enhances the rate of urea hydrolysis, which produces an ammonium ion that is sensed by the electrode. Similarly, AAO speeds up the rate of amino acid oxidation, which liberates an NH_4^+ ion which is subsequently detected. Great promise appears to exist for these enzymatic electrodes.

Higuchi, Illian, and Toussounian[60] have developed a plastic membrane electrode for analysis of organic ions. A polyvinylchloride plastic doped with dioctyl phthalate showed good response for tetralkylammonium ions down to $10^{-5}M$, with insensitivity to H^+ and K^+ ions. A nylon–phenol plastic gave selectivity for the tetraphenyl borate anion. The potentialities for electrodes able to detect organics in water open up a myriad of important applications.

The data from potentiometric monitoring with ion selective electrodes are amenable to remote telemetric recording. With the use of a remote computer, a system of linear equations could be employed which would allow for correction of the reading of a given electrode by correlating its response to the response of several other ion selective electrodes. Caution must still be employed to ensure that the electrode response is not affected by the growth of an interfering film on the sensitive surface.

3. Determination of Dissolved Oxygen

The knowledge of the dissolved oxygen (DO) content of natural and waste-water samples is extremely important in judging the extent of pollution in that water. The ability of a water system to support marine life and to assimilate and "metabolize" waste contaminants depends in great measure on the DO content. It is also important in real-time monitoring situations to have a device that can accurately reflect the DO content of a water system with a fairly rapid response. This is one area where electrochemical instruments are particularly well suited. Hoare[61] has an excellent book on the electrochemistry of oxygen which contains a good deal of material on the electroanalysis of oxygen.

(i) *Directly Immersed Electrode Systems*

The direct determination of DO by DME polarography is well known.[62] The difficulty in applying the regular DME system to continuous DO monitoring of natural waters lies in the fact that the first oxygen reduction wave is sensitive to anionic surface-active agents.[63] The overall electrode reaction for the first electron transfer is given by

$$O_2 + 2H^+ + 2e^- \rightarrow H_2O_2 \qquad E_{1/2} \text{ vs SCE} = -0.05 \text{ V}$$

This reaction is also sensitive to materials that catalytically decompose H_2O_2. The second oxygen wave

$$H_2O_2 + 2H^+ + 2e^- \rightarrow 2H_2O$$

which yields the overall reaction

$$O_2 + 4H^+ + 4e^- \rightarrow 2H_2O$$

can be used in its limiting current range of ~ −1.6 V vs SCE. However, the problem here is that one must be assured that the sample does not contain other electroreducible materials that will interfere. This problem is present in the potential region of both the first and second oxygen reduction waves; however, with increasing negative potential the number of possible interferants increases. Mancy and Okun[63] have also shown that cationic surfactants interfere with the analytical applicability of the second oxygen reduction wave, although these materials are not thought to be generally present in natural waters. An additional difficulty exists in using the DME in DO monitoring, that is, the problem of mass transport to the electrode surface. Solution flow must be carefully maintained at a constant rate to ensure reliable results.

Tyler and Karchmer[64,65] have discussed the use of a rapidly dropping mercury electrode for direct monitoring of DO. The increased drop rate of 0.25 sec/drop has the advantage of being relatively insensitive to sample agitation or flow rate. They found this electrode applicable over a pH range of 4.5–11 and at salt concentrations from 0.001 to 0.05*M*. They used a polarizing voltage of −1.6 V vs the mercury pool. At salt concentrations above 0.05*M* they observed anomalously high currents for O_2 reduction, a fact that limits the applicability of the technique.

A further extension of the rapidly dropping mercury electrode is the streaming mercury electrode. Currents obtained with the streaming electrode have been used to measure DO *in situ* in subsurface waters in Norwegian fjords.[66]

Much work has been done to investigate the mechanism of oxygen reduction on solid electrodes. This material has been reviewed by Hoare,[61] and the reader is referred to his book. However, the direct determination of DO by electroreduction of oxygen at solid electrodes immersed in the sample is complicated by both the effect of pH and the poisoning of the active surface by adsorption of other solution components. Therefore, the application of solid electrodes immersed directly in the sample is extremely limited and will not be discussed further here. The continuous renewal of surface area in a flowing mercury electrode system (either flowing or dropping) is thus an obvious advantage to direct measurement. It should be pointed out that solid electrodes do have a great deal of importance in membrane-covered cells.

The thallium electrode is one solid electrode that does not really fit into the category of those mentioned above. The reaction of thallium with DO is given by

$$4Tl + O_2 + 2H_2O \longrightarrow 4Tl^+ + 4OH^-$$

This reaction has been used by Lueck[67] to measure DO. The system consists of a flow reactor filled with thallium chips through which the sample water flows. The increase in conductivity in the sample is then related to DO concentration. The method would run into difficulties if there is a conductivity change from the formation of precipitates due to the local increase in pH. In addition, the presence of oxidizing agents able to oxidize Tl would be a source of error.

Union Carbide Instruments (White Plains, New York) also has a thallium DO electrode. This device relies on the oxidation of thallium by DO, but the principle is based on the potentiometric value of the $Tl–Tl^+$ couple measured vs an external reference electrode. The flow conditions are maintained so that the concentration of Tl^+ near the electrode is proportional to DO concentration. The advantage is that the electrode surface is constantly renewed due to the dissolution of thallium by the action of oxygen, which should, in principle, reduce surface fouling. Unfortunately, the system is affected by solution oxidants other than oxygen as well as by a nonuniform dissolution of the electrode surface, leaving pockets which rapidly lose sensitivity to oxygen.

(*ii*) *Membrane-Covered Electrodes*

The advent of the membrane-covered electrochemical cell has had much the same impact on the measurement of DO as did the glass electrode on the determination of pH. Clark[68] is generally credited with the establishment of a membrane-covered electrode system incorporating both anode and cathode within the sealed membrane compartment. The significant advantage of the membrane-covered DO electrode is that only the dissolved gaseous material can transport through the membrane to take part in the electrode reaction. Thus, ionic oxidizing agents and organic interferants are prevented from causing spurious currents. The current obtained from the reduction of oxygen is not directly affected by stirring of the sample solution, just as long as the solution in close proximity to the membrane reflects the bulk DO concentration. A typical cell patterned after the design of Mancy, Okun, and Reilley[69] is shown in Fig. 3. We have deliberately generalized the diagram

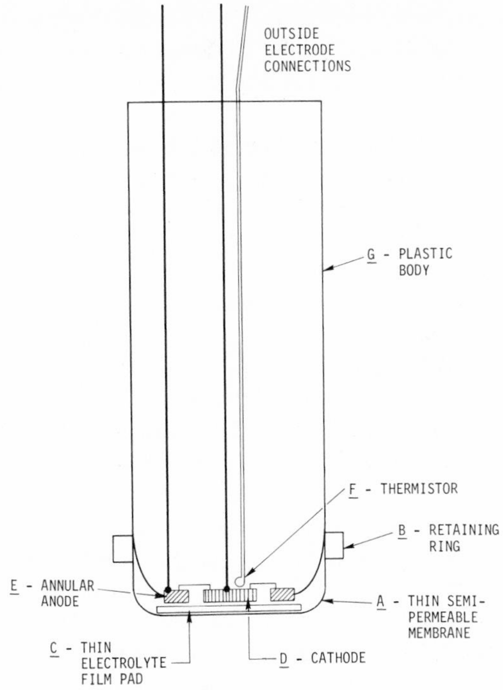

Figure 3. Schematic diagram of membrane-covered dissolved oxygen electrode.

so that it might be referred to as an example of either a galvanic or polarographic cell. The membrane *A* is stretched tight over the bottom of the sensor assembly and held in place by a retaining ring *B*. This membrane is generally polyethylene or Teflon having a thickness of ~0.5–1.0 mil. The membrane may sandwich a porous material *C*, such as lens tissue, which is saturated with the electrolyte solution. This keeps the disc-shaped cathode *D* and annular anode *E* in ionic contact. A temperature sensing device *F*, such as a thermistor, can also be inserted into the probe to allow for temperature compensation. The anode, cathode, and temperature sensor are encased in a suitable nonreactive plastic material *G*. Oxygen diffuses through the membrane and through the thin film of immobilized electrolyte before reacting at the cathode. This is the greatest source of sensitivity variation with temperature.

In general, the cathode material that has been employed is either platinum, gold, or silver, with not a great deal of difference noted. On the other hand, the anode material depends strongly on whether the cell is to be galvanic, generating its own current, or polarographic, using an external source to drive the reaction. In the galvanic cells, lead has been used most often as the anode; however, Neville[70] has used cadmium and has tested iron and zinc. Silver–silver chloride and silver–silver oxide have been employed as polarographic cell anodes.

Hersch[71] first described a galvanic cell for measuring low O_2 concentrations in gas streams. He used a lead-amalgam anode and platinum-wire cathode in an electrolyte of 24% KOH. The cathode was partially immersed in the electrolyte which formed a thin liquid film along the cathode surface. Thus, O_2 transport from the gas phase into the thin film was rapid, and good sensitivity for O_2 reduction was obtained. The work shows the efficacy of lead as an anode material.

Sawyer and coworkers[72] have investigated the polarography of several gases in KCl using a membrane-covered platinum working electrode and saturated calomel counter electrode. They found that polyethylene membranes had the best transport characteristics for gases but that Teflon lasted the longest. It is interesting to note that dissolved SO_2 gave a larger reduction wave than did oxygen, which would lead to positive errors in measuring DO in a waste water that contains appreciable dissolved SO_2 and has a low enough pH.

Carritt and Kanwisher[73] examined several aspects of the polarographic DO sensor. They selected Ag–Ag_2O as the internal reference electrode for operation in 0.5*M* KOH since they found long-term shift problems with a KCl electrolyte and Ag–AgCl reference. They also employed a thermistor imbedded in the sensor body to enable them to reasonably well temperature compensate the system over a range from 10 to 28°C. Their cathode was a platinum disc operated at −1.1 V vs the reference. The cell was covered with a 0.5-mil polyethylene membrane. A successful application was made of this system in DO monitoring of natural water in the Chesapeake Bay and in measuring the photosynthetic and respiration rate of phytoplankton.

Mancy, Okun, and Reilley[67] developed a galvanic DO sensor using a 1.0-mil polyethylene membrane, lead anode, and silver cathode in 1*M* KOH. They investigated several electrolytes including H_2SO_4 and $KHCO_3$ solutions but chose KOH because of its overall performance. They found an overall sensitivity to DO at 25°C of 1.05 μA/mg/liter with a residual current of 0.2 μA. They also determined that the sensor measured the oxygen activity rather than the concentration, but that the activity is the important parameter. The actual oxygen concentration would be determined by use of an appropriate correction factor relating the saturation O_2 solubility in the sample to that in distilled water at the same temperature and pressure. A careful analysis of the diffusion problem was also presented which showed the source of the temperature variation of the sensitivity. The Mancy electrode, as it has become known, was also applied in a number of real monitoring situations including oxygen profile in a fresh-water lake, monitoring in river water, and a study of oxygen uptake in domestic sewage.

Neville[70] described a galvanic DO sensor using a 1.0-ml polyethylene membrane with gold as the cathode, cadmium as the anode, and KCl as electrolyte. A thermistor was embedded in the assembly in intimate contact with the gold electrode. Without temperature compensation the sensitivity variation was ~4–5 %/°C. By employing the thermistor, the variation was reduced to less than 0.25 %/°C around 25°C. An important improvement was also noted in the performance by having a large anode-to-cathode area ratio of 40 : 1. This large ratio increases long-term stability and, in current use, anode-to-cathode area ratios of 400 : 1 are now being employed.[74,75]

Halpert and Foley[76] tested over 60 types of membranes and determined that 0.5-mil Teflon had both good permeability and sufficient strength to be used over a wide temperature range. They used a polarographic cell with an applied voltage of −0.7 V, with a platinum cathode and Ag–AgCl anode in saturated KCl solution. A special electrolyte containing methanol was used for low-temperature work (to −25°C).

Lipner and coworkers[77] employed a Mancy-type galvanic cell to monitor oxygen uptake rates in the biological oxidation of tyrosinase by DOPA. They used a bucking potential across the cell so that very small changes in oxygen concentration could be measured. A sensitivity in determination of O_2 uptake of 7×10^{-5} μliter oxygen per ml sample per min was obtained. This type of sensitivity is important for measurement of biochemical oxygen demand (BOD) rates.

Mackereth[78,79] has described a membrane-covered galvanic DO sensor using a large, perforated, cylindrical silver cathode and porous lead anode with a saturated $KHCO_3$ electrolyte. The probe has a sensitivity of 22 μA/mg/liter of O_2. Briggs and Viney[80] have used a Mackereth-type cell to monitor DO in river water and sewage. They report some difficulty due to $CaCO_3$ buildup on the membrane. A bridge circuit employed to temperature compensate the system seemed to work quite well.

Stack[81] has discussed the use of the Weston and Stack DO probe in a number of important environmental applications. The sensor is a membrane galvanic cell having a platinum cathode and lead-wire anode. It uses a thermistor for temperature compensation over the range of 0–50°C. In the determination of BOD, which is a measure of oxidatively biodegradable organic material, the DO sensor is used with a stirrer device, which maintains a fixed degree of agitation at the sensitive tip of the probe. Use of the probe can decrease the time required for analysis, reduce the number of dilutions, and allow for a procedure that permits periodic reaeration of the sample. An idea of the complexity of the BOD determination without use of the DO probe can be obtained from Ref. 82.

Another important area discussed by Stack[81] is the use of the DO sensor in controlling aeration in an activated sludge sewage-treatment process. Over the last several years the membrane-electrode technique has been quite popular in environmental and industrial water analysis.

In applications where power is not available to stir the sample past the membrane at sufficient rates to allow representative sampling and avoid concentration polarization, Lilley, Storey, and Raible[83] have devised a voltage-pulse chronoamperometric method. They apply a voltage step of -0.7 V for 5 sec between the cathode and a Ag–AgCl reference electrode. The current is then recorded in a range of 1–3 sec after initiation of the pulse. Since the double-layer charging current has dissipated in this time, the resultant current is due to the Faradaic process. An enhancement of sensitivity, over steady state, of a factor of three was achieved by using the current measured at 1.2 sec; 99% reequilibration was found to occur in 5 min. Thus, for a point-by-point monitor, taking data every 5 min, some benefit can be had.

Rayment[84] has discussed the temperature coefficient for polyethylene and Teflon membranes. He states that for a 1-mil thickness, the permeability increase of Teflon is 2.3 times that of polyethylene. Lucero[85] has derived expressions for the response of a membrane detector in the steady-state mode, and as a function of temperature and pressure.

With all the advantages inherent in the membrane-covered DO sensor, one must be careful to ensure that the electrode is maintained. Otherwise, the response of the membrane will soon reflect the oxygen uptake and the permeability of organic slimes that can be deposited in a real water-monitoring situation. Work is also required to find membranes that have a better selectivity for O_2 as well as an ability to resist slime growth. Presently, there are schemes, such as *in situ* ultrasonic cleaning, to eliminate this problem, but none that is foolproof. DO sensors are available from a number of commercial sources, and a listing of purveyors can be found in compilations such as Ref. 86.

4. Organic Carbon

(*i*) *Total Organic Content*

The determination of organic carbon is an important factor in water-quality analysis. In sewage treatment, the removal of organic carbon is a prime criterion of the efficiency of treatment. It is becoming more and more critical that sewage outfalls, both industrial and domestic, be analyzed and regulated for organic compounds. The types of analysis for organic content are varied and are summarized below:

(1) Biochemical oxygen demand (BOD). This measures the amount of O_2 required during biologic degradation of organic materials.

(2) Chemical oxygen demand (COD). The amount of strong oxidizing agent, such as $Cr_2O_7^{2-}$, that is consumed on reaction with organics in the sample is measured. The amount of O_2 that would be required for this oxidation is related to the dichromate equivalent.
(3) Total organic carbon (TOC). After removal of inorganic carbon, the sample is catalytically combusted, and the amount of CO_2 that is produced is detected and related back to organic content.
(4) Total oxygen demand (TOD). Once again, the sample is catalytically burned at high temperature, but now the uptake of O_2 required for combustion of nitrogenous, carbonaceous, and sulfurous matter is the measured quantity.

The classical BOD technique is long and laborious, requiring several modified Winkler titrations on serially diluted samples over a period of 1–20 days.[82] However, a number of modifications have been made in the procedure that allow for rapid monitoring of dissolved oxygen content of the incubated sample. This method is an effective application of the membrane-covered DO electrode, which has been discussed elsewhere in this chapter. The electrode is immersed in the sealed BOD bottle and the DO uptake is recorded directly. If the DO falls below 1–2 mg/liter, a level at which the rate of bacterial action is slowed, the sample may be reaerated. Considerable time is saved because guesswork in the dilution of BOD samples is eliminated.

Another application of electrochemical techniques to the analysis of BOD has been demonstrated by Clark and coworkers.[87–89] As oxygen is consumed in the biological process, CO_2 is given off. The Clark apparatus uses a glass-wool wick containing 30% KOH to absorb CO_2 in a small gas space above the liquid sample level in a BOD bottle. The resultant decrease in pressure, due to O_2 uptake and CO_2 removal, is sensed by a small manometric sensor. The sensor is part of an electrochemical cell which rests on top of the BOD bottle. As the pressure drops, the sensor actuates a relay which allows generation of O_2 at a platinum anode in the electrochemical cell. The O_2 that bubbles off the electrode is vented directly into the gas space in the BOD bottle, where it dissolves into the sample liquid. Thus, the electrolysis current used in maintaining O_2 pressure is a continuous measure of the BOD.

Steinecke[90] has discussed a similar system of measuring oxygen demand. In this case the electrochemical cell is pulsed on demand of the manometer sensor. Each pulse supplies the equivalent of 1 mg/liter oxygen. The pulses are recorded as a function of time to give the BOD. The system has been used for periods up to 5 days without requiring attention.

Several methods have been described for determining organic carbon using chemical and photochemical oxidation followed by electrochemical detection. Krey and Szekielda[91] oxidized a water sample with $Ag_2Cr_2O_7$–$K_2Cr_2O_7$ in H_2SO_4. The CO_2 given off was absorbed in an alkaline NaOH solution. The change in conductance in the alkaline solution was determined by an electrolytic conductance bridge. This conductivity change can be directly related to the amount of CO_2 absorbed. Enrhardt[92] used ultraviolet irradiation and potassium peroxydisulfate to oxidize organic matter in sea-water samples. The evolved CO_2 was also measured conductometrically. Lyutsarev[93] employed a dichromate oxidation to generate CO_2 from organic material in natural-water samples. The CO_2 was determined coulometrically after absorption in a solution of $Ba(OH)_2$, $BaCl_2$, and H_2O_2. The decrease in pH is sensed which in turn causes a current to flow at a platinum cathode placed directly in the absorbing solution. The generation of OH^- comes from the reduction of water. The integrated current necessary to restore the initial pH is related to the CO_2 given off by the sample. A silver anode in a KI solution is contained in a separate compartment. These procedures employ an initial step of removing inorganic carbon by acidification and heating or by gas sparging. Undesirable gaseous products that would cause a change in conductivity or pH are removed by scrubbing over materials such as MnO_2.

An instrument developed by Clifford[94] to measure TOD employs an electrochemical sensor. The liquid sample is injected into a combustion tube containing a platinum catalyst and operating at 900°C. A stream of N_2 gas containing a small quantity of O_2 allows for complete combustion of the sample. The uptake of O_2 is measured by a silver–lead electrochemical galvanic detector. The use of this TOD analyzer has correlated quite well with COD and BOD results.[95]

Formaro and Trasatti[96] examined the effect of adsorbed organics on the double-layer capacity and cathodic charging curves for platinum electrodes. Although the method is dependent on the extent of adsorption of the organic contaminant and of its effect on the double-layer capacity, its usefulness lies in the ability to correlate the organic carbon content of a sample whose organic composition does not change drastically, but where the concentration does vary. Essentially linear results were obtained with the variation of capacity as a function of organic concentration in a background solution of $1M$ $HClO_4$. Model compounds such as styrene, toluene, acrylonitrile, and amyl alcohol were used. It was estimated that organic carbon content as low as 0.03 ppm could be detected in industrial water. A regimen was given for the pretreatment and activation of the platinum indicator electrode.

(ii) Some Specific Organic Analyses

The polarographic analysis of nitrilotriacetate (NTA) has been discussed by Haberman[97] for river water and sewage samples. An anion-exchange preconcentration technique was used to determine NTA at low concentrations (0.0257–2.57 ppm); this step was omitted to analyze for concentrations up to 257 ppm. After removal of interferants by cation exchange, the NTA was determined by differential polarography as the In(III)–NTA complex by addition of an excess of In(III) to the sample. Interestingly, a number of normally occurring chelating agents did not interfere. As of this writing, the use of NTA in detergent formulations has been discontinued in the U.S. because of its suspected detrimental interactions with human life.

Kambara and Hasebe[98] used ac polarography to determine alkylbenzenesulfonates (ABS) in water. The ABS was extracted into $CHCl_3$ as a 1:1 complex with added methylene blue. After separation of the organic layer, the complex was returned to aqueous solution by back extraction, and the ac polarogram was analyzed at −0.1 V vs the mercury pool. Dodecylbenzene sulfonate in amounts from 0.07 to 0.28 mg was determined.

The polarographic analysis of aliphatic aldehydes and ketones was discussed by Hall,[99] who developed a technique whereby the aliphatic carbonyl was converted to an imine. The reaction of the ketone and primary amine is given as follows:

$$\begin{array}{c} R \\ \diagdown \\ C{=}O + H_2N{-}R' \rightleftarrows \\ \diagup \\ R \end{array} \quad \begin{array}{c} R \quad OH \\ \diagdown \diagup \\ C \\ \diagup \diagdown \\ R \quad NHR' \end{array} \quad \rightleftarrows \quad \begin{array}{c} R \\ \diagdown \\ C = NR' + H_2O \\ \diagup \\ R \end{array}$$

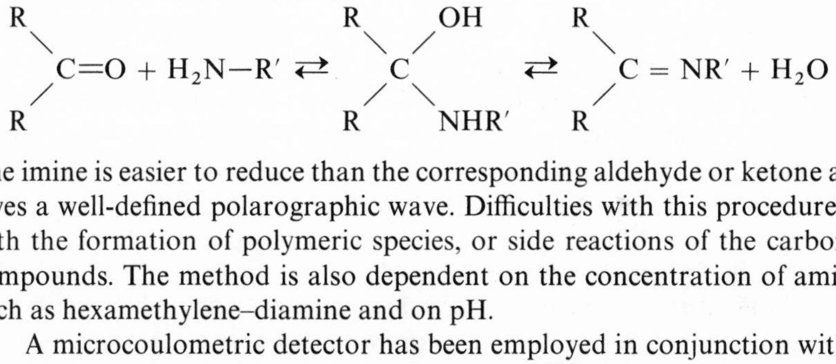

The imine is easier to reduce than the corresponding aldehyde or ketone and gives a well-defined polarographic wave. Difficulties with this procedure lie with the formation of polymeric species, or side reactions of the carbonyl compounds. The method is also dependent on the concentration of amine, such as hexamethylene–diamine and on pH.

A microcoulometric detector has been employed in conjunction with a gas chromatograph for the analysis of pesticides.[100] The gas chromatograph is used to separate pesticide fractions. The chromatograph effluent is then burned in an O_2 atmosphere in a microcombustion furnace. The chlorinated pesticide material is degraded to HCl and the gas is absorbed in the coulometric titration cell. Silver (I) is coulometrically generated to titrate the Cl^-. A pair of sensor electrodes is used to monitor the Ag^+ concentration so that the rate of Ag^+ generation is a function of the Cl^- absorbed. Sulfur-containing materials may be detected by a catalytic hydrogenation of the combustion furnace effluent which would produce H_2S. Thus, the total quantity of organic S and Cl can be determined using hydrogen reduction, and the quantity of Cl alone by bypassing that step.

The voltammetric analysis of triphenyl tin fungicide residues has been accomplished without an initial step of conversion of the tin to an inorganic

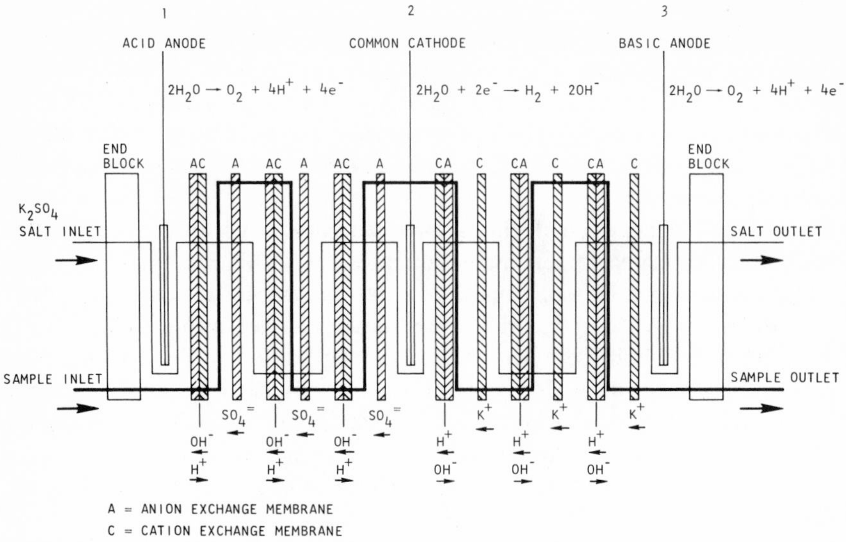

Figure 4. GGA phenol monitor electrolysis cell.

form.[101] Anodic stripping voltammetry permitted analysis of triphenyl tin compounds in concentrations as low as $10^{-8}M$. A procedure was developed for determination of phenyl tin fungicide residues in potato crops.

Anodic voltammetry at a rotating platinum electrode was used to determine cresols in industrial sewage.[102] A sensitivity of 0.5 ppm was reported for the anodic oxidation of cresols in a borate buffer.

Gulf General Atomic (GGA) has developed an instrument for continuous monitoring of phenols in waste water which uses an electrochemical cell to control pH. Martin and coworkers[103] have demonstrated that a whole class of phenols can be determined in waste water by differential spectrophotometry based on their change in ultraviolet absorption as a function of pH. In basic solution (pH ~ 12) the activated phenols show an absorption maximum near 2900 Å, while in acid solution (pH ~ 2) the absorption due to phenols at 2900 Å is very small. The heart of the GGA instrument is a compact electrochemical cell for alternating the pH of the sample stream between acid and base limits. A schematic diagram of the cell is shown in Fig. 4. The use of ion-exchange membranes permits the controlling electrodes to lie outside of the sample stream. This arrangement eliminates the generation of gas bubbles and hypochlorite in the sample, both of which can interfere with the later spectrophotometric measurement. As can been from the diagram, the "cell" really consists of two cells used in series. When electrodes 1 and 2 are connected, the sample is made acidic owing to transport of SO_4^{2-} from the recirculating K_2SO_4 electrolyte and H^+ from the water splitting at the anion–

cation exchange-membrane interface. Similarly, when electrodes 2 and 3 are connected, the sample is made basic by the transport of K^+ from the electrolyte and OH^- from the water splitting. Multiple passes increase the efficiency of pH control. The cell is arranged so that some of the acid produced during the acid cycle serves to clean the base-producing half of the cell. Sufficient salt may be kept in 1 liter of electrolyte solution for 2 weeks of operation. Overall sensitivity to phenols is based substantially on the optical cell path and selectivity of the photometer. The basic instrument package has an 0.1-ppm sensitivity with a 5-cm optical cell.

5. Chlorine Residual Analysis

To determine the effectiveness of chlorination as a disinfectant and oxidant in sewage treatment, the amount of chlorine residual must be known. The analysis, although a part of the control loop in many sewage plants, is after the fact in that there has been no successful way of determining the "chlorine demand" of waste waters.[104]

Electrochemical devices have been extensively employed as real-time residual chlorine monitors. The two techniques that have been described are the potentiometric and "amperometric." However, the "amperometric" method appears to be galvanic in that no external voltage is applied to the cell; rather, the electrolytic cell is capable of generating a current proportional to the active dissolved chlorine. Common to both techniques is the buildup of grease on the electrodes and cell compartment walls. This is a particular problem when a buffering agent is metered into the sample liquid to lower the pH to ~4. Most continuous-flow cells incorporate a quantity of grit to constantly scour the electrode surfaces.

Luck[105] and Baker[106] have discussed the potentiometric method for on-stream analysis of free chlorine. The sensor consists of a redox indicator electrode, such as platinum, and a reference electrode, Ag–AgCl or calomel. It is claimed that the sensor can be compensated for pH over a range of 5–9 and that the potential that is read is related to a composite biocidal strength of all active chlorine derivatives. Some of the reactions yielding active chlorine are listed below:

$$Cl_2 + H_2O \rightleftarrows HOCl + HCl$$

$$HOCl \rightleftarrows OCl^- + H^+$$

$$OCl^- + NH_3 \rightarrow NH_2Cl + OH^-$$

$$2HOCl + NH_3 \rightarrow NHCl_2 + 2H_2O$$

$$3HOCl + NH_3 \rightarrow NCl_3 + 3H_2O$$

The chloramines have a lower disinfectant strength than does hypochlorite.

The amperometric procedure for residual chlorine involves the following half-cell reactions:

$$OCl^- + 2H^+ + 2e^- \rightarrow Cl^- + H_2O \text{ (cathode)}$$

$$Cu \rightarrow Cu^{2+} + 2e \text{ (anode)}$$

The sensor cell consists typically of a platinum cathode and sacrificial copper anode. The hypochlorite is reduced at the cathode, and the driving force is provided by the copper anode. The current that flows in the external circuit is proportional to the concentration of hypochlorite and rate of mass transport to the cathode. The flow rate is generally controlled by the use of a constant-head, flow-chamber apparatus. If total residual chlorine is to be analyzed, including the chloramines, then the sample water is buffered to pH 4 and KI is metered in. The iodide is converted to iodine by the chlorinaceous oxidants, and sometimes by other oxidants. The liberated iodine then is the reacting species at the cathode. Systems of this type have been in use for over 20 years[106] and are described in the chemical literature[107–109] and in manufacturer's* literature. It is interesting to note that care must be taken in delivering the sample to the sensor cell. Max[108] points out that the use of a centrifugal pump breaks up enough suspended matter so that it reacts with more chlorine and drops the measured value. The use of a continuous, self-cleaning filter is also described, and the possibility of loss of chlorine to a buildup of filter cake exists as another potential source of error.

A cell for the determination of dissolve oxidants has been patented by Hersch and Deuringer.[110] It is similar to the "amperometric" cells normally employed for monitoring chlorine residual. The anode, made of silver or a type of active carbon, is oxidatively consumed. The cathode, at which dissolved chlorine or other oxidant is reduced, is made of porous carbon or platinized carbon. The completion of the galvanic circuit gives a current proportional to the rate of reaction of the dissolved oxidant.

Another continuous chlorine analyzer has been described by Takahashi.[111,112] Iron(II) is coulometrically generated from $Fe(NH_4)(SO_4)_2$ and is fed into the sample stream. After mixing, the potential of the iron Fe(II)–Fe(III) couple is sensed at a platinum–calomel electrode pair. A servo system controls the rate of coulometric generation in order to keep the sensor electrode potential constant with Fe(II) in slight excess. The coulometric current is proportional to the rate of conversion of Fe(II) to Fe(III) and therefore related to the chlorine content.

The details of some amperometric and potentiometric titrations for residual chlorine have been published[113–115] and show utility at low chlorine levels.

*Wallace and Tiernan, Inc. (Belleville, New Jersey); Fisher and Porter Co. (Warminster, Pennsylvania); Foxboro Co. (Foxboro, Massachusetts); Capital Controls (Colman, Pennsylvania).

Schwabe, Baer, and Steinhauer[116] have reviewed electrochemical methods for analysis and control of Cl_2 in cyanide detoxification, phenol in waste water, and other industrial applications.

6. Determination of Nitrogenous Materials

(i) Total Nitrogen

The determination of nitrogenous materials in water is extremely important for water pollution control. Nitrogen-containing compounds are a source of nutrient for the growth of algae. Nitrogenous compounds also present a toxicity hazard in drinking water. Kjeldahl nitrogen analyses in general are time consuming and are sensitive only to organic nitrogen and ammonia.

A method for total nitrogen that employs an electrochemical microcoulometric cell has been described in the literature.[117–119] A sample of water is injected into a hydrogen gas stream at ~600°C. It is then carried into a catalyst furnace at ~800°C, where the nitrogen compounds are reduced to NH_3. The catalyst can be granular nickel or a combination of rhodium on silica–alumina and nickel on magnesia. Halogens are converted to HX, sulfur to H_2S, and P to PH_3. These gases and any CO_2 that is formed are scrubbed out on a CaO column at 450°C. The resultant NH_3, free of acidic gases, is then bubbled into the coulometric cell, which contains a slightly acid solution of 0.04% Na_2SO_4. A glass-electrode–reference-electrode system senses the increase in pH. Acid is then generated in the cell at a platinum anode using a servo system that tries to keep the pH constant. Thus, the total charge passed to neutralize the NH_3 is a direct measure of the total nitrogen in the water sample. Albert and coworkers[117] have suggested that by varying the sample inlet temperature between ~300 and ~600°C the analysis might be split into given values for volatile organic and ammonia nitrogen and for ionic nitrogen (NO_3^- and NO_2^-). The sensitivity of this technique will give analyses down to 0.2 ppm nitrogen in less than 10 min.

(ii) Nitrate and Nitrite

Hartley and Curran[120] have described a polarographic method for the determination of NO_3^- by converting it to an aromatic nitro compound (4 nitro-2,6 xylenol). The sample is added to a nitration mixture of H_2SO_4 and HOAc to form the nitrosylium ion. The 2,6 xylenol is reacted via the process

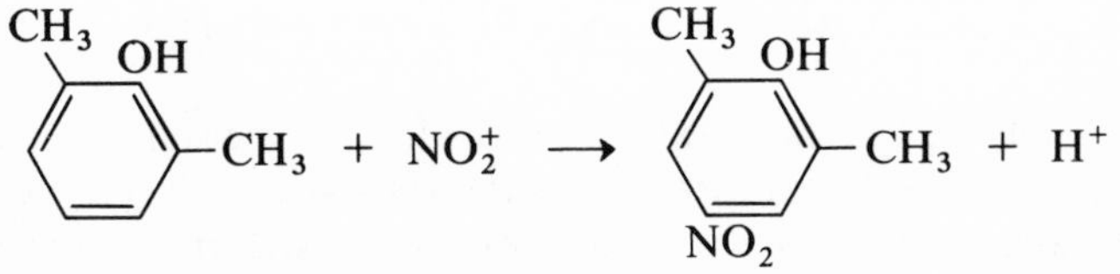

The nitro xylenol is then reduced polarographically with an $E_{1/2}$ of -0.27 V vs Hg–Hg_2SO_4. The diffusion current is proportional to $[NO_3^-]$ over a range of 1×10^{-3}–$2 \times 10^{-5} M$.

Annino and McDonald[121] developed polarographic methods for the analysis of NO_3^- and NO_2^- over a hundredfold mutual molar ratio. A pH 2 citric acid buffer was used for determination of NO_2^- by measuring the polarographic wave at -1.2 V vs SCE. Addition of NaN_3 to the acid test solution eliminated the NO_2^-, and NO_3^- was determined after the addition of uranyl ion and measurement of the height of the catalytic wave at -1.2 V vs SCE.[122]

Buck and Crowe[123] have developed a coulometric procedure for determination of NO_2^-. The procedure involves the generation of a known quantity of Mn(III), from a reagent containing Mn(II) and Fe(III), in excess of the amount of NO_2^- that is to be determined. The sample is added to the reagent solution, and the Mn(III) which remains after oxidation of NO_2^- to NO_3^- is back titrated by electrogeneration of Fe(II). A platinum indicator electrode is used to monitor the completeness of the back titration. Interference by halogen is eliminated by treatment with Ag^+.

Harrar[124] has used controlled potential coulometry at a platinum anode to determine NO_2^-. By using a pH 4.5 buffer, the loss of nitrous acid was minimized. The platinum anode did require pretreatment to secure a smooth oxidation of NO_2^- to NO_3^- at a controlled potential of $+0.95$ V vs SCE. Harrar lists a number of interferences and some ways of removing their effects; however, it would still appear that the method is not selective enough to be used directly for natural- or waste-water samples.

Manahan[125] used a combination of selective ion electrodes to potentiometrically determine NO_3^- by the method of standard addition. The reference electrode was a LaF_3-membrane F^- selective electrode and the indicator electrode a NO_3^- selective one. The method is essentially internally compensated for activity changes and shows a relative error of $<1\%$ at low ($5 \times 10^{-5} M$) NO_3^- levels.

7. Electroanalysis for Selected Materials Necessary for Water-Quality Estimation

(*i*) *Metals*

Calcium and Magnesium. A continuous polarographic determination of water hardness based on the reaction of calcium and magnesium with EDTA has been described.[126] Most common anions do not interfere; traces of iron and copper do, which would considerably limit the direct application of the method.

Chromium. A Czechoslovakian patent for the measurement of Cr(VI) in waste streams has been described.[127] It consists essentially of a concentration cell having a reference solution of CrO_3 in sulfuric acid at a carbon electrode connected via a glass frit to the sample solution. A platinum electrode in the sample solution senses the efficiency of removal of Cr(VI) in plating wastes by treatment with Fe^{2+}, SO_3^{2-}, SO_2, or $S_2O_5^{2-}$, by measuring the excess of the reductant.

Arsenic, Lead, Cadmium, and Some Other Trace Metals. Among a host of toxic inorganic materials, the ones thought to present the biggest hazards are arsenic, antimony, cadmium, lead, mercury, selenium, tellurium, and thallium, because of their use and abundance. Offner and Witucki[128] have proposed a DME polarographic system to serve as an alarm for toxic levels of these ions. Using a potential of −0.8 V vs SCE or −0.7 V for Te^{4+} all of the above-mentioned ions could be readily detected at concentrations of 5 ppm, well below their acute toxic levels. The As(V) must be prereduced to As(III) for polarographic analysis. In actual practice it would still have to be ascertained whether naturally occurring ligands would interfere with direct polarographic detection by complexation with these ions.

Maienthal and Taylor[129] have discussed the application of polarography, anodic stripping voltammetry, and linear-sweep voltammetry to the determination of trace inorganic constituents in water. They discuss procedures for determination of aluminum, arsenic, cadmium, copper, indium, iodine, iron, lead, tellurium, and zinc. By use of a comparative technique, precisions of 1% can be obtained at the 0.1-ppm level.

Arsenic was determined in natural waters over a concentration range of 0.25–50 mg/liter by conventional DME polarography.[130] Calcium oxide, citric acid, gelatin, and Trilon B were added to the water sample, giving a resultant pH of 6. The As(III) wave at −1.3 V was used to calculate the arsenic concentration either by standard addition or with a calibration curve. The effect of zinc and nickel at levels above 5 mg/liter was eliminated by use of Trilon B.

The determination of arsenic in drinking water down to a level of 5 ppb was discussed by Whitnack and Brophy.[131] Drinking-water samples were acidified with H_2SO_4, and a rapid linear-sweep voltammogram was taken at a DME after deoxygenation. The peak height of −0.97 V vs the mercury pool was evaluated by the method of standard addition. The U.S. Public Health Service recommends that the maximum permissible level of arsenic in drinking water be 50 ppb (see Table 2).

Samuel and Brunnock[132] employed square-wave polarography to analyze for copper and lead in the ppb range in crude oils. Although the method required considerable sample pretreatment, involving extraction into an aqueous phase followed by complete oxidation of residual organics, it might still have utility when applied to oily samples.

Lead in ppb concentrations was determined in Iowa City river water using square-wave polarography.[133] Organic surfactants, when present, interfered with the determination and were removed by fuming the sample with nitric and perchloric acids. The maximum sensitivity for lead was found at pH 3 in a perchlorate–fluoride supporting electrolyte.

Whitnack and Sasselli[134] have used anodic stripping voltammetry to determine copper, lead, cadmium, and zinc in sea water. Essentially no sample pretreatment was used other than deoxygenation of the sample. Concentrations of 1×10^{-9}–$3 \times 10^{-8} M$ were analyzed using preconcentration times of 10–30 min. A quartz cell was used to minimize contamination and adsorption errors.

Allen, Matson, and Mancy[135] have discussed the effect of complexation in the determination of trace metals in the natural environment by anodic stripping voltammetry. The dramatic increase in platable metal, lead, and copper, found upon acidification of a natural-water sample is a good example of the awareness that must be used when applying anodic stripping. The content of acid-exchangeable ligands can greatly affect both the determination of the trace metal and its effect on the aquatic environment.[136] The uptake rate of Cu(II) by ligands in river water was also examined. The indicator electrode that was used was the composite mercury–graphite electrode, based on the wax-impregnated graphite electrode of Matson, Roe, and Carritt.[137]

Ariel, Eisner, and Gottesfeld[138] investigated the use of "medium exchange" to determine copper in Dead Sea brine by anodic stripping voltammetry. Lead, cadmium, and zinc were determined fairly readily at a mercury electrode, even at short preelectrolysis times and rapid (220 mV/sec) anodic sweep rates. However, the copper determination was obscured by the beginning of the mercury oxidation wave. A procedure was developed whereby the copper was electrodeposited from the sample solution, but then a different stripping solution (medium exchange) was used to run the anodic oxidation. The use of an ammonium chloride–ammonium hydroxide stripping solution allowed determination of $4.5 \times 10^{-8} M$ copper in Dead Sea brine. The method appears to be applicable in a number of cases where normally overlapping peaks could be resolved.

Zinc was determined in the waste water of synthetic fiber plants by ac polarography.[139] Organics were removed by a heated evaporation of the sample in H_2SO_4 followed by another evaporation in HNO_3. The ac polarogram was measured in phosphoric acid solution.

Nickel and cobalt were determined at concentrations as low as $4 \times 10^{-8} M$ by linear-sweep voltammetry at a hanging mercury drop electrode.[140] Using a $0.1 M$ ammonia buffer containing $2 \times 10^{-4}\%$ dimethylglyoxime, nickel and cobalt could be analyzed in each other's presence at relative ratios of $\sim 20:1$. Essentially no pretreatment of the sample was required, and copper and iron did not interfere.

Uranium, Boron, and Molybdenum. Derivative pulse polarography was used to determine uranium in sea water.[114] Preconcentration was accomplished by extraction of the uranium into a solution of *bis*-(2-ethylhexyl) phosphoric acid. Elimination of other interfering ions, especially molybdenum, was brought about by a second extraction into ethyl acetate. The final pulse polarograms were related back to a sea-water content of a few micrograms per liter.

Ishibashi and coworkers[142] developed a method for analysis of uranium in sea water with a sensitivity of 1–2 ppb. The uranium was separated from the sample by solvent extraction into ether, followed by treatment with a chelating agent such as Mordant Blue 2R. Polarographic analysis of the uranium chelate gave the final answer.

A polarographic method for determination of uranium in sea and river water which does not involve a liquid–liquid extraction is discussed in Ref. 143. Uranium is concentrated on an anion-exchange column, and following several washings to remove possible interferants, it is eluted with 0.8*M* HCl.

The analysis of boron in soil and water samples has been accomplished through the conversion of boron compounds to BF_4^-.[144] Excess F^- from the reaction of HF with the prepared sample was removed by precipitation with Ca^{2+}. The analysis step measured the BF_4^- activity with a selective ion electrode. Samples containing 0.15–1500 ppm boron were analyzed with a relative standard deviation of $\leq 7\%$.

The determination of molybdenum by anodic stripping voltammetry was accomplished through the deposition of a thin film (<100 monolayers) of insoluble $MoO_2{\cdot}2H_2O$ on a mercury electrode.[145] In a pH 5 buffer, the reduction of Mo(VI) at −0.8 V vs Ag–AgCl (3*M* Cl^-) coated the electrode with $MoO_2{\cdot}2H_2O$. This film was then anodically stripped at constant current to give a stripping time proportional to the bulk molybdenum concentration. It was determined that a minimum stripping current density of 92 $\mu A/cm^2$ is required for linear response.

Mercury. The determination of mercury in natural and waste waters has recently become of great interest. Anodic stripping voltammetry would certainly appear to be a good candidate for low-level mercury analysis, 1 ppb $\cong 5 \times 10^{-8} M$. However, it must be cautioned that organomercury compounds found in the environment are not readily reduced at carbon or platinum electrodes. To ensure an accurate assay, the sample should be digested with a strong oxidant to convert all the mercury to inorganic Hg(II). A bulletin[146] published by Chemtrix Co. gives a method for stripping analysis of mercury in river water. The electrodeposition is made on a carbon cathode, such as a wax-impregnated graphite electrode, from the sample solution which has been made 0.05*M* in KCNS. It is stated that an electrolysis time of 30 min might be required to detect 1 ppb, while 1 ppm would take only

10 min. Once again, caution should be applied to any direct electrochemical analysis of mercury from "real" water systems.

A complexometric titration for mercury at low levels, 50–150 μg, was developed using *bis*(2-hydroxyethyl) dithiocarbamate (DTC) as the titrant.[147] The method uses an amperometric end-point determination that employs a rotating platinum electrode (RPE) maintained at −0.2 V vs SCE. Inorganic Hg(II) is titrated directly with millimolar DTC in a pH 9 borax, KCl buffer. The decrease in the current at the RPE is monitored as a function of titrant addition, the end point being determined graphically by extrapolation. In the case of organomercurials, the samples are first wet oxidized with a HNO_3–$HClO_4$ mixture. This pretreatment is a good one for the application of electrochemical techniques to the determination of mercury in the environment.

Anodic Stripping Voltammetry and Neutron Activation Analysis. Mark and coworkers[148–151] have discussed a technique for using anodic stripping voltammetry in conjunction with neutron activation analysis for determination of trace metals in water. Deposition of thin metal films on pyrolytic graphite, followed by neutron irradiation of a section of the graphite, was shown to be an effective way of preconcentrating and analyzing for trace metals. The use of a section of ion-exchange membrane to preconcentrate metal ions with subsequent metal deposition onto a pyrolytic graphite cathode from the ion exchanger was also helpful in eliminating some interferences found in natural waters.

(*ii*) *Nonmetallic Materials*

Fluoride. Morrow and Henry[152] have devised an "amperometric" cell for the continuous monitoring of fluoride. They found that a linear current-concentration relation could be obtained at a high-purity (99.99 % minimum) rotating aluminum anode in the limiting current region of −0.3 to −0.8 V vs SCE. It was important to buffer the test solution to pH ~4, and $10^{-4}M$ EDTA was added to remove certain interferences such as Cu(II). Calibrations were linear over the $[F^-]$ range of 0.2–2.0 ppm. In the monitoring instrument, a high-purity rotating aluminum-rod anode and concentric aluminum-alloy-ring cathode were used in essentially a galvanic configuration. Shorting of the cathode to the anode through a load resistance maintained the anode at −0.38 V vs SCE. Input sample water was first treated with an EDTA solution, passed through a flowmeter, and finally treated with buffer to bring the pH to ~4, before it flowed by the sensing electrodes. Thermistor temperature compensation held the error to less than 1 % for a sample temperature between 43 and 73°F. The Fe^{3+} and PO_4^{3-} errors could not be eliminated by reagent addition, and had to be accounted for in a calibration run.

Beyermann[153] used an aluminum anode for the determination of nanogram quantities of fluoride. With proper control of electrode pretreatment and solution pH, the current produced by a potential difference of 1.0 V between the aluminum anode and a platinum or aluminum cathode was essentially a linear function of $[F^-]$. The extent of interferences and electrode behavior limited the applicability of the technique.

Babcock and Johnson[154] have shown that the application of fluoride selective electrodes in municipal water supplies is extremely effective. They demonstrated that the probable effect of normally encountered ionic strength is small and that at typical pH's above 7 the error from $Al^{3+} + Fe^{3+}$ complexes is greatly reduced. The interference from other anions is negligible, except for OH^-, but at a pH of <9, which is very common, the effect of OH^- is minimized.

Durst[155] has applied the technique of linear null-point potentiometry to determine subnanogram quantities of fluoride. The titration is carried out in a concentration cell which uses two F^- selective ion indicator electrodes (LaF_3 membrane). The same solution, ~10 μliters, is in contact with one fluoride electrode and is connected to the titration vessel via a KNO_3 salt bridge. The fluoride concentration in the titration vessel is varied by the addition of a standard solution until an essentially zero potential difference exists between the fluoride electrode in the sample and the fluoride electrode in the titration vessel. Relative activities are made constant by the use of KNO_3.

A method for analysis of F^- at low concentrations (0.05–20 ppm) has been described by Bond and O'Donnell.[126] The method involves the shift in the polarographic half-wave potential of the uranium V–III reduction in acid solution. Bond and O'Donnell observed a linear cathodic shift in $E_{1/2}$ as the $[F^-]$ was increased from 0.05 to ~1.25 ppm. The rest of the range was obtained by use of a calibration curve. Phosphate interference was said to be not serious, owing to the use of a supporting electrolyte of 1 *M* HCl; however, F^- sensitivity was decreased. Resolution of ~1 mV in the $E_{1/2}$ determination was required. The effect of other ligands present in natural waters was not discussed.

Fluoride has been determined in the 19–190-ng range by a differential kinetic method.[157] It is based on the inhibition of the $ZrOCl_2$ catalyst, by F^-, in the reaction

$$BO_3^- + 2I^- \xrightarrow{(ZrOCl_2)} BO_3^{3-} + I_2$$

where the I_2–I^- ratio is sensed at a platinum electrode.

Cyanide, Sulfite, and H_2S. One amperometric method for the determination of cyanide in waste water involves the use of a gold indicator electrode.[158] The electrode is maintained at +150 mV vs the SCE, and the resultant anodic current is reported to be proportional to CN^- concentration (pH = 11). Periodically, the gold electrode is held at −400 mV so that its activity is regenerated. When an excess of chlorine is present, the CN^- removal is essentially complete, and the anodic current falls to a residual level. Lueck[159] has developed a potentiometric method for controlling CN^- detoxification in waste waters using a gold indicator electrode. A wiper device is used to remove deposited nonprecious metals.

Strafelda and Dolezal[160] have given a brief review of electrolytic methods for continuous analysis of sulfite-containing waters. They also have developed a polarographic analyzer for the determination of concentrate sulfite solutions in order to provide a feedback loop for the most efficient use of reagent in an SO_2-removing stack-gas scrubber. They found that the wave height for SO_2 reduction in 3.5*M* H_2SO_4 was proportional to concentration up to 0.1*M*. The automated system incorporated degassing, sample dilution, H_2SO_4 metering, a DME, and zinc-amalgam anode.

Analysis of H_2S in the operation of a sewage plant was found to be necessary from the standpoint of both corrosion and odor. Garber, Nagano, and Wada[161] adapted an electrolytic bromine coulometric titrator to determine dissolved H_2S and H_2S in the aeration supply air. Air was bubbled through the sample in a specially designed scrubber unit which permitted the stripping of H_2S from the liquid sample. The gaseous sample containing H_2S was then bubbled through the titrator electrolytic cell, where the sulfide was oxidized by electrolytically produced bromine. Mercaptans were removed by a selective scrubber. Small-chain unsaturated hydrocarbons interfered since they reacted with bromine.

Electrochemical Device for Gas Chromatographic Analysis. Lovelock, Simmonds, and Shoemake[162] have described a compact electrochemical system which can be used as a source and sink for H_2 carrier gas in a novel gas chromatographic application. The cathode is a hollow cylinder of a palladium-silver alloy immersed in an electrolyte of KOH–LiOH. The H_2 generated at the cathode rapidly diffuses into the inner wall of the cylinder, where reasonably high H_2 pressures can be built up. This net pressure of H_2 is then used as a source of carrier gas to derive a sample through a gas chromatographic column. A small amount of makeup gas is used to carry the sample into the main flow of the carrier gas. The effluent gas, mainly H_2, leaving the chromatographic column then passes through another palladium–silver tube which acts as the anode in the electrochemical cell. Rapid diffusion of the H_2 out of the gas stream to the electrode–electrolyte interface completes the electrochemical loop, since H_2 is oxidized at the anode. The remainder of the makeup gas and sample gas then flows on to the detector. Thus, the net electrode reactions are effectively

$$2H_2O \xrightarrow{+2e} H_2 + 2OH^- \quad \text{(cathode)}$$

$$H_2 + 2OH^- \xrightarrow{-2e} 2H_2O \quad \text{(anode)}$$

so that the equilibrium potential tends to zero.

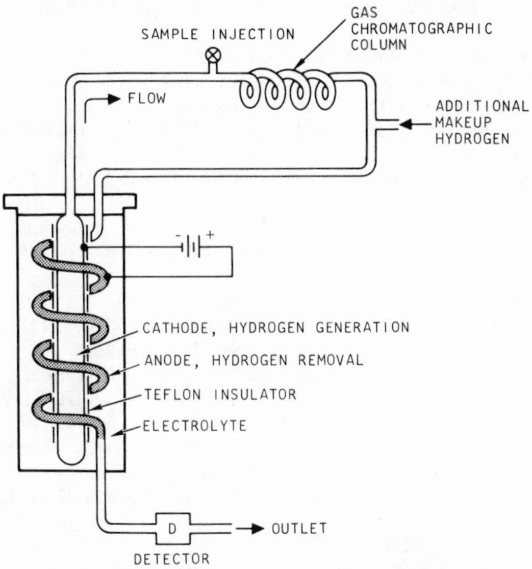

Figure 5. Electrolytic hydrogen generator and gas chromatograph.

A schematic diagram is shown in Fig. 5. The makeup gas could be H_2, generated from the same electrolyte as the main loop in a separate cathode tube. Since not much makeup gas is required, the electrolysis of 1 g of water would supply enough H_2 for 20–200 h of operation. Although the system requires further development to reduce corrosion problems and the poisoning of the electrodes by materials such as H_2S, it represents an excellent beginning of a lightweight, compact, and simple system. Hopefully, it will be pursued in both air and water pollution monitoring areas.

REFERENCES

[1]Public Health Service publication No. 956, March 1963.
[2]*Anal. Chem.*, April editions, annual reviews.
[3]*J. Water Pollution Control Federation*, June editions, annual reviews.
[4]*Treatise on Analytical Chemistry*, Eds., I. M. Kolthoff and P. J. Elving, Interscience, New York, 1963, Part 1, Vol. 4, Chaps. 42–51.
[5]*Electroanalytical Chemistry*, Ed., A. J. Bard, Marcel Dekker, New York, 1966, Vols. 1–3.
[6]W. Latimer, *Oxidation Potentials*, 2nd ed., Prentice Hall, New York, 1952, p. 46.
[7]I. G. Bowen and V. H. Regener, *J. Geophys. Res.* **56** (1951) 307.
[8]A. S. Britaev, *Tr. Tsentr. Aerol. Observ.* **37** (1960) 13; *Chem. Abstr.* **56** (1962) 6663.
[9]S. A. Poulsen and H. E. Saunders, U.S. Pat. 3,031,272, May 24, 1962; *Chem. Abstr.* **57** (1962) 1564b.
[10]A. W. Brewer and J. R. Milford, *Proc. Roy. Soc.* (*London*) **A256** (1960) 470.
[11]U.S. Pat. 3,038,848, June 12, 1962.
[12]G. M. Mast and H. E. Saunders, *Inst. Soc. America Trans.* **1** (1962) 325.
[13]L. Potter and S. Duckworth, *J. Air Pollution Control Assoc.* **15** (1965) 207.
[14]P. H. Gudiksen, P. W. Hildebrandt, and J. J. Kelley, *J. Geophys. Res.* **71** (1966) 5221.

[15]A. G. Buswell and A. Keen, *Rev. Gen. Caoutochouc Plastiques* **40** (1963) 1161; *Chem. Abstr.* **62** (1965) 1079g.
[16]P. Hersch and R. Deuringer, *Anal. Chem.* **35** (1963) 897.
[17]B. D. Epstein, E. Dalle-Molle, and J. S. Mattson, *Carbon* (in press), and references therein.
[18]G. A. Rost and J. Swartz, U.S. Pat. 3,234,117, February 8, 1966.
[19]A. F. Wartburg, A. W. Brewer, and J. P. Lodge, *Air Water Pollution* **8** (1964) 21.
[20]T. L. Duffy and P. L. Pelton, *Am. Ind. Hyg. Assoc. J.* **26** (1965) 544.
[21]B. E. Saltzman, U.S. Public Health Service 999-AP-11, D-1-D-5, 1965; *Chem. Abstr.* **65** (1966) 7889f.
[22]R. H. Hendricks and L. B. Larsen, *Am. Ind. Hyg. Assoc. J.* **27** (1966) 80.
[23]B. E. Saltzman and A. F. Wartburg, *Anal. Chem.* **37** (1965) 779.
[24]F. Schulze, *Anal. Chem.* **38** (1966) 748.
[25]A. W. Boyd, C. Willis, and R. Cyr, *Anal. Chem.* **42** (1970) 670.
[26]R. S. Ingols, R. H. Fetner, and W. H. Eberhardt, *Advan. Chem. Ser.* **21** (1959) 102.
[27]D. H. Byers and B. E. Saltzman, *Advan. Chem. Ser.* **21** (1959) 93.
[28]D. T. Sawyer, R. S. George, and R. C. Rhodes, *Anal. Chem.* **31** (1959) 2.
[29]R. Chand and P. R. Cunningham, *IEEE Trans. on Geosci. Elect.* **GE-8** (1970) 158.
[30]R. E. Rostenbach and R. G. Kling, *J. Air Pollution Control Assoc.* **12** (1962) 459.
[31]R. D. Cadle and H. S. Johnston, *Proc. Nat. Air Pollution Symp.*, Second Symp., Stanford Research Institute, Menlo Park, California, 1952, pp. 28–34.
[32]M. D. Thomas, J. A. MacLeod, R. C. Robbins, R. C. Goettelman, R. W. Eldridge, and L. H. Rogers, *Anal. Chem.* **28** (1956) 1810.
[33]R. DiMartini, *Anal. Chem.* **42** (1970) 1102.
[34]M. D. Thomas and R. J. Cross, *Indust. Eng. Chem.* **20** (1928) 645.
[35]S. G. Booras and C. E. Zimmer, *J. Air Pollution Control Assoc.* **18** (1968) 612.
[36]American Society for Testing Materials, ASTM Designation: D 1355-60 (reapproved 1967).
[37]P. Urone, J. B. Evans, and C. M. Noyes, *Anal. Chem.* **37** (1965) 1104.
[38]F. Cabot, *Ind. Res.* **12** (1970) No. 9, 70.
[39]P. A. Shaffer, A. Beiglio, and J. A. Brockman, *Anal. Chem.* **20** (1948) 1008.
[40]R. R. Austin, G. K. Turner, and L. E. Percy, *Instruments* **22** (1949) 588.
[41]Bulletin 4097A, Beckman Instruments, Fullerton, California.
[42]Chand and Cunningham, *loc. cit.* (Ref. 29).
[43]D. F. Adams and R. K. Koppe, *J. Air Pollution Control Assoc.* **17** (1967) 161.
[44]P. J. Ovenden, *J. Electroanal. Chem.* **2** (1961) 80.
[45]J. L. Roberts and D. T. Sawyer, *J. Electroanal. Chem.* **7** (1964) 315.
[46]L. A. Elfers and C. E. Decker, *Anal. Chem.* **40** (1968) 1658.
[47]S. Kaye and M. Griggs, *Anal. Chem.* **40** (1968) 2217.
[48]P. A. Corrigan, V. E. Lyons, G. D. Barnes, and F. G. Hall, *Environ. Sci. Technol.* **4** (1970) 116.
[49]R. G. Bates, *Determination of pH, Theory and Practice*, John Wiley and Sons, New York, 1964.
[50]R. A. Horne, *Marine Chemistry*, Interscience, New York, 1969, pp. 280–281.
[51]A. F. Mentink, Federal Water Quality Administration, Cincinnati, Ohio, private communication.
[52]*Ion Selective Electrodes*, Ed., R. A. Durst, NBS Special Publication 314, U.S. Government Printing Office, Washington, D.C., November 1969.
[53]G. A. Rechnitz, *Chem. Eng. News* **43** (1967) No. 25, 146.
[54]J. M. Riseman, *Amer. Lab.* **7** (1969) 32.
[55]J. B. Andelman, *J. Water Pollution Control Federation* **40** (1968) 1844.
[56]I. Nagelberg, L. J. Braddock, and G. J. Barbero, *Science* **166** (1969) 1403.
[57]G. G. Guilbault and P. J. Brignac, *Anal. Chem.* **41** (1969) 1136.
[58]G. G. Guilbault and J. Montalvo, *J. Am. Chem. Soc.* **91** (1969) 2164.
[59]G. G. Guilbault and E. Hrabankova, *Anal. Chem.* **42** (1970) 1779.
[60]T. Higuchi, C. R. Illian, and J. L. Toussounian, *Anal. Chem.* **42** (1970) 1674.
[61]J. P. Hoare, *The Electrochemistry of Oxygen*, Interscience, New York, 1968, Chap. 6.
[62]I. M. Kolthoff and C. S. Miller, *J. Am. Chem. Soc.* **63** (1941) 1013.
[63]K. H. Mancy and D. A. Okun, *Anal. Chem.* **32** (1960) 108.
[64]C. P. Tyler and J. H. Karchmer, *Anal. Chem.* **31** (1959) 499.

[65]J. H. Karchmer, *Anal. Chem.* **31** (1959) 502, 509.
[66]E. Foyn, *Chem. Abstr.* **55** (1961) 12715.
[67]W. Lueck, *Chem. Abstr.* **70** (1969) 50633q.
[68]L. C. Clark, *J. Appl. Physiol.* **6** (1953) 189.
[69]K. H. Mancy, D. A. Okun, and C. N. Reilley, *J. Electroanal. Chem.* **4** (1962) 65.
[70]J. R. Neville, *Rev. Sci. Instr.* **33** (1962) 51.
[71]P. Hersch, *Nature* **169** (1952) 792.
[72]Sawyer, *loc. cit.* (Ref. 28).
[73]D. E. Carritt and J. W. Kanwisher, *Anal. Chem.* **31** (1959) 5.
[74]Mentink, *loc. cit.* (Ref. 51).
[75]P. Kiss and J. Somloi, *Chem. Abstr.* **67** (1967) 36305a.
[76]C. Halpert and R. T. Foley, *J. Electroanal. Chem.* **6** (1963) 426.
[77]H. Lipner, L. R. Witherspoon, and V. C. Champeaux, *Anal. Chem.* **36** (1964) 204.
[78]F. J. H. Mackereth, *J. Sci. Instr.* **41** (1964) 38.
[79]F. J. H. Mackereth, U.S. Pat. 3,322,662, May 30, 1967; *Chem. Abstr.* **67** (1967) 78556.
[80]R. Briggs and M. Viney, *J. Sci. Instr.* **41** (1964) 78.
[81]V. T. Stack, Jr., *Ann. ISA Conf. Proc.* **21** (1966) 5.2-2-66.
[82]*Standard Methods for the Examination of Water and Wastepaper*, 12th ed., American Public Health Association, New York, 1965, p. 415.
[83]M. D. Lilley, J. B. Storey, and R. W. Raible, *J. Electroanal. Interfac. Electrochem.* **23** (1969) 425.
[84]R. B. Rayment, *Anal. Chem.* **34** (1962) 1089.
[85]D. P. Lucero, *Anal. Chem.* **40** (1968) 707.
[86]*Anal. Chem. Laboratory Digest.*
[87]J. C. Young, W. Garner, and J. W. Clark, *Anal. Chem.* **37** (1965) 784.
[88]W. J. O'Brien, *Dissertation Abstr.* **26** (1966) 4559.
[89]W. J. O'Brien and S. W. Clark, *Poll. Abstr.* P70-03280 (1970).
[90]H. Steinecke, *Chem. Abstr.* **66** (1967) 118715c.
[91]J. Krey and K. H. Szekielda, *Z. Anal. Chem.* **207** (1965) 338.
[92]M. Enrhardt, *Chem. Abstr.* **71** (1969) 128533j.
[93]S. V. Lyutsarev, *Chem. Abstr.* **71** (1969) 73884s.
[94]D. A. Clifford, *Chem. Abstr.* **71** (1969) 94541t.
[95]A. L. Goldstein, W. E. Katz, F. H. Meller, and D. M. Murdoch, "Total Oxygen Demand," American Chemical Society, Division of Water, Air, and Waste Chemistry Meeting, Atlantic City, N.J., September 12, 1968.
[96]L. Formaro and S. Trasatti, *Anal. Chem.* **40** (1968) 1060.
[97]J. P. Haberman, *Anal. Chem.* **43** (1971) 63.
[98]T. Kambara and K. Hasebe, *Japan Analyst* **14** (1965) 491.
[99]M. E. Hall, *Anal. Chem.* **31** (1959) 2007.
[100]L. A. Cavanagh, D. M. Coulson, and J. E. DeVries, *Chem. Abstr.* **57** (1962) 8955f.
[101]M. D. Booth and B. Fleet, *Anal. Chem.* **42** (1970) 825.
[102]O. V. Sheveleva and L. Ya. Buidakova, *Chem. Abstr.* **66** (1967) 13905m.
[103]J. M. Martin, C. R. Orr, C. B. Kincannon, and J. L. Bishop, *J. Water Pollution Control Federation* **39** (1967) 21.
[104]Q. C. Turtle, *J. Am. Water Works Assoc.* **61** (1969) 293.
[105]J. R. Luck, *Chem. Abstr.* **66** (1967) 5677b.
[106]R. J. Baker, *Ind. Water Eng.* **6** (1969) No. 1, 20.
[107]W. C. Conkling and H. R. Holland, *Water Wastes Eng.* **3** (1966) 45.
[108]R. C. Max, *Ann. ISA Conf. Proc.* **21** (1966) 2.1-1-66.
[109]H. J. Krum, *Water Sewage Works* **98** (1951) No. 9.
[110]P. A. Hersch and R. Deuringer, *Chem. Abstr.* **64** (1966) 9434g.
[111]T. Takahashi, *Chem. Abstr.* **56** (1962) 1301f.
[112]T. Takahashi and H. Sakurai, *Chem. Abstr.* **62** (1965) 9770.
[113]N. J. Nicolson, *Analyst* **90** (1965) 187.
[114]J. J. Morrow, *J. Am. Water Works Assoc.* **58** (1966) 363.
[115]K. D. Dobryshin, I. E. Flis, I. M. Vorobev, and V. A. Kokushkina, *Zh. Anal. Khim.* **21** (1966) 752.

[116]H. C. S. Schwabe, H. J. Baer, and H. Steinhauer, *Chem. Abstr.* **66** (1967) 88526x.
[117]D. K. Albert, R. L. Stoffer, I. J. Oita, and R. H. Wise, *Anal. Chem.* **41** (1969) 1500.
[118]R. T. Moore and J. A. McNulty, *Environ. Sci. Technol.* **3** (1969) 741.
[119]R. L. Martin, *Anal. Chem.* **38** (1966) 1209.
[120]A. M. Hartley and D. J. Curran, *Anal. Chem.* **35** (1963) 686.
[121]R. Annino and J. E. McDonald, *Anal. Chem.* **33** (1961) 475.
[122]I. M. Kolthoff, W. E. Harris, and G. Matsuyama, *J. Am. Chem. Soc.* **66** (1944) 1782.
[123]R. P. Buck and T. J. Crowe, *Anal. Chem.* **35** (1963) 697.
[124]J. E. Harrar, *Anal. Chem.* **43** (1971) 143.
[125]S. E. Manahan, *Anal. Chem.* **42** (1970) 128.
[126]F. Strafelda and M. Stastny, *Chem. Abstr.* **66** (1967) 68781y.
[127]*Chem. Abstr.* **66** (1967) 101106g.
[128]H. G. Offner and E. F. Witucki, *J. Am. Water Works Assoc.* **60** (1968) 947.
[129]E. J. Maienthal and J. K. Taylor, *Am. Chem. Soc., Div. Water Waste Chem., Preprints* **7** (1967) 92.
[130]L. A. Davidyuk, *Chem. Abstr.* **64** (1966) 19192a.
[131]G. C. Whitnack and R. G. Brophy, *Anal. Chim. Acta* **48** (1969) 123.
[132]B. W. Samuel and J. V. Brunnock, *Anal. Chem.* **33** (1961) 203.
[133]E. B. Buchanan, T. D. Schroeder, and B. Novosel, *Anal. Chem.* **42** (1970) 370.
[134]G. C. Whitnack and R. Sasselli, *Anal. Chim. Acta* **47** (1969) 267.
[135]H. E. Allen, W. R. Matson, and K. H. Mancy, *J. Water Pollution Control Federation* **42** (1970) 573.
[136]M. E. Bender, W. R. Matson, and R. A. Jordan, *Environ. Sci. Technol.* **4** (1970) 520.
[137]W. R. Matson, D. K. Roe, and D. E. Carritt, *Anal. Chem.* **37** (1965) 1594.
[138]M. Ariel, U. Eisner, and S. Gottesfeld, *J. Electroanal. Chem.* **7** (1964) 307.
[139]V. M. Chernov, *Chem. Abstr.* **67** (1967) 67419.
[140]E. N. Vinogradova and G. V. Prokhorova, *Zh. Anal. Khim.* **23** (1968) 711.
[141]G. W. C. Milner, J. D. Wilson, G. A. Barnett, and A. A. Smales, *J. Electroanal. Chem.* **2** (1961) 25.
[142]M. Ishibashi, T. Fujinaga, R. Izutsu, T. Yamamoto, and H. Tamura, *Chem. Abstr.* **57** (1962) 4478c.
[143]J. Korkisch, A. Thiard, and F. Hecht, *Mikrochim. Acta* **1956**, 1422.
[144]R. M. Carlson and J. L. Paul, *Chem. Abstr.* **71** (1969) 111888f.
[145]P. Lagrange and J. P. Schwing, *Anal. Chem.* **42** (1970) 1844.
[146]*The Polarogram*, No. 6, Chemtrix, Beaverton, Oregon, July 1970.
[147]S. Ehrlich-Rogozinsky and R. Sperling, *Anal. Chem.* **42** (1970) 1089.
[148]B. H. Vassos, F. J. Berlandi, T. E. Neal, and H. B. Mark, *Anal. Chem.* **37** (1965) 1653.
[149]U. Eisner, J. M. Rottschafer, F. J. Berlandi, and H. B. Mark, *Anal. Chem.* **39** (1967) 1466.
[150]U. Eisner and H. B. Mark, *Tal.* **16** (1969) 27.
[151]H. B. Mark, U. Eisner, J. M. Rottschafer, F. J. Berlandi, and J. M. Mattson, *Environ. Sci. Technol.* **3** (1969) 165.
[152]J. J. Morrow and J. L. Henry, *J. Am. Water Works Assoc.* **59** (1967) 245.
[153]K. Beyermann, *Z. Anal. Chem.* **194** (1963) 1.
[154]R. H. Babcock and K. Johnson, *J. Am. Water Works Assoc.* **60** (1968) 953.
[155]R. A. Durst, *Anal. Chem.* **40** (1968) 931.
[156]A. M. Bond and T. A. O'Donnell, *Anal. Chem.* **40** (1968) 1405.
[157]D. Klockow, H. Ludwig, and M. A. Girando, *Anal. Chem.* **42** (1970) 1682.
[158]K. Schwabe and T. Berndt, *Chem. Abstr.* **67** (1967) 11265t.
[159]W. Lueck, *Chem. Abstr.* **67** (1967) 120028v.
[160]F. Strafelda and J. Dolezal, *Collection Czech. Chem. Commun.* **32** (1967) 2707.
[161]W. F. Garber, J. Nagano, and F. F. Wada, *J. Water Pollution Control Federation* **42** (1970) R209.
[162]J. E. Lovelock, P. G. Simmonds, and G. R. Shoemake, *Anal. Chem.* **42** (1970) 969.

Chapter 7

THE PROSPECT OF ABUNDANT ENERGY*

R. Philip Hammond

Oak Ridge National Laboratory
Oak Ridge, Tennessee

I. INTRODUCTION

It has been shown that electrochemical processes can perform a wide variety of functions and can substitute for other ways of accomplishing man's tasks, frequently in a cleaner and more efficient fashion. This substitution of electric energy for other inputs is of more general significance than is indicated by the examples given. In principle, energy may be looked upon as the only key input, since everything else can be collected and recycled. From the environmental standpoint, then, the degree to which man can minimize his consumption of scarce materials through recycle will determine how well he can preserve the pleasant face of the earth. The cost of energy is a determining factor in such considerations. If energy can be made sufficiently cheap, many more substitutions of electric intensive industrial processes can occur. Figure 1 shows some examples.

Another urgent aspect relating man to his future environment is the shortage of land. As described below, sufficiently low-cost energy can make available to man's use vast areas now useless, so that the pressure on the 11% of the land now used for agriculture will be diminished.

The question of the long-term supply of energy and the possibility of lower costs are of governing importance to the kind of environment man can expect in the future. This chapter explores the basis for future energy supplies and the possibilities for substantial reductions in cost.

*Research sponsored by the U.S. Atomic Energy Commission under contract with the Union Carbide Corp.

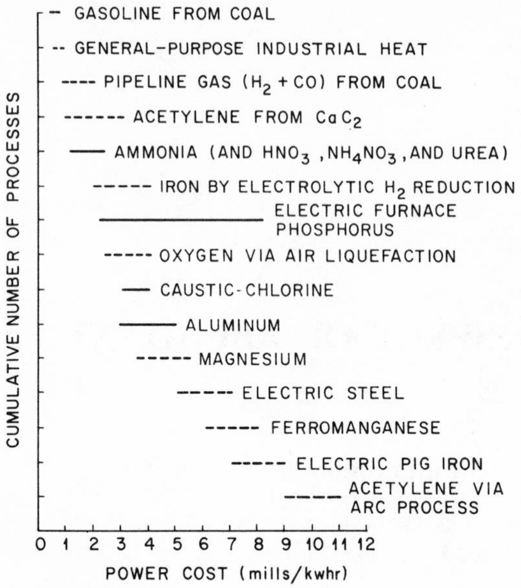

Figure. 1. Elasticity of the demand for power.

II. ENERGY SUPPLY THROUGH REACTORS

The prospect of an inexhaustible supply of energy has been recognized by some since the earliest days of the discovery of nuclear energy. One of the best papers on the subject is *Energy as an Ultimate Raw Material, or Problems of Burning the Sea and Burning the Rocks* by Alvin M. Weinberg. Burning the sea refers to the energy that can be made available by fusing the heavy hydrogen isotope deuterium into helium. Once the fusion reactor has been perfected, the deuterium in the oceans can provide an immense energy source. Half a cubic mile of water would yield the total energy used by man to date. The energy needed to separate the deuterium from the water would be a negligible fraction. Burning the rocks, on the other hand, is accomplished by a breeder type of fission reactor which is capable of releasing the energy of all the uranium and thorium in the fuel material instead of only the small portion (1 %) that is usable in ordinary reactors.

The ores which contain uranium and thorium are widespread and of varying concentrations. These regular ores contain a vast amount of energy and will last for many centuries. A still larger reservoir of energy is contained in shale deposits, some of which contain 200–500 ppm uranium. The much vaunted oil shales are usually between tenfold and a hundredfold richer in energy as nuclear ores than as combustibles. But the energy in the granites of the earth's crust far outweighs all of these. Samples indicate that the granites contain from 12 to 200 ppm of uranium and thorium. Of this, about one-third to one-half is readily extractable by crushing and leaching. The energy consumed in quarrying, crushing, and leaching is equivalent to less than 1 % of the yield. The supply of energy in both the granite and the seas is almost unthinkably large. Either source would keep us going at a profligate rate for many millions of years—as long as the solar system will last.

How close are we to tapping these great reservoirs of energy? The fusion reactor is certainly some distance away from practicality. In spite of recent strides spurred by the Russian Tokamak device, no system has yet come even close to yielding a net output of energy, and no one predicts less than a 20-year development period ahead. But the eventual outcome seems at least technically achievable. The economics are unknown.

The breeder reactors, on the other hand, have the advantage that they can build upon the enormous experience gained with military and civilian reactors. The differences between a breeder and present-day converter reactors are more of degree than of radically different principles. There is no doubt of their practicality, since operating prototypes for several different types of breeders exist, and large-scale versions are under construction in several countries. Most of these will be starting operation in the next 2–3 years, and plans for the following generation of fully commercial breeders should be possible by 1980. The year 1990 should see the more advanced breeders mentioned below, *and by 2000 the low-cost potential of nuclear energy sources should be fully apparent.* At that time authorities expect the world rate of construction of nuclear power stations will be approximately 350,000 MW capacity per year, or a 10,000-MW station completed every 10 days.

In order to compare these unfamiliar sources of energy with regular fuels one must look at the costs and consumption rates. Table 1 gives some units for energy with which we can compare large quantities and costs, while

Table 1. Units for Heat Energy

1000 Btu =	0.01 therm	= 1 ft^3 natural gas
	0.293 kW·h	
	252 kcal	$2\frac{1}{2}$ oz wheat
10^6 Btu =	1 MBtu	= 77 lb coal
	10 therms	1 house·day
		0.013 g uranium
		8 lb granite
10^{18} Btu =	1Q	= 38×10^9 tons coal
		13×10^3 tons uranium

Table 2. World Energy Consumption

Cumulative total (1 million B.C. to 1960) = $13Q$ (made up of 150 mile^3 of wood, etc., + 15 mile^3 of coal)
Present consumption rate (per year) = $0.15Q$ (45 MBtu per capita)
Expected consumption rate by 2050 A.D. = $5Q$ (300 MBtu per capita)

Table 2 shows the cumulative fuel energy consumed by man. Table 3 shows a comparison of a number of sources of energy and the quantity required to furnish $1Q$ of energy, 1 MBtu of energy, and the relative fuel costs. It is clear why reactors are displacing coal (when we build them large enough) and why we have such a strong incentive to develop breeders (see Fig. 2).

The world average use of fuel energy amounts to the equivalent of a steady rate of 1.5 kW per capita (36 kW-h per day), but as Fig. 3 shows, its distribution is far from uniform. Some backward peoples use very little more energy than their own food calories (about $\frac{1}{8}$ kW), while the U.S. averages 10 kW per person (see Table 4). The situation is paradoxical in that energy is cheap in the wealthy nations and expensive in the poor nations, as we see in Fig. 3. One may ask, is the energy the source of wealth, or is wealth necessary to use energy? It is increasingly apparent that the *basic source of wealth is the successful harnessing of energy*, first to agricultural inputs, and then to industrial products. The fact that such investments are profitable,

Table 3. Energy Sources and Costs

	$1Q$	1 MBtu	Cost/MBtu, dollars
Solar energy			
a. Ground heat	60,000 mile^2·1 yr	600 ft^2·1 day	–
b. Wheat	100×10^6 mile^2·1 crop	147 lb	4.00
Fuel oil	2.4×10^{10} tons	6.8 gal	0.40
Coal	6 mile^3	77 lb	0.20
Enriched ^{235}U	13,000 tons	0.013 g	0.15
Natural ^{235}U	0.023 mile^3 ($\frac{1}{2}$% ore)	1.3 g	0.025
^{238}U or Th			
a. in $\frac{1}{2}$% ore	0.0003 mile^3	2.6 g ore	0.0003
b. in shale	0.007 mile^3	65 g ore	0.002
c. in granite	0.3 mile^3	8 lb	0.015
Deuterium in sea water	0.055 mile^3	1 pint	0.001

Figure 2. Two forms of energy. The granite has the same weight as the coal but contains nearly 100 times as much energy.

that the first inputs of improved seed, fertilizer, and water control can sometimes offer returns of several hundred percent in a single year, is the brightest ray of hope we have that the backward two-thirds of the world can escape the trap of poverty and hunger. And if they cannot escape, there is little long-range hope for the rest of us.

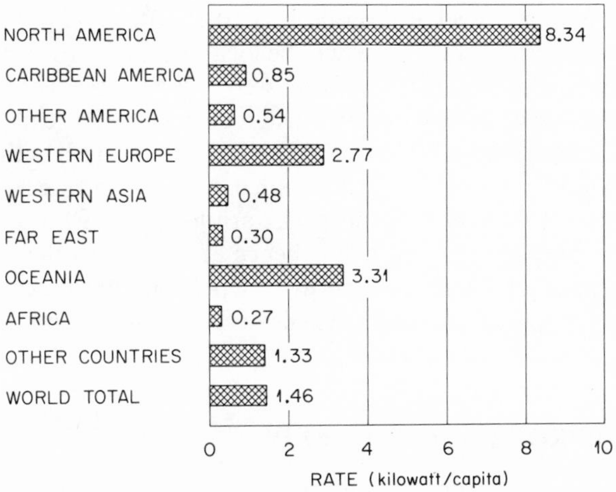

Figure 3. World energy consumption by continental areas (1966).

Table 4. U.S. Energy Use per Capita (in kW)

	Household	Commercial	Transport	Industry	Other	Total
Space heating	1.14	0.13	0.03	0.14	0.07	1.5
Other heat	0.33	0.4		2	0.6	3.3
Motive use			2			2
Electricity	0.5	0.3		1	0.3	2.1
Nonenergy uses				0.41	0.7	1.1
Total	2	0.83	2	3.5	1.67	10

It should be noted that environmental aspects of nuclear fuel supply are favorable. Replacing the present volume of coal mined each year with the same volume of rock from the richer shales and granites for use in breeder reactors would supply 10 billion persons with twice the present U.S. per capita energy (600 MBtu $\times$ 10^{10} = $6Q$ per year), or 40 times present world use.

What about the problem of the dissipation of heat? Here the ultimate limit is set by the net energy the earth receives from the sun—that is, the total heat loss of the atmosphere to space. This heat loss is 120×10^{12} kW. At present, man produces 4.5×10^9 kW, or 4×10^{-5} of the sun's contribution. With a budget of 20 kW per person (600 MBtu per year), and 10×10^9 people, the total man-made energy of 0.2×10^{12} kW would still be only 1.7×10^{-3} of the earth's natural rate of heat loss. This would increase the earth's average temperature by about $\frac{1}{10}$°C, which hardly seems intolerable, since spontaneous swings in temperature of as much as 2°C have been recorded in the geologic past. Thus, the total heat balance hardly would be affected until we reach a population level of several times 10×10^9.

Much more serious, of course, will be local heating in the vicinity of the large catalytic nuclear burners. As discussed below, it seems likely that in most cases these will eventually be clustered in what are called "nuclear parks," producing perhaps 40×10^6 kWe each, and located on the seashore. The world would require about 2000 such parks to produce the energy needed for 10×10^9 people. If the parks are located offshore, they could dissipate their heat to the ocean and thence eventually to the atmosphere.

The potential for lower cost energy is dependent on the scale of energy use. This effect of size on cost and efficiency is a general one, not limited only to energy.

III. SIZE APPROPRIATE TO NEEDS

Every tool, every process, every industry, and every human social and political institution has an appropriate size at which it functions most efficiently. There are jobs for which an abacus is superior to an electronic computer, and a skiff better than a nuclear-powered aircraft carrier, and vice versa. This question of what is the right size is highly significant to our future. J. B. S. Haldane, in his justly famous essay *On Being the Right Size*, illustrated in a delightful way that there is a best size for every animal and also for every human institution. John von Neumann, Alvin Weinberg, Harrison Brown, and others have pointed out that man's social and political institutions are frequently the wrong size for the new technological support systems upon which they now depend. It is important in a period of rapid technological change to recognize these mismatches. The nature of a satisfactory solution to a problem often changes completely with the scale of the problem. Thus, a village and a megalopolis will have entirely different approaches to government, drinking water, and waste control. There have been absurd examples of inappropriate planning even by well-known experts when they have failed to realize these effects of scale.

I should like to illustrate some of the significance of this idea of being the right size by experience with nuclear power reactors and thus lead into a discussion of energy costs. In 1962 I published two articles which showed how the weights and costs of the various components of a power reactor varied as the size was increased. For instance, to double the energy output, the core of the reactor must be doubled in volume; but the shielding which surrounds it need not be any thicker and only a little larger in dimensions. The instruments, auxiliaries, and manpower are hardly affected by doubling the capacity. The summation of many such effects means that the unit cost of large reactors is lower than for small ones and for very large ones can be much lower once we have developed the necessary hardware (see Fig. 4). Although there probably is a maximum feasible size for a reactor, it may be equivalent to a rate of power production exceeding 30,000 MW. In 1962 the largest U.S. power reactor was 180 MW, and the largest size contemplated in the AEC program was 300 MW. At present there are more than 120 nuclear power reactors on order, under construction, or operating in the U.S. But not one of 300-MW size or smaller has been competitive with coal or other fuels. Most of those ordered are 800–1100 Mw and only a few are as small as 450. The size of the network fed by the reactor determines the size which can be utilized with reliability, but, as the networks of the U.S. become more and more interconnected, larger reactors can be used and lower cost power will result.

The same size effect is true of nuclear fuel processing—the chemical recovery of the spent fuel for further use. The only commercial processing plant in the U.S. has a capacity of 1 ton of uranium per day, which is too small for good economics in the chemical business; yet it can keep up with the fuel from reactors generating a total of 10,000 MW of electricity. When we have

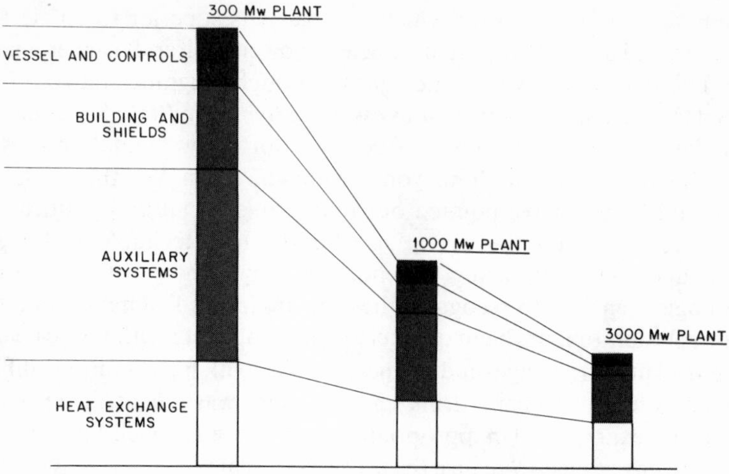

Figure 4. Unit cost of plant falls with increase in size of reactor because of the way in which the component costs vary with the size of the reactor. In breeder reactors the cost of heat-exchanger equipment is the least affected by size while auxiliary systems, instruments and shielding are very sensitive to size. Most reactors now being ordered are 800–1100 MWe although the maximum feasible size may eventually be as much as 30,000 MWe.

enough reactors on the line for 10-tons-per-day plants to be feasible, the processing costs may be only one-fifth of current costs.

The same scaling principles are true of such support industries as reactor-fuel fabricating plants. In short, the economic size of reactors and their support industries is extremely large. We do not know exactly how large, but the costs of nuclear power will continue to go down with size until the optimum is reached.

IV. POWER DISTRIBUTION

The trends toward larger reactors and larger fuel-fabricating and processing plants are connected with similar changes in power distribution methods. Individual power networks serving local regions in the U.S. have grown with the increase in population and per capita power demand of their regions, but now it is clear that there are substantial advantages in regional links of very large capacity—the so-called supergrid concept.

The supergrids will probably use very-high-voltage dc transmission lines. Developments are afoot that make it economical to put such lines underground (see Chapter 8). The larger points of connection of such a supergrid will gradually take on more and more the characteristics of energy centers: locations favorable to the generation of power where contiguous energy-producing and -consuming installations grow up. As power production increases, the inland sites will one by one become saturated from the standpoint of heat-rejection capability, or thermal pollution. Only the very largest rivers, lakes, and seacoast sites will be able to add new generating capacity. Just as for generation, one finds that the larger the quantity to be transported, the lower the costs of transporting power over long distances and that longer transmission distances become feasible as demand grows. Thus, the supergrid system will gradually change its character as its size increases, from an interconnection of semiindependent local systems to a pipeline delivery system transporting power from a few immense energy-producing centers to the rest of the U.S. The economic tendencies which will produce this trend are already well developed.

Grasping the idea that such great changes can happen rather quickly is difficult. A little perspective may be gained, perhaps, by realizing that 30 years from now the electric power industry will have doubled at least three times. (If gasoline used for cars is partially displaced by electricity, the increase may be much greater.) But three doublings mean an eightfold increase by the time that today's newest power stations will be nearing the end of their life. From the standpoint of 30 years hence, we should have little concern for what we see around us now—it will all be wiped out anyway.

V. ENERGY CENTERS

What might an energy center of 30 years hence look like? First, its size. By 2000 A.D. the U.S.–Canadian grid system, for example, is expected to have a total installed capacity of about 2 million MW. By then the location of power centers will be determined not by where we want the power but by where we can dump the surplus heat. How many places in the U.S. are suited to such a purpose? The Atlantic, Pacific, and Gulf coasts and the Great Lakes–Hudson Bay region are all that present themselves; but coastal and lake property will be priced at a tremendous premium. If reactors of today's largest size are built, there will be 2000. They would take up almost all the suitable shoreline—a sort of seawall of power plants. Therefore, it is clear that there must be fewer of them and so they must be larger than today's plants.

Even if siting rules are drastically changed, it is almost impossible to conceive of even 10 suitable land sites along each of the four coasts of North America (counting the Canadian north as a coast). Probably the largest energy centers will be located offshore on artificial islands or on floating platforms. This latter possibility is already being explored for desalination stations and appears quite feasible technically, although the

Figure 5. Floating energy center.

economics need further study. If we are fortunate enough to find 50 good locations altogether, either offshore or onshore, each must produce about 40,000 MW. Similar reasoning gives the same general size to stations in the U.K.–West Europe supergrid. If each station has four reactors to allow for failures, an individual reactor would be designed for 10,000 electrical MW, or about 25,000 thermal MW (a reactor generates heat and only a portion of the heat is converted into electricity).

I see the typical energy center in a large grid as a floating complex (Fig. 5) some 10 miles offshore. It would be located, where possible, in water at least 300 ft deep in order to reach the colder bottom waters below the mixed zone of the sea. This very cold water increases the power station's thermal efficiency. As the bottom waters are rich in marine nutrients, the 40 million m^3 of warm water per day discharged from the condensers would support a large fish population. I am indebted to colleagues at the Scripps Institution of Oceanography for this observation. A 40,000 MWe center would result in a major increase in local seafood output, according to their estimates.

Since the cost per acre of large floating platforms is considerably less than the value of most shore property, much of the regional industry would tend to be located on the platform—particularly those plants requiring sea transport and cooling water as well as power. The deep-draught vessels which are revolutionizing sea transport are especially compatible with these floating centers.

Growth in demand for electric power will almost certainly necessitate larger reactors and larger fuel factories, resulting in lower costs. The change to breeders will make the fuel consumed very plentiful and cheap. Thus, even if reactor designs and power systems retain most of their present characteristics, we can expect electric power in the future to become relatively less expensive.

Continuing technological improvements and changes are the accepted routine in industries such as aerospace, electronics, transport, chemicals, medicine, and drugs. We should be very short sighted indeed if we assume that the reactor–electric power industry will not also produce major innovations. In fact, possibilities of technical changes which may make dramatic differences in the cost and nature of nuclear power stations are already visible. Among these changes are the magnetohydrodynamic generator and the thermoelectric generator. There are possibilities of reactors containing ceramic fuel operating at white heat without having to be clad in a metal can, and of huge molten-metal reactors which consist of little but a tank. These and many other innovations exhibit clearly discernible economic savings, provided certain problems can be solved. No one knows which will succeed and which will fall by the wayside, but it is almost certain, in view of our past history, that some of them will succeed.

However, it is not necessary to go to such speculative examples. The current programs for power breeders appear very likely to result in workable, reliable power sources, both in the sodium- and gas-cooled versions, and it will not require many generations of them for their inherent advantages over present reactors to result in lower plant costs. Further substantial improvements are already waiting in the wings. The molten-salt reactor, which has operated at Oak Ridge for 11,000 h with excellent results (and circulated fuel for more than 2 years), is one type of advanced breeder which offers many routes to cost improvement. Another is the unclad metal fast breeder now under study at Oak Ridge. Both these reactors are high-gain breeders, capable of rapid output of new fuel to start up other reactors, and both use thorium as well as uranium in their fuel cycle.

Thus, there is no dearth of new ideas for the reactor portion of a power station. But what about the power conversion equipment—turbines, generators, and transformers—which constitute about 40% of the cost of a nuclear power station? Here, too, changes are afoot.

VI. POWER CONVERSION EQUIPMENT AND EFFECT ON COSTS

The weight and cost of electrical equipment have been known for some time to be affected strongly by the frequency of the ac for which it is designed. As the power industry grows towards the energy-center concept, more and more of the power produced will either be consumed by large users very close to the reactor or will be converted to dc and supplied to the supergrid. In either case, it is already clear that a reduction in conversion-equipment cost of perhaps 50% can be realized by generation with high-speed turbines driving high-frequency–say 400 Hz–generators or dc homopolar generators. Transformers and solid-state rectifiers are also less expensive at 400 Hz than at 50 or 60 Hz. Only at the points of delivery from a supergrid will converters perhaps again produce today's 50 or 60 Hz for local networks.

Unfortunately, it is not very useful to attempt to project absolute costs of electrical energy in the future. To do so one must make assumptions about the value of money and the method of financing to be used; these are neither uniform in different places nor certain in their trends.

Even an attempt to discuss purely relative changes stumbles over the different capital charge rates from different kinds of ownership and financing. But, as an example, I will show what might happen to the costs of power generation and distribution 30 years hence relative to those expected in the largest nuclear station now under construction—the Tennessee Valley Authority's 3100 MWe plant at Browns Ferry. The cost of the electricity sent out from this station will be made up of about 50% capital costs, 12% operation and maintenance costs, and 38% fuel costs. I will call the aggregate sum 1 unit. If the value of money and other such factors are held constant and only technological changes are considered, what happens to this unit cost when the station is replaced by a 40,000-MWe energy center of the breeder type?

The fuel costs of 0.38 units for the water-type reactors would fall to nearly zero for a high-gain breeder or to 0.10 for a low-gain type, so they would be, say, 0.05 units. Operation and maintenance costs, according to my estimates, would become 0.08. Capital costs would fall by about 0.20 due to scale-up effects alone, and another 0.10 from the use of high-frequency turbines and generators, so that the 0.5-unit capital item becomes 0.20. The total cost could thus be 0.33 units, or just one-third of the present cost. Innovations in reactor design, including very-high-gain breeders, could lower costs still further.

Energy produced by high-gain breeder reactors has some interesting economic properties. If such a station is in full operation at perhaps 80% capacity, it will have essentially no increase in costs if the load is increased to 95% or 100%. The costs may even be less, since the value of the extra plutonium or other bred fuel produced is likely to exceed the fabrication cost of the extra uranium or thorium consumed. All other costs remain the same. This, in fact, is a general property of high-gain breeders: the cost of additional off-peak power is near zero or even negative.

This property will have important effects on the way energy is distributed and marketed and on the way the plants are operated. Users of non-interruptible power will probably be charged only on their maximum potential peak demand and not at all for energy actually consumed. So your electricity bill will be the same, whether you leave the lights on day and night or carefully turn them off. Meter readings will not be needed and most users will get a flat-rate bill based on maximum demand or connected load.

The electricity supplier will have a strong incentive to use pumped storage units, sell interruptible power, or use other methods to increase the load factor—the percentage of the time that the generating plant is used. The transcontinental supergrids will also help to increase the overall load factor of the system by reducing the need for spinning reserves—plant held at instant readiness—since daily peaks will occur at different times in different parts of the system. Eventually, a special low-capital-cost, low-fuel-cost thermal reactor, such as a molten-salt reactor, may form the best peaking and reserve unit.

Design of turbines, motors, and power-distribution systems would also be affected by costing power only for capacity and not for energy. Lower capital costs would result when the low load efficiency is not important. Some motor drives would be designed to be left running all the time in order to avoid high starting currents. Home and industrial heating units would be universally programmed to draw power in off-peak periods only and would use stored electricity in between. Power distribution costs per kilowatt hour will be greatly reduced by the higher total loads per customer and the higher load factors of tomorrow. The result of all these measures would perhaps permit an overall system load factor of more than 90%, based on 95% availability and 95% utilization. Today's systems average less than 70% overall, so such an increase in load factor would represent a further cut in power costs. However, extra investments in reliability, control devices, peaking units, and other equipment will be needed.

Assuming that money values and other conditions remain constant, in 30 years' time the large power user close to an energy center can expect to obtain noninterruptible power at about 30% of present lowest costs and off-peak power at perhaps half of this. The average industrial or residential user will pay less than he now does for distribution and accounting costs, so that he may pay only one-fifth of his present delivered price. He may, however, use 20 times as much electricity as he now does.

VII. SOCIAL AND ENVIRONMENTAL ASPECTS OF LOW-COST ENERGY

Returning now to the social and environmental effects of low-cost energy, we mentioned above that energy has the capability of increasing the productivity of land and of actually increasing the supply of land. Dr. Perry R. Stout of the University of California at Davis has shown that input of external energy seems to be the key to escape from subsistence farming and that such inputs may yield very high rates of return on the investment. An important new factor makes this major agricultural step much easier in some of the most backward parts of the world. In the U.S., the change from the horse-drawn farm economy to the modern, mechanized, highly productive version took many years, mainly because the crops themselves had to be adapted to the new methods—for example, types responsive to fertilizer had to be found and disease-resistant strains developed. The hybrid corn revolution is the most important example. But the process may be much faster now because of the work of the Rockefeller Foundation and others in developing spectacularly improved varieties of some of man's most basic food crops—wheat, rice, and potatoes. Having these already available will make the change to new methods much more rapid, even in very backward regions. Agronomists in India have been introducing small quantities of Sonoran dwarf wheat, developed by Rockefeller in Mexico, into the Ganges Plain. The reception has been explosive—the increases in yield ranged from threefold to tenfold, and farmers who were supposedly too poor to afford it were offering $100 per pound for the seed and large premiums for fertilizer. In each new area these new strains will have to be tested. Pests and climatic conditions vary and different diseases will be endemic. Nevertheless, the techniques of selecting, improving, and adapting a crop are now well proved.

These high-yield crops require high inputs of energy in the form of fertilizer. The larger potential output and the investment in seed and fertilizer in turn justify considerable effort to ensure a reliable water supply. Thus, a chain reaction is started—the subsistence farmer who risks little and gains nothing becomes a commercial entrepreneur. He invests substantial sums in equipment, water pumps, and fertilizer. He must be concerned with such matters as markets, storage, and transportation. He is more likely to join a cooperative than before and his consumption of energy and other commodities is increased manyfold. An innocent-sounding innovation in farm practice can become a social revolution.

Although my principal concern in this article is with the period in the future when, if civilization still exists, some kind of solution must have been found for the present food–population crisis, a brief look at the nature of this crisis will show a further role that energy can play. The population explosion does not produce local famines such as have been known in the past. It is a new problem—a major, continuing, global pressure of population increase, far beyond our present food resources. I am confident that the explosion can be arrested and the population leveled off. But demographers tell us that any credible method of doing so will take time—enough time for the earth's population to have doubled at least once before it becomes constant. The situation has been discussed by many writers and is authoritatively documented by the recent report of the President's Science Advisory Committee on the World Food Problem. This report concludes in part: "The scale, severity, and duration of the world food problem are so great that a massive, long-range, innovative effort unprecedented in human history will be required to master it."

In the past, as the earth's population has grown, mankind has met the need for more food by putting more land into production. But now experts tell us that the supply of new productive agricultural land has been nearly exhausted and the population is growing faster than ever before. In the words of Roger Revelle, Director of the Harvard Center for Population Studies: "... one would expect that in the future it would be possible to continue to expand the area planted to food crops. This will be much more difficult, and probably impossible on any large scale, in the crowded countries of Asia where most of the easily culturable land has long since been placed under the plow."

Most of the hungry regions of the earth today are characterized by crowding of the farmland and its division into smaller and smaller units. Yet in none of the areas is the shortage of land itself—the average fraction of the land used for agriculture is only 11% (see Fig. 6). The shortage is of fertile, well-watered land, The rest of the land is too dry, too salty, too rocky, too wet, or the rains come at the wrong time. A tremendous fraction of the earth's land surface is arid. However, according to UNESCO, the supply of warm, fertile, accessible but arid land exceeds severalfold the total land now used for major food crops. Since much of this land lies close to the sea in latitudes potentially favorable for year round crop production, the idea of

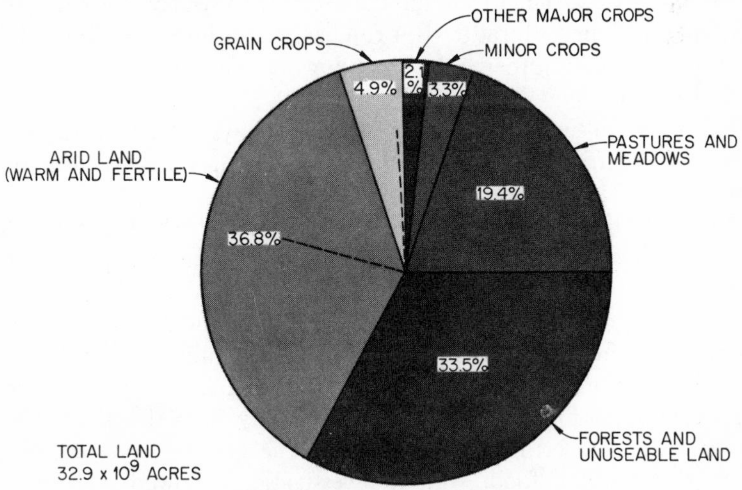

Figure 6. Present utilization of the earth's land surface.

Figure 7. Artist's concept of an agroindustrial complex of the future.

large-scale desalting of sea water to irrigate the coastal deserts has excited man's imagination. Now the prospect of unlimited energy at low cost from the atom makes such an idea not only conceivable but possible. There is no question that, if we wished to commit the necessary resources, we could in the near future make the Colorado River run backwards into the arid southwest of the U.S. or build another Nile River flowing inland from the sea to the western desert of Egypt.

But we can do such things only if the project pays—if the food and other benefits are an attractive return on the investment in water. In spite of the glamor of making the deserts green, the desalting cost has always been too high. Recently, though, several events have occurred which together give new hope. One is the success and rapid growth of nuclear energy that I have mentioned, with the trend to lower costs. Second is the rapid development of seawater evaporator equipment, both in the U.S. and in Europe. Third is the concept of the "food factory"—the idea of an agricultural enterprise geared to the optimal production of food from high-cost water. Such an enterprise is operationally integrated all the way from the water supply, energy source, and fertilizer production to the growing, processing, and marketing of finished foods. Initial investigations showed that a man's daily food supply could be produced under such conditions with less than

200 gal of water, costing only a few cents. In 1967 the AEC sponsored a study in depth of this overall concept at the Oak Ridge National Laboratory.

The study explored the technical and economic feasibility of large nuclear-powered complexes, called *agroindustrial complexes*, located on an arid coastal desert typical of the millions of square miles of such land (see Fig. 7). The complexes included dual-purpose power and desalting plants to take advantage of the inherent thermodynamic savings thus obtained. The electric power produced is consumed mostly within the complex, in captive power-intensive processes such as the manufacture of electrolytic ammonia and phosphorus fertilizers, caustic soda and chlorine, and aluminum. The study team utilized an impressive panel of expert consultants from U.S. industry, agriculture, universities, and government agencies and from foreign countries. The initial study was confined to only the essential basic processes of industrial raw materials, chemicals, and marketable farm produce. The secondary and peripheral layers of labor intensive industrial, processing, transporting, and other activity which would be part of any real project were omitted. These would create several times as many jobs as the primary complex and would produce substantial increases in economic activity without much increase in overall capitalization. Some attempt will be made to include such activity in further studies now under way but, even without it, the conclusions of the study are generally favorable and have attracted much interest.

VIII. CONCLUSION

Thus, energy in low-cost abundance can possibly open up a vast new portion of the earth's surface for man's use and can support profitable cities and farms based on water from the sea. With this possibility and others, the inevitability of 6 billion people crowding the earth by 1995 does not seem so frightening. One thing that seems certain is that nuclear power and nuclear desalting will both become big business. If only 10% of the 3 billion new people of the next 30 years are supplied with water from the sea, the money involved in constructing such stations will amount to about 1 billion dollars.

The technical achievement of abundant cheap energy will be of primary importance to the extent that it can be applied to man's pressing needs—needs that are, in fact, basically social in origin. The task of predicting exactly what these needs will be or how our technology can be applied to them is far more difficult—and yet more important—than forecasting technology itself. The difficulty arises because the social problems to be solved do not yet exist—at least not in the context to which our technology can be related.

Alvin Weinberg has pointed out some examples of what he calls "technological fixes" for social problems and has shown the importance of finding more of them. The achievement of unlimited cheap energy and the elimination of "have-not" regions should lead to a large number of such fixes, most of which we cannot hope to predict. The new technology and the social changes thus generated will undoubtedly create some new problems for mankind in the course of solving others.

There is a conclusion which seems justified, however, in spite of our ignorance of the details; namely, that as energy consumption grows larger and yet less costly, mankind will exercise increasing degrees of control over his environment. And as environmental control is essentially synonymous with wealth, so kilowatts become, in effect, the medium of exchange. If man and his technology can somehow "fix" enough of his social ills, he has the means to become wealthy indeed in the future.

The ways in which our lives will be affected by the development of low-cost, unlimited sources of power extend beyond the growth of new electrochemical industries and the appearance of a profusion of electrical and electronic gadgets. The social consequences are far more important and, of course, much harder to predict. But what is emerging is the possibility of a human society in which economic opportunity is largely decoupled from the accidents of geography. If any nation can find in the common rocks underfoot, in the air overhead, and in the waters of the sea all the major ingredients to produce unlimited supplies of energy, fertilizer, fresh water, and other necessities, then the useful part of the earth will be increased manyfold and the distinction between the "haves" and "have-nots" will tend to vanish.

REFERENCES

[1] P. C. Putnam, *Energy in the Future*, D. Van Nostrand Company, New York, 1953.

[2] A. M. Weinberg, Energy as an ultimate raw material or—problems of burning the sea and burning the rocks. *Physics Today* **12** (1959) No. 11, 18.

[3] J. B. S. Haldane, On being the right size, in *The World of Mathematics*, Ed. J. R. Newman, Simon & Schuster, New York, 1956.

[4] John von Neumann, Can we survive technology? *Fortune* **51** (1955) 106.

[5] A. M. Weinberg, "The Social Responsibility of the Nuclear Scientist," paper presented at the Institute of Atomic Energy, Sao Paulo, November 29, 1963.

[6] Harrison Brown, James Bonner, and John Weir, *The Next Hundred Years*, The Viking Press, New York, 1954.

[7] R. P. Hammond, Large reactors may distill sea water economically. *Nucleonics* **21** (1962) No. 12, 45.

[8] A. Sesonske and R. P. Hammond, "A Preliminary Evaluation of Fast Oxide Breeder Reactors for Sea Water Conversion," USAEC Report LA-2733 (1962).

[9] R. P. Hammond, R. E. L. Stanford, and J. R. Humphreys, Jr., "Mobile Fuel Plutonium Breeders," USAEC Report LA-2644 (1961).

[10] Roger Revelle, Population and food supplies: The edge of the knife. *Proc. Natl. Acad. Sci. U.S.* **56** (1966) No. 2, 328.

[11]P. Meigs, "Geography of Coastal Deserts," UNESCO, Arid Zone Research Report No. 28 (1966).
[12]R. P. Hammond, Desalted water for agriculture, in *Water for Peace*, U.S. Govt. Printing Office, Washington, D.C., 1968, Vol. 2, pp. 184–197.
[13]A. M. Weinberg, Can technology replace social engineering? *Bull. At. Sci.* **22** (1966) No. 10, 4.
[14]"Nuclear Energy Centers—Industrial and Agro-Industrial Complexes," USAEC Report ORNL-4290, Oak Ridge National Laboratory (November 1968).
[15]A. M. Weinberg and R. P. Hammond, Limits to the use of energy. *Am. Sci.* **58** (1970) No. 4, 412.
[16]R. P. Hammond, Low cost energy: A new dimension. *Science Journal* **5** (1969) No. 1, 34.
[17]P. R. Stout, Power: The key to food sufficiency in India? *Bull. At. Sci.* **24** (1968) No. 8, 26.
[18]John D. Isaacs and Walter R. Schmitt, Stimulation of marine productivity with waste heat and mechanical power. *Journal du Conseil International pour l'Exploration de la Mer* **33** (1959) No. 1, 20.

Chapter 8

THE HYDROGEN ECONOMY

D. P. Gregory, D. Y. C. Ng, and G. M. Long

Institute of Gas Technology
Chicago, Illinois

I. INTRODUCTION

Although electrical energy today is considered to be a universally convenient energy source that is instantly available at the turn of a switch, we tend to take for granted the additional availability of two other energy sources—natural gas and oil–gasoline. These chemical energy sources have two outstanding operational advantages over electricity: (1) they can be stored up in varying amounts, either within their distribution networks or in portable containers; and (2) transportation of energy over long distances is far cheaper for natural gas or oil than for electrical power. Present trends in the use of energy accentuate these differences, as the user tends to concentrate his use of power into smaller peak periods of the day, and as the intense concentration of population in local areas strains the electrical transmission network. Moreover, as society is becoming increasingly conscious of the need to protect the environment it lives in, electric power cables are being forced underground at phenomenal expense, to lie out of sight in company with the existing natural gas and oil pipelines. Today's increasing demands for electrical power are resulting in an increasing potential for atmospheric pollution resulting from the need to burn more "dirty" fuels at the electric power stations.

When we consider the day when all our fossil fuel reserves are exhausted,[10] it is at first a comforting thought that, through nuclear power stations, we shall have a clean, silent, "all-electric" society. But as we examine this concept more closely, we realize the unacceptable inconvenience of being without our natural-gas supply and without gasoline, and realize the high-cost prospects of providing all our domestic and industrial energy requirements with nuclear electricity. Perhaps the dominant problem is that the future nuclear fission stations, and fusion stations when they come along, are only efficient when they are very large, and therefore must be sited far from many of their users. In addition, nuclear power stations operate best at constant power outputs, which does not coincide with the pattern of demand. Thus, the "storability," portability, and transmission advantages of chemical energy intermediates will be required in the future just as they are now.

We have therefore established the demand for a synthetic "fuel" that can be produced at the nuclear station and that will take the place of natural gas or gasoline in our economy. Consider first an analog of natural gas. We have the choice of synthesizing a gas that is compatible with existing natural gas, or converting the gas industry to the use of an alternative material. The latter course seems to be technically preferred when we consider the possibility of hydrogen as a fuel, but would only be acceptable if it could utilize the large investment already made in natural-gas distribution systems.

Hydrogen can, in principle, be readily produced electrolytically from water (which is usually available in abundance), and can be used as a fuel that produces only water as a combustion product. The advantages to the environment are immediately obvious, and no other synthetic fuel has such a simple cycle. To provide portability, a liquid fuel derived from hydrogen must be considered. Ammonia and hydrazine, which involve nitrogen as well as water in their combustion cycle, are also attractive from the pollution aspect, but are more difficult to synthesize and are themselves toxic. To obtain a convenient liquid fuel synthetically, we are forced to consider carbonaceous fuels such as methanol. Cleaner than gasoline, synthetic methanol would still produce CO_2 as a pollutant and would require a carbonaceous feedstock for its production.

Coal, natural gas, and oil are conventionally used as important chemical raw materials. In an age without fossil fuels, alternative synthetic sources must be found for the essential heavy organic and pharmaceutical industries. The role of hydrogen as a chemical raw material must therefore be considered.

Proper planning of a systematic changeover from natural gas to hydrogen would involve the conversion of the entire gas transmission and distribution system, and the appliances, somewhat in advance of the actual time when our fossil fuels are exhausted. Today, most industrial hydrogen is produced from methane, but increasing demands for hydrogen will have to be met by production from coal, whose reserves are considerably greater than those of oil and gas. These hydrogen production plants, however, would then be phased out as nuclear–electrolytic plants are commissioned and other, perhaps chemonuclear, hydrogen sources are developed.

This chapter, therefore, sets out to study the impact on industry and society of a transition from fossil fuels to hydrogen as our basic source of stored energy. Some speculative assumptions have had to be made, among them the basic one that hydrogen will indeed be accepted as a fuel and will take the place of methane in the natural-gas industry. Having made these assumptions, we will attempt to discuss real effects and not to speculate further. We will consider future hydrogen production techniques, problems, and its characteristics in pipeline transmission; means for storing hydrogen to provide short-term energy reserves; the use of hydrogen as a conventional fuel for cooking, heating, vehicle propulsion, and the generation of electricity locally; the storage of electrical energy using hydrogen fuel cells; and the expanded use of hydrogen as a chemical raw material. In line with the theme of this book, electrochemical aspects will be emphasized and the impact of our assumptions on the pollution of the environment will be discussed.

Finally, we may well ask why we are engaging in the analysis of a nonfossil hydrogen fuel system at present. The transition from fossil to nuclear nonfossil chemical fuel is an event that will probably come into full bloom in the twenty-first century. However, the history of our energy-use pattern indicates that almost revolutionary changes could occur within 60 years. Figure 1 shows the U.S. energy sources as percentages of aggregate energy consumption from 1850 to 1965. The primary fuel in 1850, firewood, was reduced to less than 10% of the energy needed by 1910. Over the period shown, 1850–1965, aggregate energy consumption rose from about 2–54

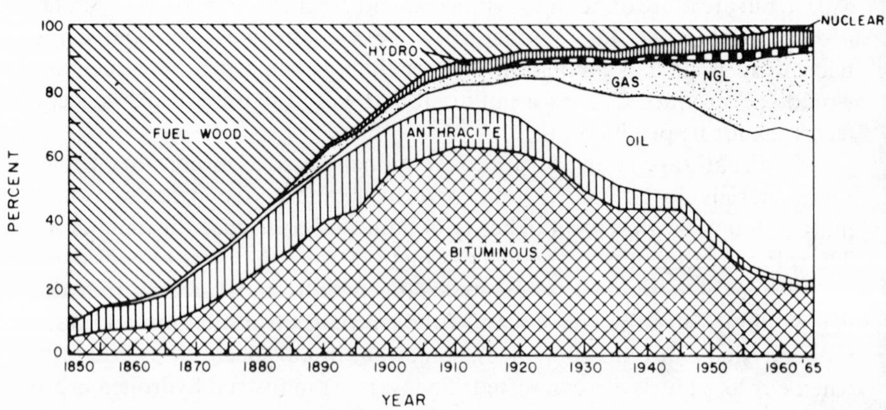

Figure 1. Specific energy sources as percentages of aggregate energy consumption: 5-year intervals, 1850–1965.[42]

quadrillion Btu/yr. The even more rapid growth rate of energy use in the last decade should warn us that the changing level and pattern of energy use between now and the early part of the twenty-first century will occur at a much accelerated rate. The insidious nature of these changes further argues for analysis and planning of alternative or replacement energy systems well in advance of their actual need. The hydrogen fuel system is not only a new energy fuel source but also a source of chemical raw materials for a vast array of synthetic products. The system and equipment changes required are considerable, and will be time consuming.

All aspects of the fuel system need to be explored so that the ultimate system will make efficient use of our energy resources and result in the minimum undesirable perturbation in the cycles of the biosphere. Therefore, we do feel that it is timely to begin to explore the many aspects of a nonfossil hydrogen fuel.

We expect the subject matter of this chapter to be controversial. We have not set out to make a clear-cut case in favor of a transition to hydrogen fuel; rather, we bring out many questions that cannot be answered today and show the continuing need for long-term planning to meet our future energy demands.

II. ECONOMIC CONSIDERATIONS

The use of nuclear–electrolytic hydrogen as a fuel will only be justified if it is economically attractive. At the present time, electrolytic hydrogen is produced competitively for industrial use as a chemical raw material, but only in areas where electrical energy is very cheap (e.g., from hydroelectric stations); even then it is far too expensive to be considered for use as a fuel. Nevertheless, if we look forward to the time when fossil fuels are exhausted and our only fuel source is nuclear electricity, electrolytic hydrogen will be competing with electricity itself as a "fuel" or heating source. Is it possible to examine the economics of such a situation?

Basically, we observe that the cost of transmitting energy over long distances in the form of a gas in a pipeline is considerably cheaper than transmitting it in the form of electrical energy in overhead cables. Thus, if we are forced to transmit energy over considerable distances, the savings in transmission cost resulting from transmitting hydrogen through pipelines can more than pay for the cost of the electrolyzer.

This situation is illustrated schematically in Fig. 2, which compares the cost of transmitting energy by long-distance electricity cables and by hydrogen gas. The figure shows the range of present-day costs of long-distance transmission of electric power at some typical voltages. The higher voltages are not normally used for distances under about 250 miles. The figure also illustrates the predicted cost of long-distance pipeline transmission of

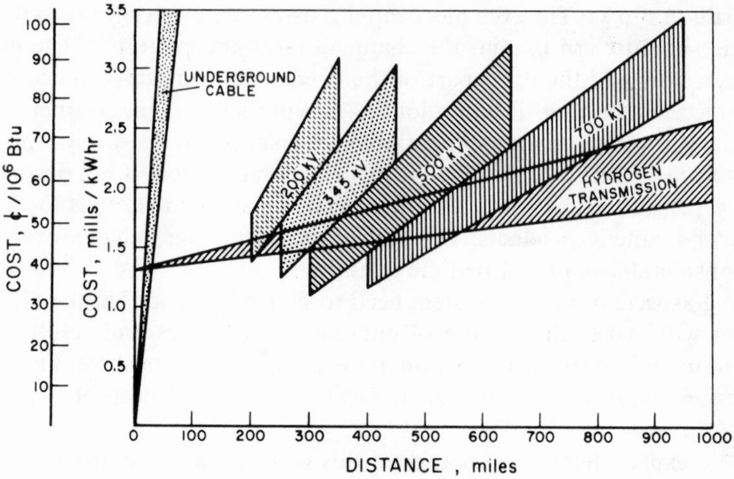

Figure 2. Relative costs of energy transmission by electricity cables and by hydrogen pipeline.

hydrogen, based on an increase of 60% over the present natural-gas transmission costs (see Section IV). The intercept of this line with the zero-distance line represents an estimate of the cost of an electrolysis plant, expressed in cents per million Btu of hydrogen produced, but excluding the cost of electricity. These figures were obtained by deducting the electricity running costs from the overall hydrogen production costs for the two advanced electrolyzers discussed in Section III and Table 1, allowing no credit for the sale of oxygen, which would be produced in such large quantities as to be almost valueless.

Making these assumptions, we see that an economic advantage exists for transmitting energy as electrolytic hydrogen over distances greater than about 250–350 miles. We may be surprised to observe that, because the basic electricity generation cost is the same for the two processes, the comparison is *not* dependent upon the availability of cheap electricity, and therefore should be valid today. (The electrolysis cost will contain a small element of electrical cost because efficiency losses are included here, but this is relatively insignificant.) Curiously, it appears that even at today's prices, electrolytic hydrogen would be a cheaper "fuel" than electricity when delivered over long distances. However, the availability of natural gas, which is both cheaper at source and cheaper to transmit than hydrogen, rules out the economic use of hydrogen as a fuel today.

Referring again to Fig. 2, we observe that the distance for which hydrogen fuel transmission becomes competitive is very dependent upon the basic electrolyzer cost and is relatively insensitive to hydrogen transmission-cost changes. It must be admitted that the assumptions made here arrive at an electrolyzer cost of 37¢/million Btu, or 12¢/1000 SCF of hydrogen, excluding power costs, and are not fully justified, which points to the need for further detailed studies of the building and operating costs of electrolyzers.

We also observe that a steepening of the electricity transmission curves (raising the transmission costs) will also have a marked effect in making electrolytic hydrogen more favorable. Trends are presently toward developing higher voltage lines, which reduce the cost, but these are economical only over very long distances indeed where gas transmission costs still have a very clear advantage. For shorter distance transmission, conservationists are pressing the electric utility companies to bury their lines undergound. By the year 2000, we expect to use 10 times the electrical energy that we are using today; thus, this social pressure to avoid an even uglier overhead network will be much increased. It costs as much as 5–10 times more to lay an underground high-voltage cable. The steepest line in Fig. 2 is an approximation of where the transmission cost might be if undergrounding long-distance electrical lines were mandatory.

This comparison has not considered the local distribution costs for energy. The average local distribution cost for electricity today is 6.6 mills/kWh, or $1.94/million Btu,[59] whereas the average local distribution cost for a natural-gas system is only about 17¢/million Btu. A strict economic comparison for our projected system cannot easily be made because we need to compare the cost of increasing the load-carrying capacity of the electricity network with the cost of converting the existing gas network to carry hydrogen.

Some other economic aspects of the use of electrolytic hydrogen are also worth considering. Even though we cannot fix a realistic value for the huge quantities of oxygen produced, which would be credited to the system's running cost, we cannot ignore the considerable production of heavy water that would result from large-scale electrolysis. Experience at the Nangal plant in India shows that 15 tons/yr of heavy water are produced in a 20-million SCF/day hydrogen plant. In other words, 1000 SCF of hydrogen will yield 0.0046 lb of heavy water as a byproduct, which, at today's price of $28.50/lb, would earn an extra 13¢/1000 SCF of hydrogen, more than enough to pay for the cost of the electrolyzer. Whether the price of heavy water would be maintained at this price by its extended use in fusion reactors remains to be seen.

Another feature of the use of nuclear–electrolytic hydrogen as a fuel is that we are able to relax the specifications for the design of the power station if it is only supplying an electrolyzer. We no longer have to maintain such a high reliability because we will have enough hydrogen storage in the pipelines to allow the electrical supply to be interrupted a few minutes at a time.

We no longer have to accurately synchronize our ac generator with all other generators on the network, and we do not need such precise control of rotor speeds or voltages. Because the cost of transmission of energy as hydrogen is so much less than by present methods, we have a greater freedom in siting the power station for other optimum needs. All of these factors should contribute to lowering the basic generation cost of electricity if it is specifically intended for hydrogen production.

Finally, we must now consider the economics of the ultimate use of the energy when it reaches its destination. If it is to be burned as heat, then the data shown in Fig. 2 are valid, for we have equated the energy content of a unit of electrical energy with the heating value of the hydrogen (3413 Btu = 1 kWh). If, however, we are to transform the energy back into electricity or into mechanical energy, we must take into account the efficiency of the device used at the destination. Even in the most favorable case of a fully developed fuel cell operating at, say, 60% efficiency, we have the situation where, for 1 kWh demanded by the customer, 1.66 kWh must be passed through the transmission system as hydrogen, and to produce this hydrogen, 1.96 kWh must be generated. Unless the nuclear generator–electrolyzer units are compelled to be built several hundred miles away from the point of use of the energy, it is hard to see how electricity → hydrogen → electricity systems will be economical. However, proponents for the siting of nuclear power stations within the cities are at present being heavily criticized by leading health physicists, who say it is wise to continue the practice of isolating large nuclear facilities away from heavily populated areas.[58] We may also want to site our nuclear–electrolytic plants on or in the ocean so we can dissipate the heat produced. Thus, we are unable to predict the average generator-to-customer transmission distance, but it is likely to be greater than it is today.

Thus, we can make an overall economic case for considering the use of electrolytic hydrogen as a fuel source in the future. Further detailed studies are already in progress and more are needed to fully justify the case. However, historically, predictions of energy costs and demands made more than 20 years into the future have been inaccurate. We seem, then, to have enough information to go ahead with more detailed considerations of the system.

III. HYDROGEN PRODUCTION

1. Introduction

Since a few decades ago hydrogen has been a heavy industrial chemical. Between 1960 and 1968, the total U.S. hydrogen production more than tripled from 3.8 to 12 billion lb[31] (722–2280 billion SCF, or 234–741 trillion

Btu).* The total world production in 1968 was 5055 billion SCF. Over 95% of the total hydrogen is used as a chemical intermediate in chemical processes to produce hydrogen-containing products such as ammonia, methanol, and refined petroleum fuels and chemicals. This portion of the hydrogen production is usually produced and consumed on site and is called proprietary hydrogen. The remaining 5% is called merchant hydrogen and is sold as an industrial gas. In the nuclear–electrochemical age when hydrogen assumes its complex energy as well as nonenergy roles, as described in the last section, the U.S. hydrogen production will have to be increased by many orders of magnitude.

The basic chemistry and physics of processes for manufacturing hydrogen from fossil fuels and water are well understood. It is in the context of the enormous quantity of hydrogen needed that the various aspects of hydrogen production must be viewed. Although we envision that both chemical and electrochemical processes will be employed in the transition period, we will examine mainly the processes involving the electrolytic and nuclear decomposition of water. Future projections of natural-gas needs will be used to estimate the quantity of hydrogen needed as an energy-providing fuel. The electrical energy required to power the electrolysis plants will be estimated. The magnitude of the hydrogen produced will compel us to also consider the vast amount of chemical byproducts such as oxygen, chlorine, and caustic soda that would be produced. Finally, we will also discuss the environmental effects of the nuclear-electrical and chemical fuel energy centers.

2. Chemical Hydrogen Manufacturing Processes

At present, the vast majority of industrial hydrogen is manufactured by several chemical processes.[15,23] In all the processes, a mixture essentially of carbon monoxide and hydrogen is made first and then further converted to a mixture of hydrogen and carbon dioxide. The carbon dioxide is finally removed to yield hydrogen of varying degrees of purity. The feed materials are gaseous or liquid hydrocarbons, coal, and water.

The chemical processes can be broadly divided into three main groups: (1) catalytic steam reforming of natural gas and hydrocarbons; (2) partial oxidation of hydrocarbons; and (3) processes using coal or coke.

The most widely used hydrogen production process is the catalytic steam reforming of hydrocarbons.[39] In the U.S. and several other countries, the most common raw material is natural gas because of its abundance and relatively low cost. Other starting materials include propane, butane, other

*Since hydrogen is being considered as an energy-providing fuel and a chemical raw material, the quantity of hydrogen will be given in mass (lb), volume (standard cubic feet), and thermal content (Btu) units in this chapter. The standard cubic foot (SCF) is the volume of gas at 60°F and 30 in. of Hg. The higher heating value of hydrogen of 325 Btu/SCF is used.

refinery gases, and petroleum naphtha. About 40% of the world's hydrogen production comes from natural gas and refinery gases. The main catalytic reaction in the natural gas (methane) process is $CH_4 + H_2O \rightarrow CO + 3H_2$. The carbon monoxide is shifted (reacted with steam) to yield more hydrogen, and the carbon dioxide byproduct is removed.

The partial oxidation of hydrocarbon processes can be further divided into two types—namely, the pressure noncatalytic and catalytic processes.[15] The most common commercial process is the pressure noncatalytic type. Natural gas and other hydrocarbon feeds (especially those that cannot be subjected to direct steam reforming) are partially oxidized by burning with oxygen or oxygen-rich gases to produce a mixture of hydrogen and carbon monoxide. Commercial, pressure, noncatalytic, partial oxidation processes have been developed by Shell[17] and Texaco.[30]

Coal- or coke-based processes were at one time widely used for hydrogen production (about 90% of the world's production before 1940). Since then, the favored raw materials have been natural gas and other hydrocarbons. By 1965 less than 25% of the world's production of hydrogen was derived from coal or coke. However, as our natural-gas reserve continues to decline rapidly, the coal- or coke-based hydrogen process will probably again be employed more widely. The most common coal or coke process is the water–gas production by passing steam over red-hot coke. The hydrogen (44–51%) and carbon monoxide (33–34%) mixture can then be further converted to hydrogen by either the catalytic water gas-steam process[15] or the steam-iron process.[15] The other type of coal- or coke-based process is the production of synthesis gas or coke-oven gas (50–55% hydrogen, 25–30% CH_4, and 4% CO) from coal and oxygen. This gas mixture is then further converted to hydrogen. Many processes have been proposed and are in varying stages of development.[15,45,51] A recent electrothermal process of decomposing coal char[21] is being considered as a source of hydrogen-rich gas in the Institute of Gas Technology's HYGAS[25] pilot plant for the production of synthetic "natural" gas (methane).

From the point of view of producing a hydrogen chemical fuel, these processes have two common characteristics. The hydrogen fuel product contains substantial amounts of CO and CO_2, and a main chemical reactant feed is hydrocarbon or carbonaceous fossil fuels.

The presence of the CO and CO_2 will affect the quality of the hydrogen fuel for both energy and nonenergy uses. The relatively inert CO_2 will dilute the heating value per unit volume of the fuel gas. In the use of hydrogen as a raw material—e.g., in ammonia production—both the CO and CO_2 are undesirable impurities. The CO content can be kept to a minimum by conversion to CO_2 and more H_2. The CO_2 must then be removed by treatment with liquid absorbents such as monoethanolamine.

The need for a fossil-fuel feed material will restrict these chemical hydrogen manufacturing processes to a limited role in the evolution of the nuclear–electrochemical age. At one extreme, these processes will be completely obsolete when all practical reserves of fossil fuels have been exhausted. In the transition period, when we have a mixture of fossil fuels and hydrogen–hydrogen-based synthetic fuels, some of these chemical processes can be effectively utilized. The natural-gas-based processes should be deemphasized because of the smaller reserves of natural gas compared with those of coal; furthermore, any synthetic hydrogen fuel will have no advantages over the natural gas fuel itself. The coal-based hydrogen manufacturing process should prove to be effective in the transition period. The transformation of coal into a hydrogen fuel will allow it to assume all the advantages and conveniences of a gaseous fuel. The coal gasification plant will probably produce the dual gaseous fuels of methane and hydrogen. The proportion of these two gases will depend on the ultimate user requirements in the area served by a particular plant. The conversion of coal to a hydrogen or hydrogen-based gaseous fuel provides an added environmental benefit; a sulfur removal step can be conveniently incorporated into the gasification process, thus eliminating the sulfur pollution potential of direct coal use. Another factor that argues for the use of a fossil-fuel hydrogen manufacturing process in the transition period is the need for time to further develop efficient and economical electrochemical and chemonuclear processes. Also, the large-scale production and use of hydrogen fuel will allow us to familiarize ourselves with hydrogen handling systems. Lastly, these interim chemical processes will buy time for the development in nuclear fission and fusion electrical energy generation technology for the full-fledged nuclear–electrochemical age.

3. Electrochemical Hydrogen Manufacturing Processes

In the nuclear–electrochemical age, a viable hydrogen production process will be one that does not require fossil fuels as either chemical raw materials or as an energy source. In almost all common conventional chemical processes, the hydrogen product is derived from the hydrogen atoms in the fossil fuel (e.g., CH_4) as well as the steam reactant. With the absence of fossil fuels, we are reduced to relying on water as our sole source of hydrogen. The energy required to decompose water into its hydrogen and oxygen constituents will be provided by nuclear fission, heat, and/or electrical energy generated by nuclear reactors.

We will first consider the water-electrolysis process that has been in commercial use for some years. Then we will discuss the direct use of nuclear heat and chemonuclear reactions in which the energy liberated by the decay of the nuclear fuel is used to break the chemical bond of water to yield hydrogen and oxygen.

Although water decomposition by electricity was accomplished as early as 1789, the definitive electrolysis of water into H_2 and O_2 was credited to Sir Humphrey Davy in 1806.[18] At present, the electrolytic process of hydrogen production is a fully commercial process. In 1963, the Norsk Hydro-Elektrisk Kvaelstofaktieselskab of Norway produced 28 billion ft^3 of hydrogen for in-plant use of ammonia production. The relatively high cost of electricity and low cost of methane combine to limit electrolytic hydrogen to a minute fraction of the total hydrogen production. The major way the cost of electrolytic hydrogen produced has been lowered is low-cost electrical energy. With the arrival of the nuclear–electrochemical age, the availability of cheap electrical energy from large fission and fusion reactors is assumed. Even in the transition period, the reserve of natural gas and other fossil fuel will be sufficiently depleted so that the increased methane cost may render the electrolytic process competitive.

The overall chemical reaction of the electrolysis of water is

$$2H_2O(\text{liquid}) \longrightarrow 2H_2(\text{gas}) + O_2(\text{gas})$$

In alkaline solutions, the cathodic and anodic reactions are

$$2H_2O + 2e^- \longrightarrow 2H + 2OH^- \quad (\text{cathode})$$

$$2H \longrightarrow H_2$$

$$2OH^- \longrightarrow 2OH + 2e^- \quad (\text{anode})$$

$$2OH \longrightarrow H_2O + \tfrac{1}{2}O_2$$

The theoretical equilibrium decomposition voltage of water at 25°C and atmospheric pressure is 1.229 V. The standard free enthalpy for the reaction is $-56{,}900$ cal/mole. At the process operating temperature of 60–80°C, the equilibrium voltage can be reasonably assumed to be 1.25 V. The resistive loss of each electrolytic cell is kept to a minimum by the use of a 20%-NaOH or 30%-KOH solution. An acidic electrolyte that can serve the same purpose is seldom used because of corrosion problems. The electrode materials should have activation overpotentials as low as possible and should also be cheap and noncorroding in the alkali solution. Iron, which has a relatively low hydrogen overpotential, is commonly used as the cathode. Cobalt, nickel, or a nickel-plated-steel cathode are the standard materials for the anode. Nickel has the added advantage of becoming easily passivated so that it is resistant to attack by nascent oxygen.

Two types of commercial electrolysis cells have been used. They are the unipolar tank-type and the bipolar filter-press-type cells. The construction details of these cells are described by Hampel[18] and Mantell.[28] Between the electrodes are diaphragms whose function is to prevent the mixing of hydrogen and oxygen. The dc passing through the cells can be from 3000 to 15,000 A. For the unipolar cell, the maximum operating current density is about 70 A/ft^2. The bipolar cell can sustain current densities as high as 150 A/ft^2. The cell operating voltage is about 2.0–2.25 V. The current efficiency is 96–100%. The cells are operated at 60–70°C and essentially atmospheric pressure. A number of cells have been operated at elevated pressures of up to 200 atm and at a temperature of 95°C. The effect of increased pressure and temperature is to reduce overpotential and the gas bubble size at the electrode. The overall benefit is, therefore, a decrease in the operating voltage.[28] The feedwater is distilled water with NaOH or KOH additive, and the makeup water required is from 6.5 to 8 gal/1000 SCF of hydrogen. The power requirement of existing commercial hydrogen electrolyzers is about 125–160 kWh/1000 SCF of hydrogen.

At present, the relative high cost of the electrical energy required for conventional electrolysis processes limits electrolytic hydrogen production to plants where special low-cost electricity is available or to special situations in which increased hydrogen cost can or has to be tolerated. The plant in Norway already referred to is an example where cheap hydroelectric power is available.

At Trail, British Columbia, cheap hydroelectricity and the need for the oxygen byproduct in a metallurgical plant make the electrolysis process practical. With the availability of cheap natural gas, recently added hydrogen production capacities at both of these sites are provided by the steam-reforming process.

The main parameters that affect the performance of the electrolysis are cell design, electrode structure, electrolyte concentration, operating pressure, and temperature. These parameters determine both the current density and voltage at which the cell may be operated, which are interdependent. For a given set of cell design and operating conditions, the cell voltage increases with the increase of current density. The cell voltage is related to the energy efficiency or operating costs of the process, and the current density is inversely proportional to the area and number of cells and therefore the capital costs of the electrolysis plant. An optimal electrolysis plant design requires the proper trade off of the cell voltage, current density, and the other process parameters.

The low performances of present electrolytic cells are mainly caused by high cell internal resistance, high electrode activation overpotential, and high polarization at the electrodes, for example, caused by gas bubble accumulation. These problems are identical to those encountered in the development of electrochemical fuel-cell devices. In the last decade, extensive research, particularly for military and aerospace applications, has advanced techniques to solve these problems. This fuel-cell technology is directly transferrable to the water-electrolysis process. The operation of a fuel cell (described in some detail in Section VI) produces water by electrochemically combining hydrogen and oxygen. The electrons released are captured as usable electrical energy. It is the exact reverse of an electrolysis process in which electricity is used to decompose water into its hydrogen and oxygen constituents. In general, application of fuel-cell technology to an electrolytic cell will allow the increase of operating temperature and pressure and the increase in the key parameter of operating current density at an efficient cell voltage. The result is reduced electricity consumption and electrolyzer size or lower operating and capital costs. Two of these advanced water-electrolysis processes[1,8,34] which are based on fuel-cell technology are of particular interest. The Allis-Chalmers system[1,8] is in an advanced stage of development with single- and multicell test data available. The General Electric system[34] is only conceptual at present.

The basic electrolysis cell unit of the Allis-Chalmers system consists of two uncatalyzed, porous, nickel electrodes separated by a thin asbestos matrix that contains the aqueous KOH electrolyte. The porous electrodes permit the product gases to pass through the electrodes. Thus, both concentration polarization and cell internal resistance are reduced.

A schematic diagram of the Allis-Chalmers basic electrolysis cell is shown in Fig. 3. This cell is capable of operating at a current density of up to 1600 A/ft^2 at 200°F. The cell voltages are 1.95 and 2.20 V at current densities of 800 and 1600 A/ft^2. At 250°F the predicted cell voltages at these two current densities are 1.78 and 1.99 V. A current density of up to 4600 A/ft^2 at a reasonable cell voltage was achieved experimentally. This represents an improvement of up to 23 times over the 200 A/ft^2 operating current density of a conventional electrolysis cell. Based on these single- and multicell data, water-electrolysis systems capable of producing 4400 lb H_2/day (or 305 million CF/yr) at 800 and 1600-A/ft^2 current densities and 44,000 lb H_2/day (or 3050 million CF/yr) at 1600-A/ft^2 operating current density were studied. A simplified block diagram of the electrolysis system is shown in Fig. 4.

The operating temperature and pressure of the reference system of 4400 lb of H_2/day at 800 A/ft^2 are 250°F and 300 psig. The overall power requirement is 115 kWh/1000 SCF of hydrogen. This represents an overall energy efficiency of about 83%. Detailed economic analyses of the capital and operating costs of the system are presented in the papers by Costa and Grimes[8] and Mrochek.[34]

The General Electric system (Fig. 5) is based on a high-temperature vapor-phase electrolysis cell. The electrolyte is solid zirconia (ZrO_2) doped with controlled amounts of calcia (CaO), yttria (Y_2O_3), and ytterbia (Yb_2O_3). A half-cell assembly is shown in Fig. 5. The operating temperature is from 500 to 800°C. The simple cell stack may consist of a $\frac{3}{8}$-in. diam tube of

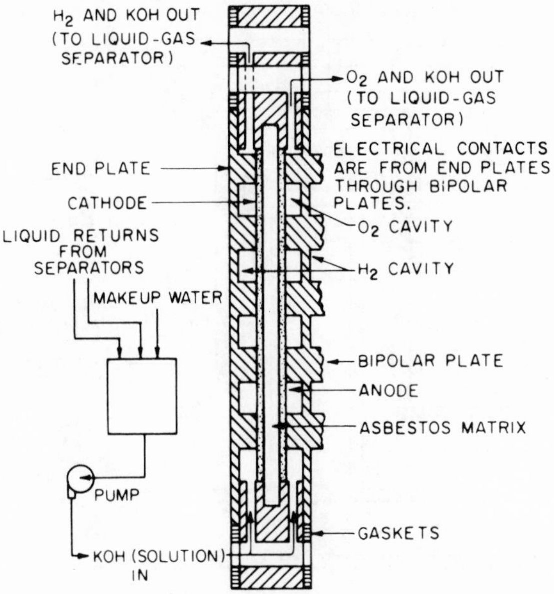

Figure 3. Schematic design of end cell in Allis-Chalmers bipolar water-electrolysis cell. (From Allis-Chalmers Manufacturing Co.)

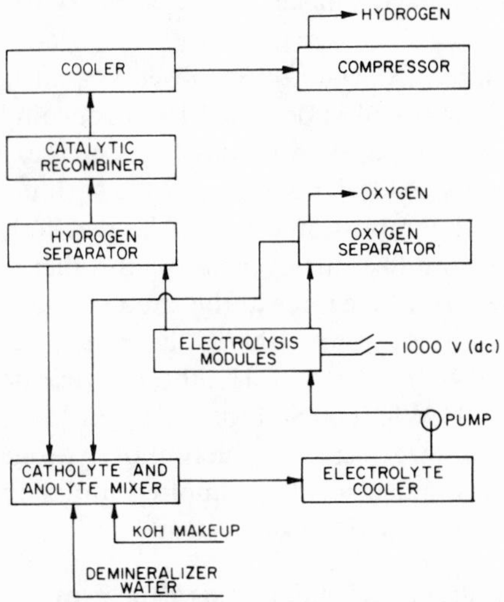

Figure 4. Simplified block diagram of the electrolysis system.

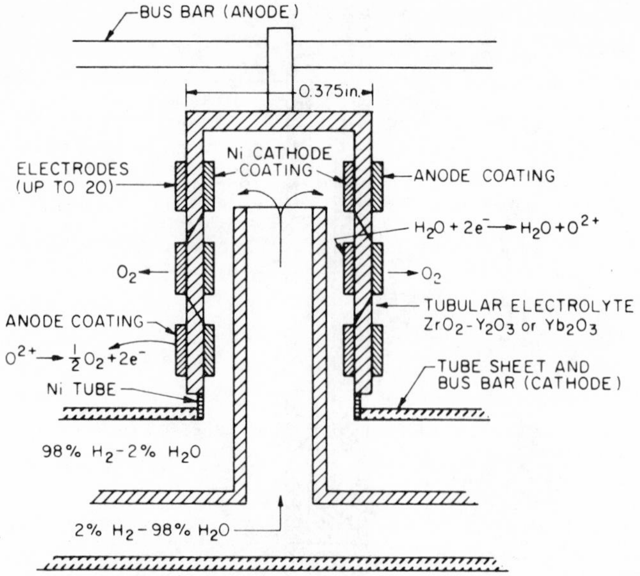

Figure 5. Schematic design of single tube in proposed General Electric steam–hydrogen electrolysis cell. (From General Electric Co.)

zirconia with about 20 bands of electrodes bonded to the inner and outer tube wall. The electrodes are nickel and a complex oxide, praseodymium cobaltate.

The so-called modified water-electrolysis system based on this high-temperature vapor-phase cell is designed to produce hydrogen while consuming the oxygen byproduct. The anode of the cell is exposed to a reducing gas such as carbon monoxide. The source of the carbon monoxide is a producer gas from partial combustion of coal. The reaction of CO with the O_2 at the anode reduces the O_2 partial pressure and therefore simultaneously reduces the back electromotive force at the anode. It has been estimated that at operating conditions of about 1000°C, a current density of 3260 A/ft^2, and a coal addition of 48 lb/1000 SCF H_2, the electrical power requirement is reduced from 80 to 20 kWh/1000 SCF of hydrogen.[34]

Our discussion of large-scale production of hydrogen will be based on both conventional water-electrolysis technology and the proposed advanced Allis-Chalmers and General Electric systems.

4. Chemonuclear Hydrogen Manufacturing Processes

The water-electrolysis process will certainly be utilized increasingly as we approach and eventually arrive at the nuclear energy–electrochemical age.

It is an off-the-shelf process and the technology needed to further improve its performance is largely available. However, the overall energy efficiency of converting the energy liberated by the nuclear fuel decay to the thermal content of the hydrogen is severely limited by the Carnot efficiency of the electricity generation step. In addition, the capital costs of the electricity generation plant must be accounted for in the manufacturing cost of the hydrogen. These added inefficiency and capital costs can be eliminated if we can develop a direct nuclear process to produce hydrogen without the intermediate step of heat–electricity generation. Two main approaches of direct nuclear production of hydrogen from water may be considered—nuclear heat decomposition of water and fissiochemical reaction.

At present, direct nuclear heat decomposition of water seems unlikely to be more efficient than water electrolysis. However, fissiochemical reactors have been investigated because their essential directness, simplicity, and possible cost reduction are substantial incentives to develop solutions to the formidable technical and economic problems that exist.

The first alternative to electrolysis is the use of nuclear heat directly to promote the decomposition of water into hydrogen. The nuclear heat source can contribute toward the production of hydrogen from water by reaction with a hydrocarbon raw material such as coal or natural gas. The basic hydrogen production process would be conventional; the nuclear heat merely provides the process heat requirement. In the transition period from the fossil-fuel age to the electrochemonuclear age, this process may contribute toward the supply of hydrogen, but as the reserve of fossil fuel approaches exhaustion, the raw material for hydrogen production will be largely restricted to water only.

The use of nuclear heat to promote the thermal decomposition of water alone to hydrogen and oxygen does not appear to be practical at present. Funk and Reinstrom[12] performed a detailed thermodynamic analysis on the energy requirements for the production of hydrogen from water. They compare the energy requirements of water electrolysis to a one-step direct water thermal decomposition process and a multistep process. In the multistep process the water first reacts with a compound (e.g., a metal) to form an oxide and hydrogen. The oxide is thermally regenerated into the original compound and oxygen. Similarly, a hydride intermediate can be used. Funk and Reinstrom[12] found that the work required for the one-step process could be reduced by increasing the operating temperature. But the reduction is very modest if an upper temperature limit of 1100°C is imposed by limitations in the materials of construction.

For the multistep process, the thermodynamic work can theoretically be reduced to zero by operating reactions with positive entropy changes at high temperatures and with negative entropy changes at low temperatures.

Hypothetical intermediate compounds were specified in terms of operating temperatures, changes of fuel energy of reaction and formation, and the absolute entropy change that accompanies the addition of one oxygen or hydrogen atom to the compound. However, the realization of a practical multistep process that is more efficient than water electrolysis is frustrated by the failure to find a chemical compound which will comply with the specifications derived. The compounds examined included monoxides and hydrides of lead, mercury, nickel, cadmium, lithium, and vanadium. Thus, the general conclusion is that the direct use of nuclear heat to produce hydrogen does not presently appear to have a significant efficiency advantage over water electrolysis. Recently, a novel hydrogen production process using nuclear heat was proposed by de Beni and Marchetti.[55] They provide little information beyond the fact that it is a three-step process (at 100, 250, and 730°C) involving $CaBr_2$ and Hg.

Instead of allowing the energy liberated by the decay of ^{235}U, ^{239}Pu, or ^{233}U atoms to degenerate into heat energy, as in a conventional fission reactor, can we utilize this energy to decompose the molecules of starting materials into the required products, in our case, water into hydrogen and oxygen? The concept of a fissiochemical reactor was first proposed by Harteck and Dondes[56] in 1956. Experimental research and system design studies[54] were conducted on a variety of reaction systems. The main interest had centered on the fixation of nitrogen in the air to produce NO_2. A succinct review of the technical and economical problems of the fissiochemical reactor was presented by Juppe[57] in 1969. His conclusion is that the formidable technical problems of fuel design and product decontamination, while not inherently prohibitive, still need substantial research and development before the practicality of the fissiochemical reactor can be determined. The experimental radiation chemical yields of reactions such as nitrogen fixation and carbon dioxide and water decomposition fall far below the theoretical maximum possible.

The energy liberated by the decay of the ^{235}U, ^{239}Pu, or ^{233}U fuels consists mainly of electron, neutron, and gamma radiation and the kinetic energy of the fission fragments. If a power reactor is used for chemical production, only about 8% of the total fission energy can be available for chemical production using the conventional fuel design. The majority (about 84%) of the fission energy is in the form of the kinetic energy of the fission fragments. It is the utilization of this portion of the fission energy that is of interest in a direct nuclear water-decomposition process. Ordinarily, this energy is released as heat within the fuel elements. A key fact that governs the design of a chemonuclear reactor and the treatment of its product is the extremely short range of these fission fragments; the average fission track lengths in gases and condensed systems are 2–2.5 cm and 5–25 μ, respectively, and the width of the fission track is only 100–300 Å. Thus, the starting material, i.e., water, have to be in intimate contact over a large surface area with the radioactive uranium or plutonium fuels themselves. The practical engineering consequence is the need to design a fuel-element configuration with a large surface area. The experimental chemonuclear fuel-element designs that have been tested are honeycomb structures, glass-fiber substrates, and fluidized beds. The most advanced design is the honeycomb structure, which seems to have the correct combination of reactivity and mechanical strength.

A second consequence of the need to have intimate interface between the starting material and the radioactive fuels is the contamination of the products by radioactive fission products, uranium fuel particles, and radioactive species produced by neutron activation. Although advanced nuclear-fuel processing technology can be adopted to decontaminate the products, the equipment involved would probably be very complex, elaborate, and expensive. The need of this added product-treatment subsystem may eliminate the potential simplicity and reduced capital costs of the chemonuclear process when compared with the water-electrolysis process.

In addition to the problems of fuel configuration and product decontamination the chemonuclear reactor suffers from low radiation chemical yield. Experimental data to date indicate that efficiencies of only about 0.5% can be achieved for water decomposition. Unless the radiation chemical yield can be significantly raised, the chemonuclear reactor will not be able to compete with conventional processes. The radiation yield is measured by the *G* value, which is defined as the number of molecules formed per 100 eV of radiation absorbed. The *G* value is inversely related to the energy requirement (e.g., kWh/lb) for the process. For endothermic reactions a maximum theoretical *G* value can be calculated. This maximum G value for decomposition of water into hydrogen and oxygen is 338.8 molecules/100 eV. However, the experimental value for the same reaction is only 1.75 molecules/100 eV. The thermal efficiency expressed as the ratio of experimental to theoretical *G* value is 0.5%. The need to improve upon this experiment *G* value by theoretical and experimental research is obvious.

As well as having a low efficiency, the decomposition of water in a chemonuclear reactor yields an explosive mixture of hydrogen and oxygen. The separation of these two gases will be extremely difficult. Chemical scavenging of either of the two will most likely induce an explosion of the unseparated mixture. The palladium-foil-diffuser separating technique will also produce a similar effect.

In a conventional steam–hydrocarbon or steam–metal hydrogen generation process, the oxygen of the water is effectively separated from the hydrogen by chemically binding it with carbon (as CO_2) or a metal (as oxide). In this form mechanical or chemical separation of the hydrogen is relatively easy.

In the water-electrolysis process the hydrogen and oxygen are physically produced at two locations and can be easily separated by the use of a diaphragm or other similar device. An equally expedient scheme must be invented to separate the hydrogen and oxygen gases from the chemonuclear reactor. This would, however, undoubtedly further add to the system complexity and capital costs of the fissiochemical plant.

A novel hydrogen production system that would combine the use of the nuclear reactor as a heat source and a fissiochemical reactor was proposed by Von Fredersdorff.[60] The hydrogen is produced (1000–1400°F) by the action of steam on iron at elevated temperature as in the conventional steam–iron process:

$$Fe + H_2O \rightarrow FeO + H_2$$

$$3FeO + H_2O \xrightarrow[\text{heat}]{\text{nuclear}} Fe_3O_4 + H_2$$

The iron oxides are then regenerated by carbon monoxide:

$$FeO + CO \rightarrow Fe + CO_2$$

$$Fe_3O_4 + CO \rightarrow 3FeO + CO_2$$

The departure from the conventional steam–iron process is the source of the carbon monoxide. Instead of a water–gas mixture of hydrogen and carbon monoxide from a fossil fuel such as coal, we obtain the CO by fissiochemical decomposition of carbon dioxide:

$$CO_2 \xrightarrow[\text{decomposition}]{\text{fissiochemical}} CO + \tfrac{1}{2}O_2$$

The maximum G value for this reaction is 48.1 molecules/100 eV, and the experimental G values for gamma radiation and fission fragment kinetic energy are 0.13 and 1.9, representing efficiencies of only 0.2 and 3.9%. The oxygen byproduct from the CO_2 decomposition is recovered by the use of a metal–metal oxide cycle.

$$M\ (\text{metal}) + \tfrac{1}{2}O_2 \rightarrow MO$$

$$MO \xrightarrow[\text{heat}]{\text{nuclear}} M + \tfrac{1}{2}O_2$$

The major advantage of this hybrid fissiochemical–thermal process is the elimination of the hydrogen–oxygen separation problem described above. The steam–iron portion of the process is well known and has been commercially used to produce hydrogen. Future research and engineering work should include (1) investigation of the fissiochemical decomposition of carbon dioxide to increase its efficiency and (2) system analysis of the total hybrid system.

In view of the immaturity of the direct fissiochemical processes the following discussion of large-scale hydrogen production will be based on the electrolytic hydrogen process only.

5. Large-Scale Production of Hydrogen

The 1968 hydrogen production in the U.S. was about 2.28 trillion SCF. Practically all the hydrogen is consumed as nonenergy chemical raw materials. As indicated previously, this amount of nonenergy hydrogen consumption will grow rapidly as we approach the nuclear–electrochemical age.[32] In addition, the use of hydrogen as an energy-providing fuel will also begin to build to a substantial level. We will therefore attempt to estimate the magnitude of the future hydrogen production in the U.S. by examining the projected energy and nonenergy hydrogen needs. We will also consider the impact of the huge quantity of hydrogen to be produced on water-electrolysis technology, the electrical energy requirements, the economics, and possible environmental effects.

In the nuclear–electrochemical age, hydrogen and its derivatives will ultimately replace natural gas, petroleum products, and other fossil chemical fuels. In the transition from fossil to nonfossil chemical fuels, because natural gas is the most readily replaced fossil fuel, it will probably be the first one replaced by hydrogen. In estimating and discussing the magnitude of the hydrogen production, we will occasionally make references to the natural-gas requirements.

The yearly electrical energy (trillion kWh) and electrical power (million MW) needs as a function of the yearly hydrogen production are shown in Figs. 6 and 7. The upper curve in each figure is based on the power requirement of present commercial electrolyzers (150 kWh/1000 SCF H_2). The lower curve is based on the proposed General Electric hybrid coal-electrolysis process. The electrical energy requirement is assumed to be 50 kWh/1000 SCF H_2. The balance of the energy requirement in the process is derived from the coal. The middle curve is based on the advanced Allis-Chalmers electrolysis process using 115 kWh/1000 SCF H_2. At the abscissas of Figs. 6 and 7 the thermal equivalents of natural gas in trillions of cubic feet are shown to put the hydrogen rate of production in realistic perspective.

The rapid growth and change of our technological society make the prediction of future energy needs unusually difficult and hazardous; even the forecasting of the requirements of natural gas, which has an established

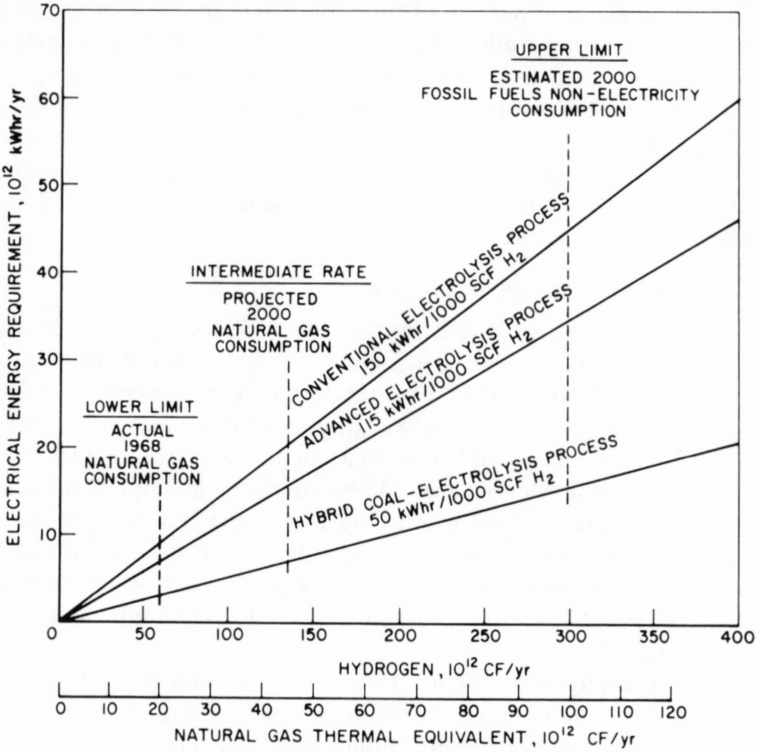

Figure 6. Electrical energy requirements needed to produce projected hydrogen demands.

usage history, is subjected to influence by unknown factors. For hydrogen chemical fuel the added uncertainties of what percentage of fossil fuels will be replaced by hydrogen and when this process will begin must also be considered. Therefore, instead of predicting a specific quantity of yearly hydrogen production at a specific year, we will examine a spectrum of possible production rates. The three signposts that underpin this hydrogen production-rate spectrum are shown in Figs. 6 and 7.

The lower limit is chosen as the same energy content as the actual consumption of natural gas in 1968 (19.9 trillion CF). In the year 2000, it has been predicted[27] that this natural-gas consumption will have risen to 45 trillion CF; the corresponding energy content of this consumption level is used as an intermediate limit. The upper limit is based upon the assumption that all fossil-fuel energy sources, other than those used for electricity generation, are replaced by hydrogen by the year 2000. This yearly consumption

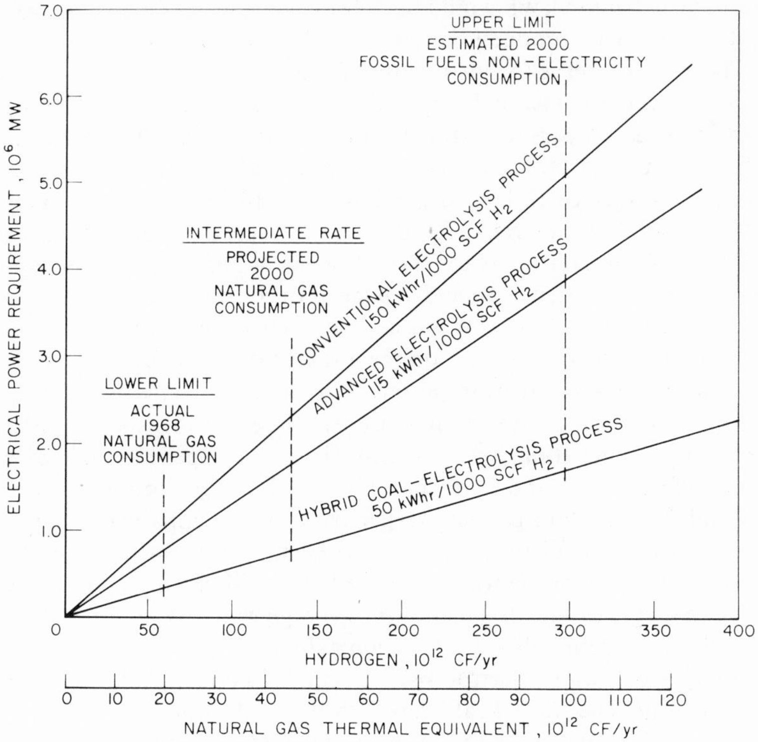

Figure 7. Electrical power requirements needed to produce projected hydrogen demands.

rate is equivalent to the thermal content of 98 trillion CF of natural gas and is calculated from data published by Starr.[44]

If we consider the ultimate total replacement of fossil fuels by hydrogen, by the year 2000, enormous quantities of hydrogen are required, and the corresponding electricity demand is huge. About 300 trillion CF of hydrogen (789 million tons or 107×10^{15} Btu) per year are required; the electricity demand to produce this amount is 15–45 trillion kWh at a rate of 1.7–5.1 million MW. The projected U.S. electrical energy productions for the years 1975 and 2000 are "only" 1.96 trillion[42] and 10.9 trillion[24] kWh. In other words, if a substantial transition from fossil fuel to electrolytic hydrogen is to take place by the year 2000, the electrical energy output will have to be two to five times the projected figure.

Even if we consider our lower limits based on 1968 figures, the situation is alarming. If the energy content of the 1968 consumption of natural gas had been provided as electrolytic hydrogen, using even the most advanced technology, 3 trillion kWh would have been consumed. Yet in that year the total U.S. electricity production was only 1.436 trillion kWh.[49]

These figures serve to illustrate the vast discrepancy between the energy capacity of the present gas industry and that of the present electrical industry, and the enormous needs for extra electrical plant capacity if we are to make the transition to electrolytic gas in the foreseeable future. Fusion power stations are projected to have capacities up to 10,000 MW.[48] The number of these power stations required to cover the spectrum of hydrogen production rates ranges from 34 to 500 stations over and above those required to meet normal electricity demands. In terms of conventional power stations, with a maximum of 1000 MW on a single generator shaft, an inordinate number of new generating units, on the order of 1000–5000, are required.

Present commercial water-electrolysis processes have makeup water needs of about 8 gal/1000 SCF of hydrogen. Based on this specification, the water requirements at the lower, intermediate, and upper rates of hydrogen production in Figs. 6 and 7 are 480, 1080, and 2400 billion gal/yr, or 480, 1080, and 2400 million gal/day, respectively. These apparently high rates only represent a small fraction of our present water consumption, about 400 billion gal/day for the whole U.S. and 866 million gal/day for Chicago. If the load is spread over 100 plants, the average water consumption at each plant will range from only 4.8–24 million gal/day.

A large-scale water-electrolysis plant can be thermally integrated with a desalting plant. The waste heat from both the nuclear generating plant and the electrolysis plant can be used in the distillation units of the desalting plant. In return, a small fraction of the fresh or distilled water product can be used as starting feedwater for the electrolyzer. Engineering feasibility studies of dual nuclear generating and desalting plants have been conducted by the Metropolitan Water District of Southern California[2] for a plant producing 150 million gal/day of fresh water and 1800 MW gross electrical power output. Alternatively, sea water or concentrated sea water can be used in an electrolysis process to produce hydrogen gas, chlorine gas, and caustic soda (NaOH). The power requirement per unit volume of hydrogen in this process will double, and the need for and utilization of the chlorine and caustic soda byproducts will also have to be considered.

In the conventional electrolysis process using distilled water (with NaOH or KOH additive) the major byproduct is oxygen. For every ton of hydrogen, 8 tons of oxygen are produced. At the lower, intermediate, and upper hydrogen production rates in Figs. 6 and 7, the oxygen byproduct output rates are 1.34, 3.02, and 6.72 billion tons/yr or 3.7, 8.3, and 18.4 million tons/day, respectively. These oxygen production rates represent several hundred times the present U.S. oxygen consumption rate. In 1965, only 7.5 million tons of oxygen was made. In 1961, the total U.S. oxygen plant capacity was 22,000 tons/day, and that for the whole world was only 72,300 tons/day. Therefore, in addition to the electrical energy supply, the enormous surplus of oxygen must be considered if practical large-scale hydrogen production by water electrolysis is to become a reality. One of the schemes is the depolarization of the excess oxygen electrode of the electrolyzer by SO_2.[20] In this case we may merely be exchanging an excess oxygen problem for one of excess sulfuric acid. Besides, it should do no harm to discharge the excess oxygen directly to the atmosphere.

The economics of large-scale hydrogen production depends on a complex set of interdependent factors such as electrical energy costs, waste-heat recovery, feedwater pretreatment, and byproduct credits. A detailed price analysis of a nonfossil hydrogen fuel would have to wait until some of these plant specifications and design parameters are chosen. However, some preliminary and simplified estimates can be made. Table 1[34] presents the estimated manufacturing costs for hydrogen based on the advanced systems proposed by Allis-Chalmers and General Electric.

The hydrogen prices range from 25.2 to 33.6¢/1000 SCF H_2. On a thermal basis the prices are \$0.78–1.03/million Btu. Present costs of hydrogen from conventional fossil-fuel-based processes range from 30 to 70¢/1000 SCF H_2 (\$0.92–2.15/million Btu). Current natural-gas prices vary from

Table 1. Estimated Manufacturing Costs for Hydrogen from Two Advanced Electrolysis Processes[a]

Item	Porous electrode cell	High-temperature vapor-phase cell
	Cost, ¢/1000 SCF H_2	
Utilities[b]	29.5	22.0
Maintenance and operating supplies	4.2	4.1
Labor at \$4/h	0.5	0.9
Overhead at 60% of labor	0.3	0.6
Fixed charges[c]	7.6	6.1
Gross manufacturing cost	42.1	33.7
by-product O_2 credit	(8.5)[d]	(8.5)[e]
Net manufacturing cost	33.6	25.2
	Capital investment, \$$10^6$	
	11.8	9.5

[a]Based on hydrogen production, 40 × 10^6 SCF; hydrogen delivery pressure, 1700 psig; fixed charges, 9%.
[b]Electrical energy at 2.5 mills/kWh.
[c]Depreciation, 6.7%; local taxes and insurance, 2.3%.
[d]Oxygen delivered at 300 psig, \$4/ton.
[e]Oxygen delivered at atmospheric pressure.

20¢/million Btu for domestic gas at the wellhead to 70¢/million Btu for imported liquefied natural gas at the port of entry. The major cost item in Table 1 is the electrical energy cost, which represents 60–70% of the gross manufacturing cost. If we should use distilled-water feed from a desalting distillation plant, the cost of the feedwater would only be 0.4¢/1000 SCF H_2 (based on desalted-water costs of 50¢/1000 gal). The byproduct credit for oxygen of $4/ton used in Table 1 contains some uncertainty, since a surplus will be produced. Another unknown and potentially significant byproduct credit is the possible coproduction and sales of heavy water from water-electrolysis plants.[18]

The large-scale production of hydrogen by water electrolysis and the replacement of fossil fuels by hydrogen fuel introduce complex perturbation in the mechanisms and rates of the natural energy, water, oxygen, hydrogen, and other cycles of the biosphere. Detailed analyses of the impact on the biosphere should be included in any design of large electricity–hydrogen plants in the nuclear–electrochemical age.

The absence of fossil-fuel combustion at these nuclear electricity–hydrogen plants will eliminate the conventional chemical contamination of the air and water. For the fusion-type reactor, the radioactive emission would also be minimized. In terms of volume of radioactive materials and the length of time that they would be retained by the human body, a 5000-MW fusion plant presents about the same degree of radioactive hazard as a 5-kW fission plant. However, the nuclear electricity–hydrogen plant does have a high potential for thermal pollution. Technical and economic requirements will increase the nuclear fusion generating plant capacity to 5000–10,000 MW[48] compared to the present fossil-fueled plant maximum size of about 1000 MW. If we assume a thermal efficiency of 40%, the amount of waste heat at each 10,000-MW fusion generating plant would be 15,000 MW or 0.512×10^{11} Btu/h.

This very high rate of waste-heat discharge at one site severely limits the location of these plants if we are to avoid excessive thermal pollution. The only practical location for such stations is on the sea coast or on offshore platforms because air or inland-water cooling techniques to remove the waste heat have limited capacity. The only practical heat sink with sufficient capacity appears to be the water in the ocean. Not only do we have an enormous total amount of water for cooling purposes (about 3.7×10^{20} gallons), the constant motion of the ocean water will distribute the waste heat over a wide area and thus limit the temperature rise of the water surrounding a plant.

As has already been suggested in a previous section, sea water could be used as the water source for electrolysis. The sea water can either be used as a brine or as distilled water from a desalination plant integrated to the electricity–hydrogen complex. The desalting plant would also help reduce the thermal pollution of the complex by utilizing part of the waste heat for distillation. The total water withdrawal for electrolysis of 2 billion gal/day appears negligible when compared to an average natural evaporation rate from the ocean surfaces of 3.15×10^5 billion gal/day. Effluent water contaminated thermally or by chemicals should be properly treated before it is returned to the sea. The huge oxygen byproduct of any electrolysis plant may be used in sewage treatment,[41] aluminum scrap recovery,[5] and other new processes. The availability of large amounts of inexpensive oxygen may serve as an incentive for the development of new industrial and anti-air- and anti-water-pollution processes. The use of oxygen in any industrial process will in itself help solve our pollution problems as oxides of nitrogen from air combustion will be reduced. If the cost of air is included in future manufacturing pricing systems, a manufacturer will be under pressure and have incentives to consider the use of oxygen.

IV. HYDROGEN TRANSMISSION, STORAGE, AND DISTRIBUTION

If hydrogen is to be used as a nationwide or regional fuel to the extent that natural gas is today, the method of transportation from source to user that would evolve for hydrogen would probably be the same as for natural gas now. The two gases are similar enough in most properties so that it is entirely feasible to use the same technology (or even the existing systems) of transmission pipelines, intermediate storage, and urban distribution networks used for natural gas. The same gas-utility organizations would also appear ideal to carry on with hydrogen distribution. Indeed, the old manufactured gases, distributed for some 80 years in most major U.S. cities prior to the widespread availability of natural gas, contained a high percentage of hydrogen.

1. Transmission of Hydrogen by Pipeline

Cross-country pipelines to carry hydrogen gas from source to market can be constructed from the same designs and materials and installed in much the same manner as today's natural-gas transmission lines. Offshore oil- and gas-well technology would also serve well to bring hydrogen ashore from ocean-sited plants manufacturing hydrogen by electrolysis, should the use of such plants develop.

Because of hydrogen's lower heating value [325 Btu/SCF vs natural gas at 1000+ Btu/SCF (both gross dry)], it might appear at first glance to require significantly larger pipelines to carry the same amount of thermal energy. It will certainly require the evident increase in number of standard cubic feet transported—i.e., $\frac{1000}{325}$ or roughly triple the present amount. However, because of hydrogen's significantly lower specific gravity [0.0695 vs natural gas at 0.60 (as an average)], we have a nearly compensating increase of approximately 2.5 times the flow capacity.[50]

Hydrogen does not compare as favorably, however, when compression costs are added. As cross-country pipelines generally need pressure boosting stations every 100 miles or so, such costs are a significant part of the total cost of transporting gas. Here three times the quantity—i.e., volume of gas compressed—means roughly three times the horsepower would be required. This translates into larger investments in compressors, their facilities, operating personnel, and fuel.

The end result of adding pipeline and compression costs to yield a total gas transmission cost shows hydrogen some 60% more expensive than natural gas to pipeline on an equal-energy basis, presuming the same pipeline (diameter, wall thickness, strength, internal smoothness, etc.), length, and compressor station spacing (but not the same size stations).[50] As transmission costs are not the major portion of final costs at the point of use, this 60% premium would not appear prohibitive. When one considers that hydrogen plants can certainly be built closer to our major coastal markets (presuming ocean sites) than today's or tomorrow's gas fields, the total transmission costs will probably be considerably less than those for natural gas as a consequence of the shorter transmission distances.

It is also possible to transport hydrogen in other forms. Liquefied hydrogen is perhaps the most obvious form, with a commercial history already established in the space program and industry. Technically, this form would be adaptable to transport by either pipeline or in unit containers such as ships. Again, judging by the evolved methods of transporting natural gas, liquefaction appears to be economically justified only when transport of large quantities over great water distances by ship is involved. So far, no cross-country liquefied natural-gas pipelines have emerged, indicating unfavorable economics for the present technology of transport in this form.

Also possible is the transport (and storage) of hydrogen in various chemical forms such as ammonia, and methanol. All of these can be easily synthesized from hydrogen and easily thermally broken down to yield hydrogen. Whether or not hydrogen would be transported in these forms would be determined by the costs of putting the hydrogen into such forms and freeing it at the market end rather than by the mere transportation costs.

Presuming pipeline transport as a gas under conditions previously stated, hydrogen transmission may be expected to cost 60% more than natural gas. Today's costs for transmission of natural gas are about 1.1–2.4¢/million Btu-100 miles. This leads us to estimate hydrogen transmission costs of about 1.8–3.8¢/million Btu-100 miles. When compared to the cost of transmitting the same amount of energy as electricity (14–19¢/million Btu-100 miles in 345 kVac transmission systems), hydrogen has an advantage of 6:1 and natural gas has an advantage of 10:1.[11]

Adjusting these figures to account for differences in utilization efficiencies between gas and electric appliances will not change the relative positions of the costs. Indeed, indications are that heating appliances using hydrogen gas can be almost 100% efficient as no venting would be required most of the time and then only to prevent overhumidification from the combustion product, water vapor.

2. Storage of Hydrogen

Like most gases, hydrogen may be stored in any of several forms. In the quantities which hydrogen's use as a major energy form would dictate, natural gas can again provide a guiding experience. Experience with natural gas, for example, indicates possible underground storage if the proper geology could be found in or near the market area. Depleted oil and gas wells are the most numerous form of this kind of storage geology; porous, permeable, water-bearing sandstone overlain by impermeable cap rock is another. This latter formation is called an aquifer. Mined caverns are a third form of underground storage. However, hydrogen's smaller molecule makes it possible for it to leak upward through certain cap rock that might readily retain heavier gases such as natural gas. This would need to be considered for all underground storage.

Liquefaction for storage (and/or transport) is another natural-gas precedent. In fact, hydrogen is currently delivered and stored as a liquid in the space program and for other large industrial users. The volume reduction achieved by liquefaction of hydrogen is greater than that of natural gas, which is a plus. Expressed as a liquid–gas ratio (liquid at boiling point, gas at 70°F and 1 atm), hydrogen is 1:850[7] as compared to a natural gas at 1:630.[9] Lower temperatures must be achieved, however, to change hydrogen gas to a liquid: −434.6°F at 1 atm compared to natural gas at −260°F. Thus, liquefaction of hydrogen would cost more money.

Storage as a gas under pressure is the most common current form of hydrogen storage. Cylinders or tube trailers at the user's location contain the gas. Larger pressure vessels, with spherical and other shapes, located both above or below grade could also follow from natural-gas precedents. None of these heavy, metal-shelled forms of gas storage, however, is economical for the larger volumes needed for a major energy source in or near a major market area because of their higher cost compared with the relatively thinner walled metal tanks or concrete tanks used for liquefied gas storage. They compare even less favorably with underground formations.

As was true with its transportation, hydrogen can be stored in various chemical forms. Should the production of hydrogen, by whatever future methods evolve, provide it economically in one of these chemical forms which also proves to be safe, compact, readily transportable, and storable, the overall economics dictate such a divergence from natural-gas practice. However, because of the adverse economics and physics involved,[9] this is unlikely and seems no more probable than storage of natural gas in any of its chemically variant forms by absorption, adsorption, as a hydrate, etc.

3. Distribution of Hydrogen

Probably the major problems with hydrogen distribution are the safety concerns and existing codes rather than the technology itself. As stated earlier, hydrogen in mixed-gas form has been distributed for over a hundred years and, in many cities of the world, still is distributed as "manufactured" or "city" gas. Carbureted water gas, blue gas, coke oven gas are all forms of "city gas" and all contain substantial percentages of hydrogen; water gas is 50% hydrogen. However, urban distribution of straight hydrogen as either a gas or a liquid would undoubtedly call for tighter piping systems and more diligent patrolling and maintenance than have proved acceptable in city-gas or natural-gas distribution operations. As a consequence, mixing of other gases with the hydrogen prior to entering urban areas may prove to be the most practical alternative. These mixing gases should narrow hydrogen's flammable limits, increase the energy required to ignite it, increase flame visibility, and, ideally, add to the heating value of pure hydrogen.

A possible design for distribution is the "pipe within a pipe," with pressurized hydrogen gas occupying the inner or carrier pipe and with an inert gas such as nitrogen or helium filling the annular space between inner and outer pipe. An alternative would be to sweep the annular space with air to keep possible small hydrogen leaks from building up concentrations to the lower flammability limit. Either system could be equipped with monitoring alarms to summon corrective actions. Existing low-pressure (inches water column), relatively large-diameter mains in our older urban gas distribution systems would lend themselves to this "pipe within a pipe" design by accepting the insertion of the smaller diameter, relatively higher pressure hydrogen carrier pipe or tubing.

The technology of distribution of hydrogen can be very similar to today's distribution of natural gas once the above-cited added precautions are considered. As mentioned under transmission, we would suffer from hydrogen's lower heating value and thereby need to distribute three times as many cubic feet of hydrogen to match the thermal capacity of a given system of natural gas. But, again, we have a nearly compensating increase of

2.5–2.7 times in flow capacity because of hydrogen's significantly lower specific gravity. Leakage through a given size hole would be similarly increased, however. As gas distribution systems rarely deal with compression, no cost penalties of increased compressor horsepower for hydrogen's added volume would hurt them directly, save through increased gas costs charged them by the transmission companies. A distinct benefit to distribution would be hydrogen's tendency to heat up slightly upon expansion, which would eliminate the need for pipeline heaters at the "city-gate" stations, as if often necessary with natural gas, in order to prevent frost damage due to gas cooling after expansion.

V. HYDROGEN-FUELED EQUIPMENT

1. Introduction

Hydrogen, produced in central plants from electrical energy and distributed to user premises, must be considered in competition with electricity, and is likely to be relatively expensive. The use of hydrogen as a fuel will be contemplated where there are economic or convenience advantages. The capital and operating costs of transmitting hydrogen gas or the construction of a system (i.e., underground) will, in many instances, prove to be less expensive than those of transmitting electricity. However, the efficient use of this gas for conventional fuel applications demands certain changes in the appliances we are conventionally using.

Basically, hydrogen fuel will be applied for two types of applications: (1) to generate heat for heating, cooking, etc., and (2) to generate mechanical energy. In the first case, the hydrogen would be burned in an appropriate burner, and the heat used in conventional ways. Appliances for conversion of gas to useful heat fall into a wide range of sizes, from food-warming devices to large industrial furnaces. All of these uses depend upon some form of burner. In general, we may anticipate that hydrogen burners will be smaller than natural-gas burners; the combustion chambers might also be smaller. Thus, hydrogen-fueled appliances can be smaller than present ones and will produce clean products of combustion. Appliances for conversion of gas into mechanical energy are either internal or external combustion engines. The latter depend upon burners again, and hence the above comments apply. Internal combustion engines of the piston type and gas-turbine type that are designed to operate on hydrogen as a fuel appear to be attractive propositions because of their clean exhaust characteristics.

2. Burner Design

In conventional practice, gas used for heating or cooking is burned in a flame-type burner where high temperatures are developed locally and the heat of the flame is transferred to its destination by radiation or conduction. The product gases, which may contain H_2O, CO_2, CO, and sulfur compounds, according to the gas used, are ducted away in a flue for high-output devices (furnaces, etc.) or are allowed to escape to the surrounding atmosphere for low-output appliances.

The burner is designed to produce a stable flame.[47] Most burners are equipped with means to mix air with the gas prior to burning (primary air) by an inspiration technique. The velocity of the gas–air mixture is arranged so that it travels toward the open end of the burner at a rate higher than the flame-propagation rate in the mixture. This prevents "flashback" or ignition of the gas inside the burner.

At the burner mouth, the gas velocity at the rim is zero, so the flame attaches itself to the rim, and propagates itself across the gas–air mixture in a cone, such that the component of gas velocity normal to the cone surface equals the burning velocity. Outside the cone, the combustion proceeds with secondary air which diffuses in from outside the flame. If the gas velocity is decreased below the burning velocity, the flame flashes back. If the gas velocity is increased beyond another critical value, at which the gas flow velocity exceeds the burning velocity at all points, the flame is "blown away." Thus, a conventional burner has limits within which it can operate.

Figure 8 shows flame velocities for various mixtures of some typical fuel gases with air. Hydrogen has a uniquely high flame velocity, particularly in the region of less than stoichiometric air contents such as would prevail in the primary gas–air mixture. Thus, a burner designed for methane, for example, would not operate with hydrogen, and a different design would have to be employed. However, since data exist for the design of stable-flame

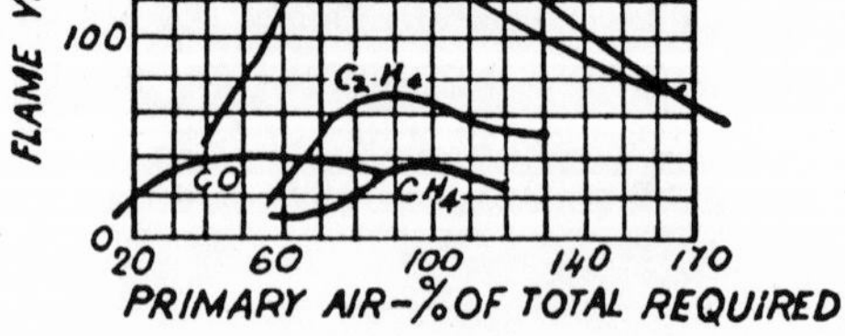

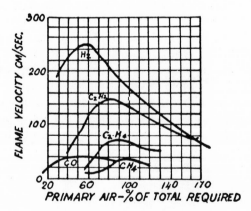

Figure 8. Burning velocities of combustible gases mixed with air.[47]

burners operating on many gas mixtures, including pure hydrogen,[16] the design and construction of conventional heating and cooking appliances that use hydrogen present no severe design problem. Figures 9 and 10 show burner design data for methane and hydrogen, indicating that hydrogen burners are more prone to flashback than blowoff.

Perhaps the most significant difference in the use of hydrogen as fuel is the absence of a need for a flue; as the product gases consist only of water vapor, no CO_2, CO, or sulfur pollution can occur. Domestic space-heating systems might operate without a flue, utilizing the humidification potential of the exhaust gas instead of a separate humidifier. Such a heating system would be more efficient than a conventional fueled furnace because all the

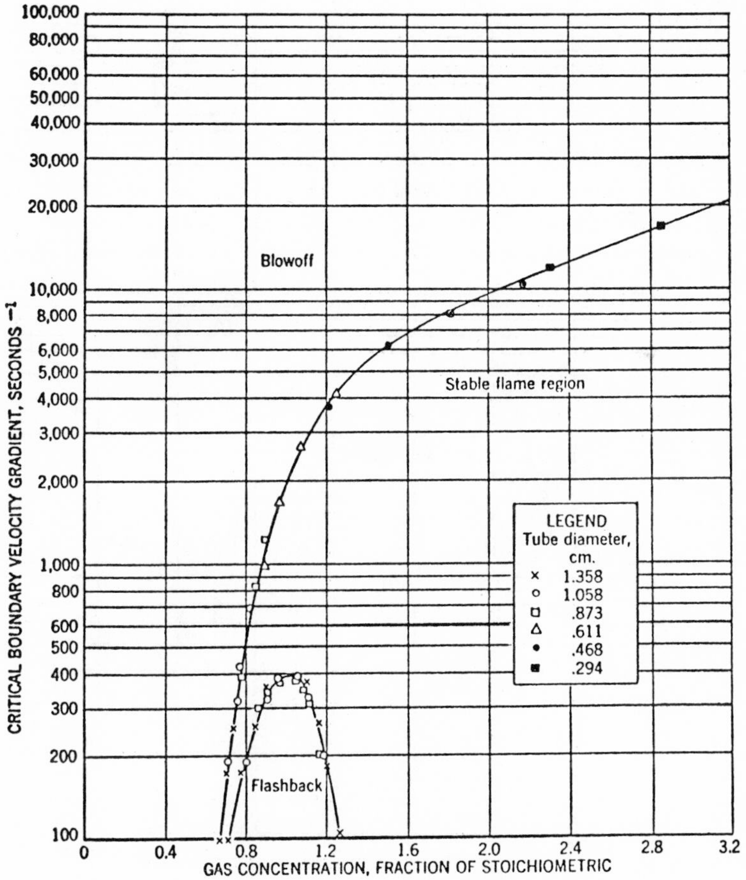

Figure 9. Flame-stability diagram for a fuel containing 100% CH_4.[16]

heat of combustion is discharged into the house. However, on the coldest days, the heating requirement will be so high that excessive humidity will result, unless part of the exhaust is ducted outside. For example, at a 35,000-Btu/h heating load, about 20 gal/day of water is produced. Table 2

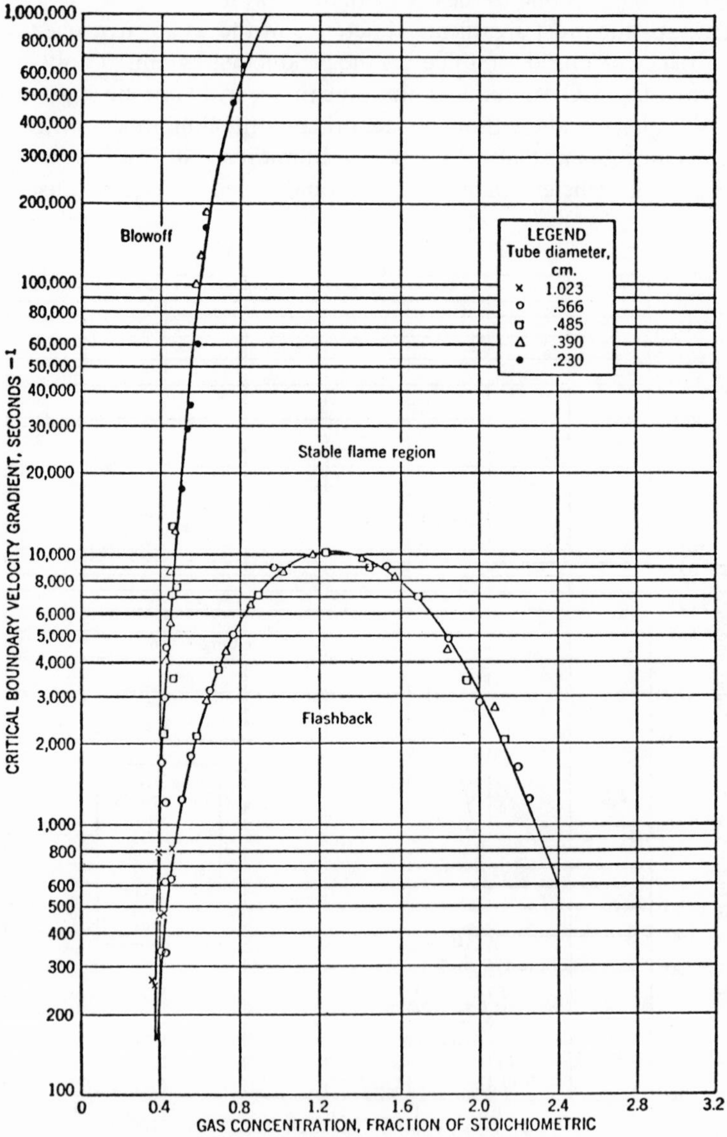

Figure 10. Flame-stability diagram for a fuel containing 99.7 % H_2 and 0.3 % O_2.[16]

Table 2. Air Consumption and Combustion Products of Hydrogen and Methane

				Combustion products		
Gas	Volume of gas, CF	Heating value, Btu	Air required, CF	Water vapor, CF	CO_2, CF	N_2, CF
Methane	1	1013	9.528	2	1	7.5
Hydrogen	1	325	2.382	1	0	1.88
Methane	0.98	1000	9.3	1.95	0.98	7.3
Hydrogen	3.07	1000	7.3	3.1	0	5.8

shows the relative volumes of air consumed by hydrogen and methane, together with the relative volumes of combustion products. Since the calorific values of the two fuels are vastly different, it is of interest to see the relative volumes of reactants and products for a given heat output, as shown in the lower part of Table 2.

3. Catalytic Burners

Various gas research projects over the past decade have demonstrated the advantages of using catalytic burners for cooking and heating appliances.

In a research bulletin published by the American Gas Association[14] in 1963, the following six advantages were listed for properly designed catalytic burners.

(1) The burners generally operate with little or no visible flame mantle.
(2) Catalytic burners generally transfer more useful heat through infrared radiation than do ordinary flame-type burners.
(3) A wider span of operating radiant emission temperatures is available, through choice of catalytic burners, than is available from the flames of more conventional burners.
(4) The radiant source temperature is generally distributed uniformly over the entire surface of a properly designed catalytic burner.
(5) The shapes of catalytic burner surfaces may be tailored to fit the needs of the application.
(6) In certain applications, catalytic burners can be placed close to the materials or heat-transfer surfaces being heated, thus reducing the time required for preheating. (As discussed later, preheating is necessary for catalytic burners operating on natural gas.)

These advantages stimulated further efforts to harness catalytic combustion for gas appliances. In the early and mid-1960's, research progressed on the assumption that, to be economically practical for wide application, any catalytic burner should operate with the most commonly available gaseous fuel, natural gas. This assumption imposed a serious limitation on the technology because of the difficulty in catalyzing natural gas. Consequently, practical catalytic burners for natural gas still require a pilot, or electrical preheating, to initiate combustion. The use of hydrogen, however, eliminates the need for an ignition device and also makes the design of a catalytic burner a simpler proposition.

Catalytic burners fall into two types. The low-temperature type consists of about 70% porous ceramic plate, which might be made of coerderite or mullite. The internal surface of the porous plate is coated with a catalyst, of which platinum and the other precious metals are best, but cheaper materials are possible. Air is supplied entirely as primary air, mixed with the gas, and fed through the porous plate. Combustion occurs at quite low temperatures within the pores of the plate, and no flame appears. The whole structure of the porous plate is heated by the combustion process, and the temperature can be controlled from barely warm to its maximum value by adjusting the flow rate of gas. At low output, such a burner may operate at, say, 300–500°F, where useful heat is produced, but no fire hazard exists. In fact, gasoline can be poured over an operating catalytic burner without igniting. The high-temperature type of catalytic burner requires a backing for the catalytic surface. Figure 11 shows the essential contruction of a typical burner, which can operate up to high temperatures and produce radiant heat. In this case, some combustion takes place on the plate surface with the aid of secondary air.

A hydrogen-fueled catalytic burner of the low-temperature type would be ideal for use in a cooking appliance. A smooth-top cooking range, using a flat ceramic surface heated underneath by catalytic burners, would have

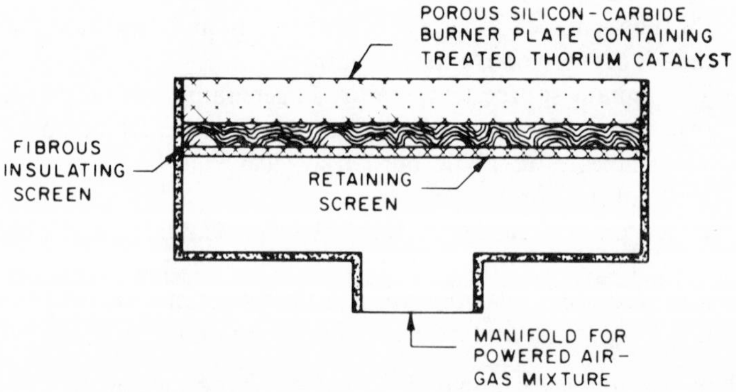

Figure 11. Contemporary, high-temperature catalytic burner.

an almost infinite turndown ratio, to provide rapid cooking or slow warming. Radiant burners would also provide cooking services. One specific advantage of the low-temperature burner is that since no part of it ever reaches a local high temperature, an appliance such as a warming tray could be fabricated from attractive decorative materials such as plastics and aluminum.

Hydrogen-fueled radiant burners would also be attractive for room space heating. Needing no flues, they could be mounted in any location to provide spot heating. Since the heat is generated at low temperatures, the heaters could be covered, for example, with decorative plastic laminate finishes.

So far, we have considered domestic heating and cooking. The appliances described would be competing with conventional electrical appliances because we assume electricity to be universally available at an attractive price. Apart from the possibility that hydrogen might be supplied to the more distant consumer more cheaply than electricity, it is good to be able to offer the choice of heating and cooking techniques to the user as we do today.

However, the generation of large quantities of heat in commercial and industrial applications merits special consideration. At present, space heating in large commercial premises is almost exclusively derived from fossil fuels, not from electricity, while large-scale industrial furnaces used, for example, in the steel, glass, pottery, and chemical industries, are normally fossil-fuel fired. Even if electrical energy were economically more attractive, a gas-fired device may, for many reasons, be preferred. For example, combustion gases are often used to provide inert or reducing atmospheres in conjunction with the process operating within the furnace. Nevertheless, the design and other technical problems of really large-scale hydrogen-fueled flame-type and catalytic burners have not been tackled, and present an area where future design work is needed. No insuperable difficulties are anticipated, but, for safety reasons and for the reasons outlined in the discussion of burner design, major reconstruction of the plant and equipment may be necessary, and this will not be done overnight.

4. Internal Combustion Engines

One of the dominant factors in the choice of fuel for a piston engine is the compression ratio at which the engine can be run. In general, high compression ratios favor higher outputs and higher thermal efficiencies. The permissible compression ratio is determined not only by the fuel characteristics but also by such factors as mixture ratio, size and speed of engine, combustion chamber design, ignition timing, etc. The important fuel characteristics are the flame speed and the limits of flammability.

Hydrogen has very wide limits of flammability when mixed with air (4.1–74 % H_2 in air), enabling an engine to be operated far away from its stoichiometric mixture of 29.6. The very high flame speed of hydrogen in comparison to other fuels will lead to a marked tendency for "knocking." However, for all fuels, flame speeds are low near the limits of inflammability, and a fuel–air mixture that has an inherently high flame speed can be used at relatively lean mixture ratios without combustion requiring too much time.

Therefore, although hydrogen–air mixtures have a great tendency to knock and also to backfire, this can be eliminated if the fuel–air mixture is adjusted to contain at least 25% excess air.[6,35] Indeed, King *et al.*[22] reported that hydrogen has been used in a laboratory test engine at a compression ratio of 10:1 without combustion knock, provided no fluffy carbon deposits were present. In an engine specifically designed for, and confined to operation on, hydrogen, engine deposits cannot build up from the combustion of fuel, but we must consider the fate of the lubricating oil in the engine, which is probably a prime source of engine deposits. Perhaps in our nuclear age we shall have to use noncarbon-based lubricants for reasons other than their short supply. A modified Briggs and Stratton 4-hp engine was run at Oklahoma State University,[37] using a special fuel-injection system, on hydrogen as a fuel. Preliminary emission measurements indicated that oxides of nitrogen pollutants in the exhaust were at an extraordinarily low level.

Other than lubrication, no significant problems seem to be confronting the use of hydrogen in a conventional piston-type engine. Indeed, such an engine operating on pure hydrogen could be a high-performance, high-compression-ratio engine, with a clean water-vapor exhaust and an extended life resulting from the absence of deposit buildup.

Many modern engine applications are met by the use of gas turbines; aircraft engines are, of course, an outstanding example of their use. The design of a gas turbine to operate on hydrogen as a fuel presents no insuperable problems and generally follows the factors already outlined for burner design. Because the flame speed is so much higher than for hydrocarbons, the problems of flame "blowoff" are easier to overcome. In a conventional gas turbine, gas streams through the combustion chamber are caused to be turbulent and are slowed down to appropriate values to prevent blowoff. For hydrogen the required degree of slowing will be less. Again, lubrication problems become important because present lubricants are almost always hydrocarbon based.

While we are considering aircraft propulsion, we might consider what alternatives we will have for a propulsion unit in the "nuclear" age. It seems doubtful that storage batteries and motors will ever be light enough for an electric aircraft; we can only consider the aid of a synthetic liquid fuel—ammonia or methanol, or hydrogen—operating a gas turbine. Considerable objections have been raised to the use of hydrocarbon-fueled turbine aircraft in the upper atmosphere because of pollution considerations; it is not known how long pollutants will survive at high-altitude levels. Unless these objections are overruled, we can eliminate the use of methanol, and probably also ammonia, on similar grounds. Hydrogen-fueled aircraft remain the only choice.

The use of liquid hydrogen as an aircraft fuel, however, would greatly increase the demand for this material. The National Air Pollution Control Association estimates[37] that a single hypersonic transport flying 5000 miles a day at Mach 6 would consume 100 tons/day of liquid hydrogen, more than half the present world's production.

Since electrolysis will make oxygen as well as hydrogen available, we can consider the use of a hydrogen–oxygen gas turbine running from cryogenic storage. This offers the advantage that the energy involved in the "cold" of the cryogenic gas can be used. The gases can be readily vaporized under pressure, and fed directly to the burners. This eliminates some of the energy taken up within the gas-turbine engine by the compressor.

In summary, the design and operation of "appliances," ranging from domestic heaters and cookers to large static and mobile engines, to run on hydrogen appear to have no serious technical obstacle. However, few appliances can simply convert from present fuels to hydrogen; thus, a difficult transition period will have to exist while existing equipment is being modified or replaced.

VI. LOCAL ELECTRICAL GENERATION FROM HYDROGEN

Although we look forward to nuclear energy, and hence electrical energy, as our prime source of power, supplemented by some portable synthetic chemical fuels, local on-site electrical generation will still be needed in many applications. Such generation will be needed for precisely the same reason that today we supplement main electricity supplies with small engine-driven sets, total energy systems, and emergency power supplies. The convenience of electrical energy can then be combined with the economics and logistic advantages of "portable" fuels. Although it may appear nonsensical to use electricity to make a synthetic fuel, then convert this fuel back to electricity, the principle is sound as long as highly efficient conversion devices are used and as long as the economics of fuel transportation vs electrical transmission costs are favorable.

Just as hydrogen may be produced by the electrolytic decomposition of water, the process in reverse is capable of generating electrical energy by the electrochemical reaction of hydrogen with oxygen. This is the principle of the fuel cell, which, although it was demonstrated as early as 1839 by Sir William Grove, has remained a classic problem for development into a commercially attractive unit until the present time. However, over the past 10 years or so, much effort has been applied to the development of hydrogen–oxygen fuel-cell systems for space use or of fuel-cell systems that would generate economic electricity from cheap, conventional fossil fuels such as hydrocarbons and atmospheric air and that would be able to compete with the well-developed motor–generator equipment presently available. Even though the problems are great, it is a far easier task to develop a fuel cell to run on pure hydrogen and oxygen, or on hydrogen and air; in fact, fuel-cell systems which powered the *Gemini* and *Apollo* spacecraft were of the hydrogen–oxygen type.

Of course, very little development effort would be required for electrical power to be readily generated by a rotating machine driven by a hydrogen-fueled internal combustion engine. However, this is a heat engine concept and is therefore subject to the inefficiency characteristics of the Carnot cycle; it is unlikely that overall efficiencies greater than 20–30% could be achieved. The fuel cell, on the other hand, is an isothermal device, not subject to Carnot cycle limitations, that can achieve overall efficiencies of 50–80% using a clean hydrogen fuel. Since our electrolytic hydrogen will be "synthetic" and relatively expensive, this increased efficiency is of vital importance to the economics of the concept and also results in a minimum of thermal release to the environment.

The construction and operation of a fuel cell has been adequately described in detail elsewhere.[4,26] In summary, a fuel cell is defined as an electrochemical cell in which electrochemical reactants are continuously fed from outside the cell and the reaction products are continuously removed. A fuel cell consists of two electrodes and an electrolyte. The electrodes are inert components which themselves are not consumed or altered in any way, but which provide the reaction sites for an electron-transfer process to occur between the reactant and the ionic species in the electrolyte. The electrolyte serves to carry the ionic species from one electrode to the other and to prevent the direct mixing and reaction of the reactants themselves.

The hydrogen–oxygen cell operates best with either a strongly acidic or strongly alkaline electrolyte[53] because the electrode potentials are sensitive to pH, and these electrolytes combine high conductivity with good buffering action to keep the pH constant. At the hydrogen electrode, hydrogen forms hydrogen ions in an acid electrolyte, or reacts with hydroxyl ions in an alkaline medium.

$$H_2 \rightarrow 2H^+ + 2e^- \quad \text{(acid)}$$

or

$$H_2 + 2OH^- \rightarrow H_2O + 2e^- \quad \text{(alkaline)}$$

At the oxygen electrode, reactions occur which may be simply represented as follows, although the actual processes are far more complex[40] than this:

$$\tfrac{1}{2}O_2 + 2H^+ + 2e^- \longrightarrow H_2O \quad \text{(acid)}$$

$$\tfrac{1}{2}O_2 + H_2O + 2e^- \longrightarrow 2OH^- \quad \text{(alkaline)}$$

Figure 12 shows a simple schematic diagram of a hydrogen–oxygen fuel cell.

Notice that these reactions involve reactants or products in three different phases; for example,

$$\underset{\text{(gas)}}{H_2} \longrightarrow \underset{\text{(liquid)}}{2H^+} + \underset{\text{(in solid conductor)}}{2e^-}$$

The electrodes therefore have to provide a reaction site for a three-phase reaction, which requires the use of properly designed, highly porous, solid structures within which a stable gas–liquid interface may be established. This can be done by utilizing the surface-tension effects of the electrolyte, either with a biporous electrode whose fine pores are flooded and whose coarse pores are kept free of liquid by the applied gas pressure, or by using hydrophobic agents that keep some of the electrode pores free of electrolytes and exposed to gas. The electrode reactions require catalysis, and this is provided with an "electrocatalyst"[46] applied to the porous electrodes. Electrocatalysts must perform the conventional role of chemical catalysis, but, in addition, they must be electronic conductors, must be resistant to corrosion by electrolyte, and must not be poisoned by species present either in the electrolyte or in the gas supply.

Although the electrochemical reaction in a fuel cell is relatively efficient, energy losses result from various sources. First, only the free energy of reaction ΔG is converted to electrical energy, the remainder of the ΔH, heat of reaction, being released as heat. Second, as increasingly high currents are

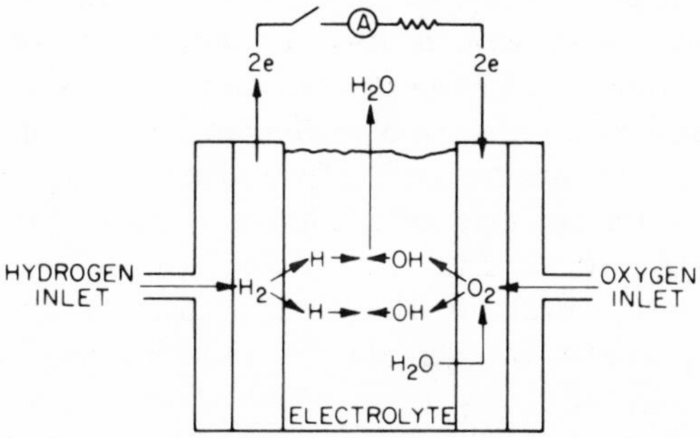

Figure 12. A simple fuel cell.

drawn from the cell, the electrodes "polarize"; that is, they shift their potentials closer to one another. This "lost voltage" manifests itself as a heat release at the electrodes. The cell components have conventional ohmic resistance, which results in further heat release. A small amount of direct reaction of hydrogen and oxygen can occur because of the reactants dissolving in the electrolyte, resulting in further heat generation. Depending on the current drawn from the cell, the efficiency varies from about 90% to as little as 50% at maximum power levels. These characteristics contrast with conventional "engines" in producing the highest efficiency at lowest power levels.

The need to remove heat from the electrochemical cell results in the selection of an operating temperature somewhat above ambient. As the designed temperature is raised, heat rejection becomes easier, reaction rates increase, and the need for expensive catalysts diminishes, but corrosion and stability problems increase and the system loses its capability of instant start-up. Successful fuel-cell systems are a compromise between these conflicting trends.

To keep the cell operating continuously, reactants must be fed at the appropriate rate, and products must be removed appropriately. For pure hydrogen and oxygen feeds, the cell is self-regulating, requiring only a constant-pressure gas feed. The electrodes will only consume the amount of gas they need. For an air-fed cell, air must be pumped at a rate proportional to the current drawn, and the excess air must be vented to remove the nitrogen from the reaction zone; thus, control becomes more difficult. Removal of product water is by evaporation from the electrolyte, which is partly self-regulating as the water vapor pressure increases as the water content rises. To use this mode of control requires a tolerance in the cell for electrolyte volume changes, which strains the stabilization of the gas–liquid interface at the electrodes. Water can be evaporated into the vented air stream, but in a hydrogen–oxygen cell an alternative removal method, such as recirculation of the hydrogen through a condenser, is required.

The fuel-cell system thus consists of a multiple stack of electrochemical cells, connected in series to generate the required voltage. Each cell is fed with hydrogen and oxygen–air supplies at identical rates, and equipped with appropriate means for water removal. The cell stack is equipped with a heat rejection system, a temperature control system, a fuel supply system, oxidant supply system, water removal system, and electrical output. A control device coordinates all these operations. The power output is a variable low-voltage dc and is conditioned to the required output by, for example, inversion to ac with voltage stabilization.

Even a simple hydrogen–oxygen fuel-cell system is a complex, sophisticated piece of chemical engineering, requiring far more than just the electro-

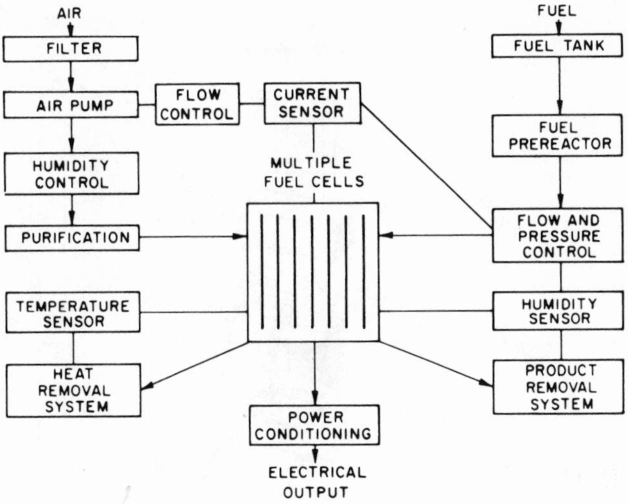

Figure 13. Typical fuel-cell system.

chemical component. A schematic diagram of a typical fuel-cell system is shown in Fig. 13. As an example of an acid and an alkaline hydrogen–oxygen system, we can briefly survey the construction of the *Gemini* (General Electric) fuel cell (Fig. 14), and the *Apollo* (Pratt & Whitney Aircraft) fuel cell (Fig. 15).

The GE cell uses an ion-exchange resin as the electrolyte in the form of a thin sheet of a sulfonated polystyrene, behaving as a strong acid. The electrodes are of platinum black, embedded in the surfaces of the sheet, in contact with cooled current-collector plates made of titanium or tantalum. The plates carry both coolant tubes and wicking devices to remove the water formed on the membrane and transport it to an external reservoir. The *Gemini* system contained 96 cells, each 8 × 7 in. arranged in three modules of 32 cells. The 1-kW unit was 25in. long, $12\frac{1}{2}$ in. in diameter, and weighed 68 lb. In addition, auxiliary pumping and control systems were provided for the supply of hydrogen, oxygen, and coolant.

The Pratt & Whitney cell, used in *Apollo*, was developed from the hydrogen–oxygen alkaline system pioneered in Cambridge by F. T. Bacon. The electrodes are of the biporous type and are made of porous nickel in two layers of different porosity. One electrode is fed with hydrogen from cryogenic storage and remains in the metallic form. The other, fed with pure oxygen, is preoxidized prior to assembly in the cell, converting the internal surface of the sintered nickel structure to nickel oxide. The electrolyte is 85% KOH, and the cell is operated at 200–250°C. Water is removed by

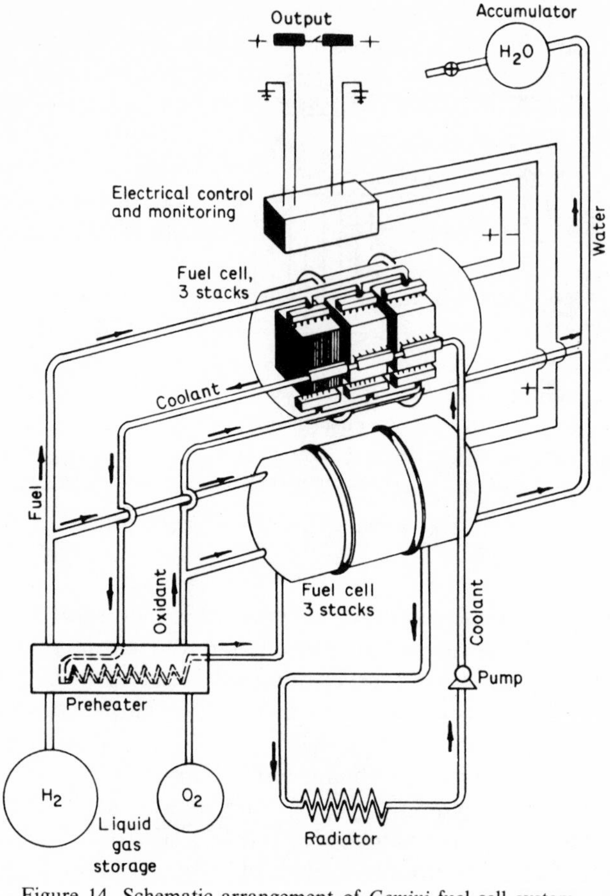

Figure 14. Schematic arrangement of *Gemini* fuel-cell system. (From A. B. Hart and G. J. Womack, *Fuel Cells: Theory and Application*, Chapman and Hall, London, 1967.)

evaporation into the hydrogen stream, which is recirculated through a liquid-cooled condenser. The *Apollo* system consists of three independent units, each composed of 31 cells in series, about 10 in. diameter. Each 2-kW unit weighs about 200 lb.

Both of these hydrogen–oxygen fuel-cell systems were specifically designed for space operation, where fuel efficiency and reliability are most important. Similar units, designed to be operated on hydrogen and air for terrestrial use, are likely to have quite different characteristics.

How, then, would hydrogen fuel cells be used in our future nuclear age, and what effects would they have on our environment? Because a fuel cell operates with relatively few moving parts, it will be quiet—almost silent—in operation, and can therefore be installed close to the user without inconvenience. Because it has a higher efficiency than conventional engines, less heat would be released—therefore, also less thermal pollution. The exhaust products are water or steam, which are harmless as long as they are not released to cause local air saturation of water.

In industrial or commercial areas where continuous electrical supply is mandatory, standby hydrogen fuel cells will replace standby diesel generators. The cell can be fed either from the hydrogen gas main or from pressurized tanks for a guaranteed service.

Premises where private electrical generation from engine–generator sets are now proved economical will install hydrogen–oxygen or hydrogen–air fuel-cell generators, running off the piped gas supply. Overall system economics are bound to change, but it is likely that the economics of total energy systems, in which a high proportion of the energy content of a single fuel is usefully employed in heating, cooling, or power generation, will become even more favorable for hydrogen than for conventional fuels use. Such total energy systems will incorporate quiet, almost silent fuel cells, and can conceptually be designed in smaller units than is possible with conventional engines.

If a cheap, reliable natural-gas-fueled fuel-cell system were available today, it would find use in two extreme conditions met in the electrical supply industry. In the more remote locations, provision of electrical power

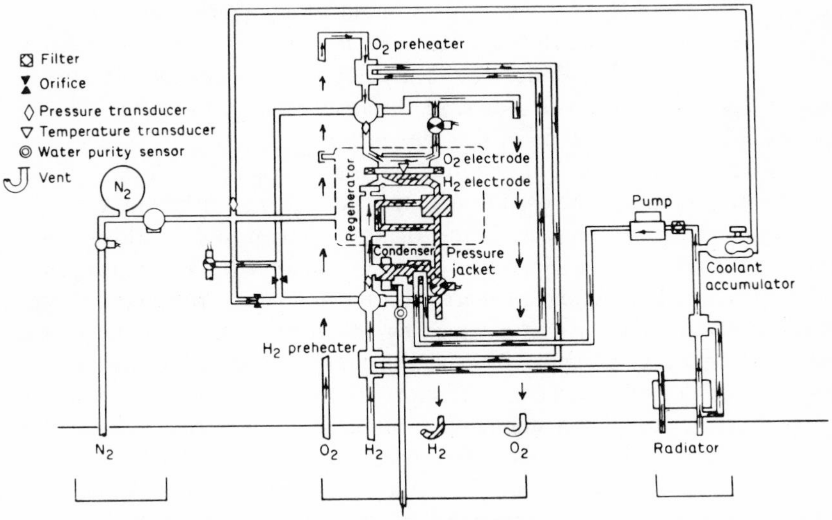

Figure 15. Schematic flow diagram of *Apollo* fuel-cell power plant, model PC3A-2. (From A. B. Hart and G. J. Womack, *Fuel Cells: Theory and Application*, Chapman and Hall, London, 1967.)

lines or overhead cables is expensive. In many such cases it is economically justifiable to provide an underground gas supply, but not electricity. At the other extreme, in highly congested city areas, electrical supplies must be run underground for amenity reasons, and the provision of high power levels requires expensive, water-cooled cables. High-capacity gas lines are far smaller, and easier to install. If we substitute hydrogen for methane in the gas lines, cheap, reliable fuel-cell systems become an order of magnitude closer to reality, and would find economic justification for these applications in a very short time. The London Electricity Board in England, discussing its problems[33] in reinforcing the electricity supply to the Board's medium-voltage distribution network, has considered the use of 500-kVA fuel cells interspersed with its transformer stations and states that the development of fuel cells for this purpose is a problem of utmost importance to it. Although all fuel cells would have a low pollution effect if they were located throughout the city, our hydrogen-fed cells would be even more socially acceptable than conventional transformer stations. Similar electrical distribution problems exist in most of the world's large cities.

One of the severe problems of the electrical supply industry is the need to match generating capacity with fluctuating demands during the day. Not only must seasonal trends and daily use patterns be anticipated, but the balance between supply and demand is so fine that even the influence on electricity demand of popular TV programs must be anticipated. (In Britain, the electrical demand always surges up immediately after a popular sports program or a "Miss World" telecast, as the majority of the population revert to their normal cooking, etc., habits all at once.) The advent of an all-nuclear generating age will not ease this problem, as the large nuclear stations cannot be started up or shut down rapidly.

Hydrogen-fueled generators (fuel-cell or internal-combustion-engine generators) will have to be used to take the place of the "spinning reserves" and diesel or gas-turbine auxiliary plant at the power stations. Such generators can satisfy the short-term variations in load, but the daily cycling variations really need a form of energy storage that can be recharged electrically. Electrochemical systems are being investigated for this use[19]; the regenerable hydrogen–oxygen fuel cell appears a promising concept. Such a unit could be located anywhere in the supply network, from the power station to consumer premises, but the closer it is to the consumer, the greater the saving in load-carrying capacity of the transmission lines.

During low-demand periods, the system will be run as an electrolyzer, producing high-pressure hydrogen in tank storage and oxygen that is either similarly stored or vented to the atmosphere. During peak-demand periods, the system would operate as a fuel cell to boost local power supplies. The prime disadvantage of a secondary battery for such an application is its finite capacity. Unpredicted and lengthy power demands, for example, during excessively cold weather or in the event of a supply breakdown, would result in complete discharge of the battery and consequent total failure of supply. The reversible hydrogen–air system, however, may be connected to the hydrogen pipeline supply for just such an emergency; therefore, it represents a storage device with, in emergency conditions, an infinite discharge capacity. The use of an on-line storage battery located near the consumer, capable of supplying the peak-demand loads, not only enables more even continuous operation of the central power station, which results in an even and predictable thermal disposal load, but also requires rather lighter loading in the transmission cables, which are therefore easier to render inconspicuous, for example, by burying them underground. Even though no high-rate storage battery is 100 % efficient, widespread use of any kind of storage device would result in a thermal disposal problem at consumer locations.

Perhaps the ultimate goal of the fuel-cell researcher has been to develop an economic electric vehicle. The use of hydrogen as a vehicle fuel presents problems.[29] For this purpose we are forced to consider methanol as the ultimate synthetic fuel. Most efforts to the present time have concentrated on the need to use a liquid hydrocarbon fuel for vehicle fuel-cell propulsion; however, this has met with no success because hydrocarbons are able to undergo direct electrochemical oxidation only inefficiently and under severe operating conditions, while the generation of hydrogen from hydrocarbons in a vehicle is not practicable. Nonetheless, several more successful attempts have been made to produce methanol fuel cells, both of the direct-oxidation type and the hydrogen-generation type.[52] A methanol fuel-cell-powered vehicle is undoubtedly a simpler proposition than a hydrocarbon-fueled one, but it is still fraught with difficulties. A methanol cell will still produce CO_2 as an effluent, and may also produce aldehydes and other products of incomplete oxidation under transient conditions. The outstanding advantage over secondary-battery-propelled vehicles—the only other way we can expect to travel locally in the "nuclear age"—is that the fuel-cell system may be recharged in as short a time as it takes to fill the methanol tank, in contrast to even the $\frac{1}{2}$ h or so (quoted by optimists) for fast charging a battery, which would prove unacceptable to most users. This alone will probably force the development of a methanol–air fuel cell for vehicle propulsion in the next few decades.

In conclusion then, we can see that in our hydrogen age, a demand for local electricity generation may still exist as it does today. Today, this is met by fossil-fueled engines, and work is in progress to develop fossil-fuel-fed fuel-cell systems because of their advantages in efficiency, cleanliness, quietness, and low maintenance requirements. With hydrogen readily available, hydrogen–air fuel cells may be considered with a resulting order of magnitude reduction in the size of the technical problems. Conventional generating methods, using hydrogen, may be unacceptable because of their low efficiency with an expensive fuel and because of their social inconvenience of noise, although chemical pollution would be almost absent. Hydrogen fuel cells will be favorable for use in local generation, total energy, standby supplies, and electricity storage systems. Methanol fuel cells may be produced for vehicle propulsion. The problems of both are still great, but not insuperable. The incentive to develop them will only come when we know that hydrogen and methanol will become our basic fuels. If we wait too long, so that the fuels are available before the proper means of using them are, they will be economically useless. Proper advanced planning is the only solution.

VII. HAZARDS AND SAFETY ASPECTS OF HYDROGEN

Hydrogen is widely used in industry; indeed, a code of safety has developed for its use.[36,38] Hydrogen is undoubtedly an extremely dangerous and hazardous material unless it is handled correctly, and accustomed users of hydrogen have developed a healthy respect for it. However, it is curious to observe the differences that seemingly exist between the rigidly laid down codes of practice for industrial hydrogen and the confident and informal way in which a 40–50% hydrogen mixture is treated in those areas that still use manufactured gas in a domestic supply. In this section we will outline the hydrogen hazards and attempt to explain the differences that prevail between handling industrial hydrogen and domestic manufactured gas.

Hydrogen is colorless and odorless, and is not detected by the senses. Although nontoxic, it can cause suffocation by excluding oxygen. It is lighter than air, and tends to rise if released, but just after evaporation from its liquid state (i.e., below −418°F), it is heavier than room-temperature air and tends to fall. When mixed with air or oxygen, it has inflammable and explosive limits over a very wide range of mixture ratios.

	Flammability limits	Explosive limits
Hydrogen in air	4.0–75.0%	18.3–59.0%
Hydrogen in oxygen	4.5–94.0%	15.0–90.0%

Moreover, ignition of flammable mixtures occurs with a very low energy input (0.000019 J in air)—that is, from an invisible spark. This required energy input is one-tenth that required to ignite a methane–air or gasoline–air mixture. However, although very rapid combustion of a flammable hydrogen–air mixture can occur, it requires a powerful ignition source to detonate an unconfined explosive mixture. For this reason, a hydrogen plant is often installed out of doors so that escapes result in unconfined gas mixtures. Although hydrogen is prone to leakage because of its low viscosity and molecular weight, its rapid diffusion rate in air helps it to disperse rapidly. A spill of 500 gal of liquid hydrogen in open air will have diffused below the explosive limit after about 1 min.

Safety of hydrogen handling is therefore built around three basic axioms: adequate ventilation, leak prevention, and elimination of ignition sources. It is necessary to observe *all three* for safe handling.

Among the requirements of the 1970 National Fire Code for Gaseous Hydrogen Systems,[38] it is interesting to note that for a system in which gaseous hydrogen is delivered to consumer piping, consumer systems must be located above ground, not beneath electrical power lines, or near piping or storage of flammable liquids, not within 5 ft of adjoining property, or not within 15 ft of public sidewalks. Clearly, these restrictions are unacceptable to the concept of piped consumer hydrogen to every house. Therefore, it is necessary to examine them to see if they are realistic. It appears that in countries where manufactured gas is still in use (up to 50% hydrogen) the codes of practice are less strict than are laid down for pure hydrogen in the U.S.A. Why is this?

One major difference between pure hydrogen and manufactured gas is that the latter contains up to 30% methane. (A typical analysis of "coal gas" is hydrogen 50.6%, methane 29.7%, ethane 3.2%, nitrogen 4.9%, carbon monoxide 7.3%, carbon dioxide 0.1%, other hydrocarbons 3.0%.) The lower flammability limits of methane and hydrogen in air are similar (5.3 and 4.1%), but the difference between their upper limits (14 and 74%) is vast. Mixtures of hydrogen and methane exhibit the dominating influence of methane; a 50–50 mix will have an upper flammability limit of only 22.6% in air. A typical flammability limit for "coal gas" is from 5.3 to 30.9% in air.

Thus, we observe that the flammability range of manufactured gas in air is only about one-half that for hydrogen, but it is still wide, and the lower limit is about the same. Compare this with the lower flammability limit for gasoline vapor in air, which is as low as 1.5% (7.6% upper). It would appear, then, that as far as leakage is concerned, gasoline vapor is far more hazardous than methane, coal gas, or hydrogen, which are all about equal.

Because of its low molecular weight and low viscosity, hydrogen will diffuse through materials, and leak through small cracks and holes, at a far higher rate than correspondingly larger molecules such as methane and other hydrocarbons. However, it is hard to see why pure hydrogen should be significantly more hazardous to handle than a mixed hydrogen gas, such as "coal gas," and, therefore, it is hard to see why the existing codes for its handling are so strict.

Perhaps we should recall that in the early days of automobiles, the hazards of carrying several gallons of gasoline were considered prohibitive. In England, a man on foot carrying a red flag had to walk in front of each vehicle, but familiarity with the problem has enabled such regulations to be relaxed. Similarly, it is anticipated that reasonable and proper codes of practice will evolve for the routine handling of hydrogen in acceptable conditions.

VIII. HYDROGEN AS A CHEMICAL RAW MATERIAL

We have considered a very much extended use of hydrogen as a new fuel for the future. Hydrogen, however, is already a heavy chemical product, being produced in enormous quantities, and its application as a chemical raw material is well known. Let us consider how this hydrogen is used today, the impact of changes in the source of hydrogen from fossil fuels to nuclear plants, and the indications of the needs for still larger supplies of hydrogen in the future.

The world production of hydrogen has increased sharply over the past 30 years. In 1938, the estimated production was 2.5 billion SCF, and by 1968 it had climbed to an estimated 2154 billion SCF. A contingency forecast of possible demand in the year 2000 indicates that hydrogen consumption will be 15,500–52,530 billion ft^3.[31] Most of this hydrogen is produced very close to its point of consumption, so that it does not appear to the "man in the street" as a very evident raw material, as do coal, oil, etc., but the quantities being produced and used up today are quite enormous.

Most of this hydrogen is presently being produced from fossil fuels. In the U.S. it comes predominantly from the steam reforming or partial oxidation of natural gas, but in Europe coal and oil are the raw materials. It finds its way mainly into the fertilizer industry by way of ammonia—some 66% of the 1938 production and 55% of the 1968 production being accounted for in ammonia synthesis. About 20% of the production is used in oil processing —in hydrodesulfurizing and hydrocracking, etc.—leading to fuel and petrochemical products. About 5% is used in synthetic methanol production, and an increasing amount, at present close to 20%, is used in the hydrogenation of vegetable and animal oils to produce edible fats. New and increasing uses for hydrogen today include the gasification of coal and shale, for example, by the IGT HYGAS process, and the direct reduction of iron ore to produce high-grade iron and steel.

Consider the impact of (1) a shortage of fossil fuels and (2) a more ready supply of electrolytic hydrogen on this sector of the chemical industry. Pure electrolytic hydrogen, if the price is right, will be quite suitable for all of the processes listed above. One adjustment will have to take place in the ammonia plant: at present hydrogen produced by partial oxidation of hydrocarbons, with subsequent removal of carbon monoxide and dioxide, is produced as a mixture with atmospheric nitrogen, which is then ready for conversion to ammonia. Some of the more recent plants use air liquefaction to produce oxygen for partial oxidation, and nitrogen for mixing with hydrogen for the ammonia plant. Our electrolytic hydrogen supply will be clean and pure, and will require a large supply of pure nitrogen for ammonia synthesis. Since the electrolysis of water will also produce plenty of clean oxygen as a byproduct, there seems to be less merit in running an air liquefaction plant. This change would affect the overall economics and requires further study because the quantities of ammonia produced are very large indeed.

A recent study[3] of the economics of a nuclear-powered electrolytic hydrogen–air liquefaction ammonia plant indicated that costs were comparable, based on 1.0–1.6 mills/kWh nuclear electricity and 25–44¢/million Btu gas (see Fig. 16). In this study, the importance of an integrated facility was stressed to take advantage of the byproduct oxygen produced (both from the air liquefaction plant and the electrolyte). For example, nitric acid

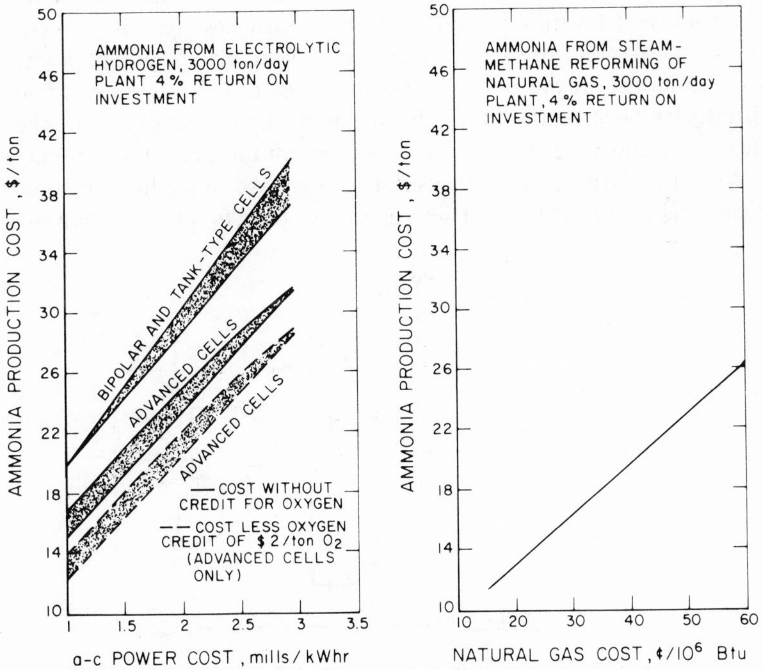

Figure 16. Comparison of ammonia production costs.[3]

produced by combustion of some of the ammonia with this oxygen is then used in the fertilizer plant complex to produce nitrate fertilizers, by reaction with either ammonia or phosphate rock.

In our nuclear age, the use of hydrogen for hydrocarbon and coal processing will presumably have vanished, but plenty will still be required for foodstuff processing. No problems can be foreseen in the use of high-pressure electrolytic hydrogen for hydrogenation of animal or vegetable oils.

Real problems, however, will face the iron and steel industry. Without fossil fuels, the blast furnace becomes inoperable. The new H-iron process developed by Hydrocarbon Research Inc. is but one of several iron-ore reduction processes using hydrogen instead of coal or coke as the reducing agent. It appears[43] to be most suitable for operation with electrolytic hydrogen because it is a recycle process capable of achieving complete utilization of the hydrogen. Figure 17 is a schematic diagram of the H-iron process, using methane as a source of hydrogen. Almost half the plant is eliminated if electrolytic hydrogen is substituted. By use of this type of technology, we can visualize a completely "clean" steel works with only water vapor as the exhaust product. Of course, the problems are severe, but this type of process is already well advanced, and no insuperable problems are foreseen. In the transition period, when blast furnaces are still in operation, the ready availability of oxygen from the nuclear–electrolytic plants would encourage its use in the basic oxygen furnaces now being adopted by the steel industry because the plume from one of these is simpler and cheaper to clean than those from the 10–12 open-hearth furnaces that it replaces.[34]

Without a large source of fossil fuels, it is hard to see how the pharmaceutical, plastics, lubricant, and detergent industries, for example, can survive.

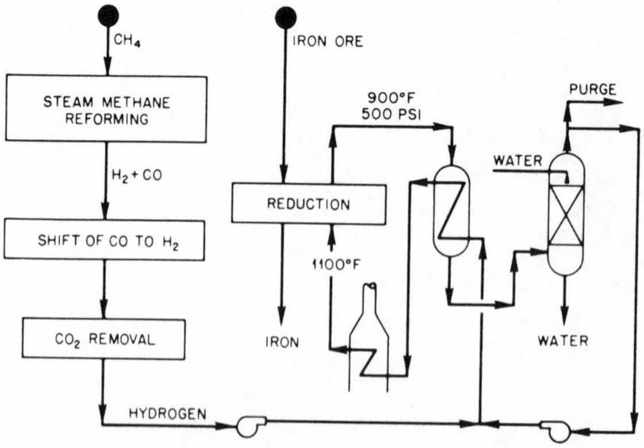

Figure 17. Schematic diagram of H-iron process for supplying 95% reduced iron to electric steelmaking furnaces.[43]

Yet it is impossible to conceive of a highly developed society living without all the products of the heavy organic chemical industry.

A carbon-based raw material is essential for all these classes of products. Synthetic methanol, or even synthetic methane, might become a very important factor in our new hydrogen age.

Methanol is produced by the reaction of carbon monoxide and hydrogen, in turn produced from a hydrocarbon feed. Without hydrocarbons, we must ask ourselves from where are we to obtain the required carbon monoxide? Several speculative processes may be considered, starting from the plentiful supply of carbonate in the form of limestone in the earth's crust. Before considering these, however, we must take into account the ultimate destination of all this carbon. Whether used as fuel or through petrochemical products finally to the urban incinerator, we finally finish up pumping more CO_2 into an atmosphere already overloaded with this product because of our having burned up the whole world's supply of fossil fuel. Clearly, the limestone → CO_2 → organic chemical route is a short-sighted policy leading to a form of suffocation.

The alternative is obvious—to extract the existing CO_2 from the atmosphere and convert it to an organic material. Natural vegetation does this in a delightfully decorative way, but the process is too slow for the pace at which mankind is moving; we are reconverting the products of millions of years of tree-growing back to CO_2 in a few lifespans. What is needed is a "synthetic, high-speed tree."

We can consider, in principle, "scrubbing" the atmosphere with huge quantities of alkali, treating the resulting carbonates to produce carbon dioxide gas, and reacting this with some of our hydrogen over a catalyst to form methane (the well-known "methanation" process). This methane would surely be too expensive for use as a fuel, but is a precious precursor of methanol and other organic chemicals. In practical terms it may be feasible to do just this because we envision the production of large amounts of alkali from the electrolysis of sea water in the electrolytic production of hydrogen. The economics of this process, which seems incredible at first, may be favorable. This represents another important use of our hydrogen "fuel" supply as a chemical raw material. It seems a pity, though, to use electrolytic hydrogen as the source of energy of the methanation reaction, which is an exothermic process and leads to still more thermal pollution.

Perhaps there is an electrochemical way of reducing CO_2 or carbonate to methane or carbon monoxide? For years, electrochemists have been striving to oxidize hydrocarbons to CO_2 in directly fueled fuel cells. Some success has been achieved, but at too low an efficiency and too high a cost of catalysts to be able to compete with conventional engines. However, are these electrochemical processes truly reversible? Giner[13] showed that carbon

Table 3. Comparative Data on Hydrogen and Other Fuels

	H_2	CH_4	CH_3OH	NH_3
Molecular weight, g/g·mole	2.016	16.04	32.04	17.03
Freezing point, deg C	−259.20	−182.5	−97.8	−77.7
Heat of fusion, cal/g	13.9	14	0.74	79.38
Boiling point, deg C	−252.77	−161.5	64.7	−33.4
Heat of vaporization, cal/g	106.5	121.9	262.79	327.7
Critical temperature, deg C	−239.9	−82.1	240	132.3
Critical pressure, atm	12.80	45.8	78.5	111.3
Critical density, g/liter	31	162.5	272	235
Liquid density, g/liter (temp. deg C)	71 (−252.77°)	425 (−252.77)	792(20)	674(−33.4)
Vapor density, g/liter (temp. deg C)	1.3 (−252.77°)	1.8 (−252.77)	–	0.89 (−33.4)
Gas density, g/liter	0.082 (25)	0.7174 (25)	–	0.71 (21)
Molar volume at STP, liter/g·mole	22.420	22.360	0.0396	22.094
Liquid–Gas expansion ratio	865	650	–	–
Heat of combustion (net) at 25°C, kcal/g·mole	57.8	191.8	152.6	75.6(15°C)
Specific heat at STP, cal/mole·deg C	6.89	8.16	18.3	8.7
Limits of flammability in air, %	4–75	5–15	6–50	15–28
Limits of flammability in oxygen, %	4–95	5–61	–	15–79
Stoichiometric mixture, %	29.50	9·47	12.24	21.81
Maximum flame temperature (air at 18°C), deg C	2200	1980	–	–
Autoignition temperature in air, deg C	571	632	470	651
Thermal conductivity, cal/cm·sec·deg C × 10^4	3.8 (0°)	0.702 (0°)	4.995 (20°)	0.478 (0°)
Viscosity, μP_2	84.11 (0°)	102.4 (0°)	8080 (0°)	92.6 (0°)
Diffusivity in air, cm^2/sec	0.611 (0°)	–	0.162 (25°)	0.198 (0°)
Solubility in water, cm^3/100 g solvent	2.1 (0°)	3.3 (20°)	∞	117,000 (0°)

dioxide can indeed be electrochemically reduced by chemisorbed hydrogen to form a "reduced CO_2" complex. Here is a glimmer of hope for a process that could produce both fuel and organic chemicals from a nuclear–electrochemical plant.

ACKNOWLEDGMENTS

This chapter was prepared with the help of services provided by the Institute of Gas Technology. The authors wish to acknowledge the help of Dr. H. R. Linden, Dr. R. B. Rosenberg, R. J. Dufour, G. G. Yie, H. C. Maru, Mrs. S. M. Grom, Miss A. C. Roess, Miss M. K. Field, and others at IGT.

REFERENCES

[1]Allis-Chalmers Manufacturing Company, *Design Study of Hydrogen Production by Electrolysis*, Publication No. *ACSDS0106643*, Milwaukee, October 1966.

[2]Bechtel Corporation, *Engineering and Economic Feasibility Study for a Combination Nuclear Power and Desalting Plant, Phases I and II, TID 22330*, U.S. Atomic Energy Commission, Division of Technical Information, Washington, D.C., 1965.

[3]R. E. Blanco *et al., Chem. Eng. Progr.* **63** (1967) 46; **63** (1967) 49.

[4]J. O'M. Bockris and S. Srinivasan, *Fuel Cells; Their Electrochemistry*, McGraw-Hill, New York, 1969.

[5]J. E. Browning, *Chem. Eng.* **75** (1968) 88.

[6]A. F. Burstall, *S.A.E. Proc.* **22** (1927) 365.

[7]Compressed Gas Assoc., Inc., in *Handbook of Compressed Gases*, Reinhold, New York, 1966, pp. 94–100.

[8]R. L. Costa and P. G. Grimes, *Chem. Eng. Progr. Sym. Ser.* **63** (1967) No. 71, 45.

[9]B. E. Eakin, in *American Gas Association Operating Section Proceedings—1960*, CEP-60-5.

[10]M. A. Elliott, Paper No. ASME-NAFTC-1 presented at the North American Fuel Technology Conference, Ottawa, May 31–June 3, 1970.

[11]Federal Power Commission, *1970 National Power Survey*, Part 3, Washington, D.C., 1970.

[12]J. E. Funk and R. M. Reinstrom, *I & EC Process Design Develop.* **5** (1966) 336.

[13]J. Giner, *Electrochim. Acta* **8** (1963) 857.

[14]J. C. Griffiths, C. W. Thompson, and E. J. Weber, *A.G.A. Res. Bull.* **96** (1963) 18.

[15]A. T. Grisenthwaite, *Trans. Inst. Chem. Eng. (London)* **34** (1956) 235.

[16]J. Grumer, M. E. Harris, and V. R. Rowe, *U.S. Bureau of Mines Report Invest. No. 5225*, Washington, D.C., 1956.

[17]L. W. der Haar and J. E. Vogel, *World Petrol Congr. Proc. 6th*, Sect. 4, 383–94, Frankfurt/Main, 1963.

[18]C. A. Hampel, Ed., in *The Encyclopedia of Electrochemistry*, Reinhold, New York, 1964, pp. 1156–1160.

[19]A. B. Hart, *Design Eng.*, February (1970) 71.

[20]W. Juda and D. M. Moulton, *Chem. Eng. Progr. Sym. Ser.* **63** (1967) No. 71, 59.

[21]V. J. Kavlich, B. S. Lee, and F. C. Schora, paper presented at the 3rd Joint Meeting of the Instituto de Ingenieros Quimicons de Puerto Rico and the American Institute of Chemical Engineers, San Juan, Puerto Rico, May 17–20, 1970.

[22]R. O. King, W. A. Wallace, and B. Mahapatra, *Can. J. Res.* **26F** (1948) 264.

[23]R. E. Kirk and D. F. Othmer, in *Encyclopedia of Chemical Technology*, 2nd Ed., Interscience Publishers, New York, 1966, Vol. 11, pp. 338–379.

[24]H. H. Landsberg and S. H. Schurr, *Energy in the U.S., Uses and Policy Issues*, Random House, New York, 1960.

[25]B. S. Lee, paper presented at the American Power Conference, Chicago, April 21–23, 1970.

[26]H. A. Liebhafsky and E. J. Cairns, *Fuel Cells and Fuel Batteries*, John Wiley and Sons, New York, 1968.

[27]H. R. Linden, paper presented at the Institute on Exploration and Economics of the Petroleum Industry, International Oil and Gas Educational Center, The Southwestern Legal Foundation, Dallas, March 4–6, 1970.
[28]C. L. Mantell, *Electrochemical Engineering*, McGraw-Hill, New York, 1960.
[29]C. Marks, E. A. Rishavy, and F. A. Wyczalek, S.A.E. Paper 670176 (1967).
[30]C. P. Marun and W. L. Slater, *World Petrol. Congr. Proc. 6th*, Sect. 4, 373–82, Frankfurt/Main, 1963.
[31]P. Meadows and J. De Carlo, in *Mineral Facts and Problems*, U.S. Bureau of Mines, Washington, D.C., 1970.
[32]G. A. Mills and J. S. Tosh, Paper No. ASME-NAFTC-4 presented at the North American Fuel Technology Conference, Ottawa, May 31–June 3, 1970.
[33]A. G. Milne and J. H. Mattby, *Proc. Inst. Elec. Eng.* **114** (1967) 745.
[34]J. E. Mrochek, in W. W. Grigorieff, Ed., *Abundant Nuclear Energy*, U.S. Atomic Energy Commission, Washington, D.C., 1969, pp. 107–122.
[35](U.S.) National Advisors Committee for Aeronautics, *NACA Rep.* No. *535*, Washington, D.C., 1935.
[36](U.S.) National Aeronautics and Space Administration, *NASA Technical Memorandum TMX-52454*, Washington, D.C., 1968.
[37](U.S.) National Air Pollution Control Administration, *NAPCA Contract* No. *EHS70-103*, Washington, D.C., 1970.
[38]National Fire Protection Association, Standard NFPA No. 50A, in *National Fire Codes*, Boston, 1969–70, Vol. 2.
[39]R. M. Reed, *Trans. Amer. Inst. Chem. Eng.* **41** (1945) 453.
[40]A. C. Riddiford, *Electrochim. Acta* **4** (1961) 170.
[41]W. F. Schaffer, Jr., *USAEC Report ORNL-TM-1629*, Oak Ridge National Laboratory, Oak Ridge, Tenn., 1968.
[42]S. H. Schurr and B. C. Netschert, *Energy in the American Economy 1850–1975*, John Hopkins Press, Baltimore, 1960.
[43]A. M. Squires, in W. W. Grigorieff, Ed., *Abundant Nuclear Energy*, U.S. Atomic Energy Commission, Washington, D.C., 1969, pp. 181–196.
[44]C. Starr, in *Proceedings of 4th Intersociety Energy Conversion Engineering Conference*, Washington, D.C., 1969, p. 1072.
[45]G. R. Strimbeck *et al.*, in *American Gas Association Operating Section Proceedings—1952*, pp. 778–817.
[46]S. Srinivasan, H. Wroblowa, and J. O'M. Bockris, *Advances in Catalysis*, Academic Press, New York, 1967, Vol. 17, pp. 351–418.
[47]M. W. Thring, *The Science of Flames and Furnaces*, John Wiley and Sons, New York, 1962, pp. 156–167.
[48]D. J. Rose, *Nucl. Fusion* **9** (1969) 183.
[49]U.S. Bureau of Census, *Statistical Abstracts of the U.S.*: *1970*, 91st ed., U.S. Government Printing Office, Washington, D.C., 1970.
[50]C. G. Von Fredersdorff, in *American Gas Association Operating Section Proceedings—1959*, CEP-59-18.
[51]C. G. Von Fredersdorff and E. J. Pyrcioch, in *American Gas Association Operating Section Proceedings—1952*, pp. 685–701.
[52]K. R. Williams, *Advan. Sci.* **22** (1966) 617.
[53]K. R. Williams and D. P. Gregory, *J. Electrochem. Soc.* **110** (1963) 209.
[54]M. Beller, L. G. Epel, and M. Steinberg, *Chem. Eng. Progr. Sym. Ser.* **63** (1967) No. 71, 31.
[55]G. de Beni and C. Marchetti, *Euro Spectra* **9** (1970) 46.
[56]P. Harteck and S. Dondes, *Nucleonics* **14** (1956) 22.
[57]G. Juppe, *Euro Spectra* **8** (1969) 39.
[58]K. Z. Morgan, *Elec. World* **174** (1970) 132.
[59]U.S. Congress, Joint Economic Committee, *The Economy Energy, and the Environment*, 72, by the Environmental Policy Division Legislative Reference Service of the Library of Congress, U.S. Government Printing Office, Washington, D.C., 1970.
[60]C. G. Von Fredersdorff, private communication, 1959.

Chapter 9

HYDROMETALLURGICAL TREATMENT OF SULFIDE ORES FOR ELIMINATION OF SO_2 EMISSIONS BY SMELTERS

T. A. Henrie and R. E. Lindstrom

Bureau of Mines
U.S. Department of Interior

I. INTRODUCTION

The growing affluence of our society, coupled with population increases and military requirements, has placed unprecedented demands on an already burgeoning minerals industry and intensified problems created by the discharge of waste products into the environment. Among the serious offenders in this regard are base-metal pyrometallurgical operations which emit sulfur oxide gases into the atmosphere during the processing of sulfide ores. Increasing concern over the problems facing man in his environment requires that efforts be made to eliminate, as much as possible, such sources of contamination.

II. HYDROMETALLURGICAL PROCEDURES FOR THE RECOVERY OF METAL

Considerable research has been conducted on hydrometallurgical procedures for recovery of metal from sulfide ores. This has been well summarized, by Forward and Warren in 1960,[1] and later in publications on copper hydrometallurgy[2] and recovery of sulfur from sulfide ores.[3] In recent research, an electrooxidation procedure developed by the U.S. Bureau of

Mines[7,8] has been shown to be a potentially effective hydrometallurgical method for treating sulfide ores. The technique involves crushing and grinding of the ore, forming a pulp with brine solution, and electrolyzing the pulp. Oxidation of sulfide minerals occurs as a result of the controlled production of hypochlorite during electrolysis of the brine solution containing the finely ground ore. The anode reactions involved in forming the oxidizing species are shown in the following equations:

$$2Cl^- - 2e \rightarrow Cl_2 \quad (1)$$

$$Cl_2 + H_2O \rightarrow OCl^- + 2H^+ + Cl^- \quad (2)$$

The hypochlorite ion formed as a result of the anode reactions oxidizes the sulfide mineral according to the following reaction:

$$MS + 4OCl^- \rightarrow MSO_4 + 4Cl^- \quad (3)$$

Subsequent recovery of the metal values from solution is accomplished by precipitation with an active metal, such as zinc, as shown by the following equation:

$$M^{2+} + Zn^0 \rightarrow M^0 + Zn^{2+} \quad (4)$$

Similar oxidation of sulfide minerals by the addition of alkali-metal hypochlorites and chlorine has been considered by several investigators.[4,5] More recently, Parks and Baker[6] were issued a patent for a process employing the leaching of cinnabar with sodium hypochlorite and carbon adsorption of the mercury from solution.

The electrooxidation approach was developed by the Bureau of Mines, largely as a result of laboratory and pilot-mill studies and observations on the oxidation of gold ores that contained quantities of $Au(CN)_2^-$—adsorbing carbonaceous material in the form of carbon and humic acid-type hydrocarbons. It was observed that mercury present in the gold ore in the ppm range was oxidized and extracted as a soluble salt, and subsequent investigations have shown that several sulfide minerals behave in a similar manner.

Process details as presented are based on extensive laboratory and pilot-mill studies conducted principally on mercury, silver, and molybdenum ores, with parallel studies on antimony and copper ores. The flow diagram in Fig. 1 shows the essential details of the process and the sequence of operations as developed in laboratory and pilot-plant experiments.

Pilot-plant experiments were conducted in the installation shown in Fig. 2. Equipment consisted of an 18-in. by 3-ft rod mill, four rubber-lined 55-gal electrolysis tanks fitted with marine-type agitators, a 55-gal drum, agitated digestion vessel, five 3 by 3-ft thickeners, a 1 by 3-ft rotary vacuum filter, a precipitation unit, and requisite pumps, piping, and valves. Each electrolysis

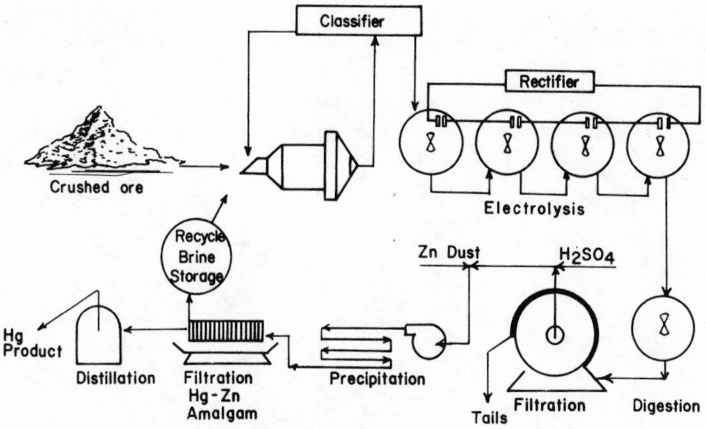

Figure 1. Conceptual flow diagram for electrooxidation of mercury sulfides or similar ores.

tank was fitted with a 4 by 20-in. multiple-electrode unit. Direct current was supplied by rectifier. Mill capacity varied between 1.4 and 1.8 tons/day, depending on the ore being processed.

Direct current electrolysis was conducted on ore that had been wet ground in salt solution to 90–96% passing 65 mesh. The sodium chloride served as an effective electrolyte constituent, as well as a chemical reactant. The hypochlorite ion, being a product of electrolysis of a brine solution, is the oxidizing species that reacts with the sulfides to form sulfates and various soluble metal salts. Electrolysis of a variety of cinnabar ores in neutral 4–10% salt solutions has been shown to oxidize the sulfide to mercuric sulfate as shown in the following reaction[8]:

$$HgS + 4OCl^- \rightarrow HgSO_4 + 4Cl^- \quad (5)$$

which is hydrolyzed to form the basic sulfate:

$$3HgSO_4 + 2H_2O \rightarrow HgSO_4{\cdot}2HgO + 2H_2SO_4 \quad (6)$$

These mercury compounds have limited solubility in relatively neutral solutions; however, they react with the sodium chloride electrolyte to form the soluble tetrachloro mercury complex

$$HgSO_4 + 4Cl^- \rightarrow HgCl_4^{2-} + SO_4^{2-} \quad (7)$$

and

$$HgSO_4{\cdot}2HgO + 12Cl^- + 2H_2O \rightarrow 3HgCl_4^{2-} + SO_4^{2-} + 4OH^- \quad (8)$$

Figure 2. Pilot-plant electrooxidation installation.

Electrolysis of silver sulfide and other minerals, such as argentite, pyrargyrite, stephanite, polybasite, proustite, galena, and sphalerite, in neutral salt solution results in formation of the soluble tetrachloro complex of silver by the following reaction:

$$AgS + 4OCl^- \rightarrow AgCl + SO_4^{2-} + 3Cl^- \quad (9)$$

Silver chloride is subsequently dissolved in large excess of chloride ion, principally as the tetrachloro complex, according to the reaction

$$AgCl + 3Cl^- \rightarrow AgCl_4^{3-} \quad (10)$$

Electrooxidation of MoS_2 in molybdenite ores is readily accomplished at pH 7 in 4–10% salt solutions, forming the soluble molybdate ion. The reactions involved are nearly identical to those observed during the addition of hypochlorite, wherein the molybdenite is converted to molybdate according to the following reaction:

$$MoS_2 + 9OCl^- + 6OH^- \rightarrow MoO_4^{2-} + 2SO_4^{2-} + 9Cl^- + 3H_2O \quad (11)$$

Copper sulfide responds to electrooxidation in 4–10% salt solutions, forming chloro complexes. For example, chalcocite oxidizes to cupric salts and the sulfate according to the following reaction:

$$Cu_2S + 5HOCl \rightarrow 2CuCl_2 + SO_4^{2-} + Cl^- + H_2O + 3H^+ \quad (12)$$

Excess chloride ion in the system results in formation of chloro complexes of copper, such as

$$CuCl_2 + 2Cl^- \rightleftarrows CuCl_4^{2-} \quad (13)$$

The use of acid systems employing both sulfate and chloride appears to be of advantage, particularly for treatment of chalcopyrite mineral.

Present practice of roasting antimony sulfide, Sb_2S_3, with emission of SO_2 gases can also be circumvented by electrooxidizing the stibnite in 4–10% salt solution to form Sb(V) compounds as shown in the following reaction:

$$Sb_2S_3 + 14OCl^- + 3H_2O \rightarrow Sb_2O_5 + 14Cl^- + 3H_2SO_4 \quad (14)$$

III. ELECTRODE SYSTEMS IN ELECTROLYSIS

Two types of electrode systems were evaluated for efficiency in electrolysis, one utilizing an iron cathode and a lead dioxide-coated titanium anode and another using an all-graphite system. Laboratory experiments demonstrated that higher concentrations of salt were required if graphite electrodes were used, as compared with lead dioxide-coated anodes. Good results required about 8–10% salt for the graphite–graphite electrode system as compared with 3–4% salt in experiments using the iron–lead dioxide–titanium electrode system. For example, in comparable experiments electrooxidizing cinnabar ores in 4% salt solution with iron–lead dioxide–titanium electrode systems, and in 10% salt solutions with graphite–graphite electrodes, it was shown that 94–95% mercury extraction can be achieved with either electrode system. However, power consumption was about 20% higher in the 4% salt solution than in the 10% salt solution.

The pH of the electrolyte was generally allowed to follow the natural pH obtained on contact of solution with the ore. This usually fell in the pH 6–7 range and is consistent with efficient production of hypochlorite in the electrolysis of salt solution. At lower pH values, hypochlorous acid decomposes to give off chlorine according to the following equilibrium:

$$HOCl + HCl \rightleftarrows H_2O + Cl_2 \quad (15)$$

Conversely, insoluble metal salts may be formed if the pH is too high. Effective dissolution of the molybdate ion required pH values generally above 7.5.

Agitation of the pulp was necessary during electrolysis so that the oxidizing species produced at the anode was effectively mixed with the mineral particles for chemical reaction. The electrodes were immersed in the pulp and positioned so that the agitator continuously forced the pulp between the electrodes. Hypochlorite ion reacted as rapidly as it was formed in the agitated cell, thus maintaining hypochlorite at a low concentration so that undesirable oxidation reactions were minimized.

Since the ore particles in the brine solution tend to inhibit the conduction of current through the cell, it is necessary to maintain anode–cathode spacing as close as possible so that effective transport of material between the electrodes and adequate electrical conduction can be obtained. It was found that electrode spacing of about $\frac{1}{4}$–$\frac{1}{2}$ in. for electrodes about 3–4 in. in width gave the most satisfactory results. The electrode design and spacing in the cell are shown in Fig. 3. In the larger scale operations, the cell vessel was contained in a 55-gal drum with agitation provided by a 10-in.-diam impeller. This particular cell design consisted of four graphite cathodes positioned between five graphite anodes. All electrodes were made from $\frac{3}{4}$-in. graphite sheet and were 3–4 in. wide. Electrodes were immersed 20 in. in the brine pulp. Continuous flow of the pulp through four electrolytic cells placed in series was effective for treating the ores. A fifth barrel with an agitator was placed at the end of the circuit to insure that the chemical reaction was complete before

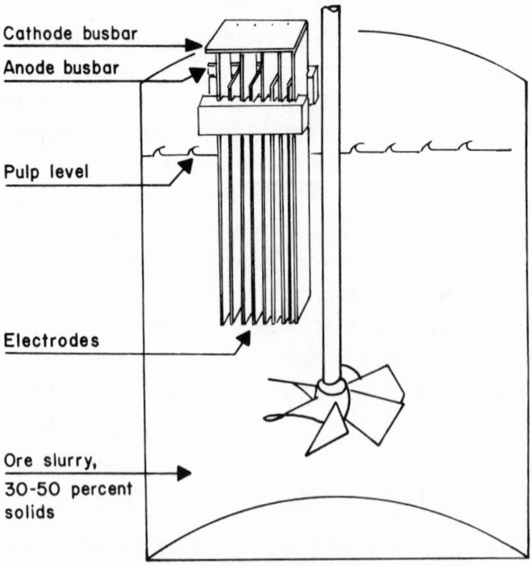

Figure 3. Electrooxidation cell and agitation vessel.

the pulp entered the liquid–solid separation sequence. The current required was determined by the rate of flow of the particular ore being treated. Retention time in the electrolytic cells was 4–6 h. The power consumption for electrolysis depended on the ore being treated. The mercury ores investigated required from 17–60 kWh per ton of ore, with a mean of 35 kWh required to obtain favorable extraction from a variety of ores containing 1–4 lb of mercury per ton. Molybdenum extraction from molybdenite ores consumed 160 kWh of power per ton. However, oxidation of molybdenite requires nine equivalents, corresponding to 20 kWh per pound of molybdenum recovered. Power consumption in extracting silver was dependent upon the refractory nature of the ore and ranged from 40 to 70 kWh per ton of ore. However, a 20 % salt concentration is required to maintain favorable solubility of the tetrachloro complex. Several of the ores investigated contained elemental sulfur, which remained largely in the elemental form during the electrooxidation procedure and reported to the solid tails in the processing sequence.

It was found that the rate of reaction in oxidizing the mineral increases with temperature; however, the stability of hypochlorite decreases with increasing temperature so that the optimum operating temperature is in the range 25–40°C.

The technology of liquid–solid separation for recovery of a pregnant solution closely follows conventional practice used in countercurrent-slimes circuits. If the ore contains considerable quantities of clay, countercurrent decantation or washing is usually required to achieve low soluble mercury values in the mill tails. Liquid–solid separation by filtration is generally favored whenever permitted by the character of the ore. Whereas soluble mercury losses to the tails are approximately equal to those obtained with countercurrent decantation, the sodium chloride losses encountered in filtration are only $\frac{1}{2}$ of the losses in a countercurrent decantation circuit. Since salt losses in the tails can represent an important economic factor, it is necessary to provide reservoirs in which tailings can be collected.

Recovery of metal values from pregnant solutions is accomplished by conventional means. For example, precipitation of mercuy from the pregnant chloride solutions readily takes place by contact with an active metal without the degassing and clarification required by the classical Merrill Crowe process for gold precipitation from cyanide solution. Precipitation is rapid, effective, and essentially complete at a pH of 2.5 by contacting the pregnant solution for 2 min with 200-mesh metallic zinc dust in the amount of 1.5 lb Zn/lb Hg. Copper is also effective for mercury precipitation; however, precipitation on copper appears to be surface dependent so that a much greater surface area is required than with zinc. The gold and silver dissolved during electrolysis follow the mercury through the system and are coprecipitated with the

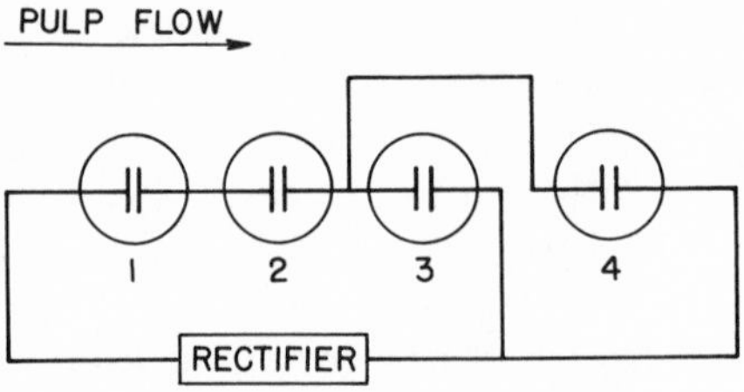

Figure 4. Pilot-plant cell circuitry.

mercury. Barren brine solution, resulting from the mercury precipitation step, is recycled through the system. The resultant precipitate consists of a zinc–mercury amalgam with a nominal composition of 40% mercury and 60% zinc, including minor amounts of lead, silver, gold, copper, etc. Essentially complete mercury recovery from the amalgam is readily accomplished by distillation at 600°C and condensation of the vapors. The zinc calcine is washed with ammonium hydroxide or hydrochloric acid to remove the portion of zinc oxidized during distillation of the mercury and then is recycled to the precipitation unit.

Solvent-extraction or ion-exchange techniques can be utilized to recover the molybdate ion from molybdenum leach circuits. The metal then is recovered from solvent strip solutions by any of several methods as a high-purity premium product.

Copper ion in electrooxidation leach solution would respond favorably to solvent extraction by well-known LIX procedures and subsequent cementation or electrolysis.

Data from one of several pilot-plant operations on mercury ores are presented in detail to show overall results, to delineate the problem areas, and to illustrate the potential of the procedure. The ore not only responded favorably to electrooxidation, but settled rapidly so that either filtration or countercurrent decantation could be used for liquid–solid separation. The ore consisted primarily of quartz, calcite, and dolomite. The dolomite and calcite were intermixed with the quartz in amounts varying from 95% quartz to less than 50% quartz. Many small grains of oxidized iron minerals were noted, but very little clay was present. The cinnabar occurred as isolated grains throughout the host rock. Some very fine veinlets of cinnabar were found to be distributed throughout the silicified calcite and dolomite. The ore had a specific gravity of 2.8. This ore was ground in 10% salt solution to 90% minus 65 mesh. Analysis of the screen fractions showed that mercury content of the various fractions ranged from 0.8 lb of mercury per ton in the plus 65-mesh fraction to 2.5 lb of mercury per ton in the minus 200-mesh fraction. Mercury content of the heads averaged 2.5 lb per ton. The pulp density at the rod mill was maintained at 64% solids and was diluted to 44% solids for agitation during electrolysis. Salt content of the solution was 10%.

Electrolysis was conducted with graphite cathode–graphite anode systems consisting of four cathodes 2.5 in. by $\frac{3}{4}$ in. thick and 24 in. in length, interspaced with three anodes 2.5 in. by $\frac{3}{4}$ in. thick and 24 in. in length, positioned $\frac{1}{2}$ in. apart and immersed in the pulp to a depth of 20 in. Figure 4 shows the circuitry used in connecting the electrode systems to the rectifier. The cell in barrel 3 is connected in parallel to the cell in barrel 4, and these cells are connected in series with cells in barrels 1 and 2. Under these conditions cell voltages are in the 4–4.4-V range, and the rectifier reads 12.6 V and 120 A. This series-parallel arrangement allows the operator to design the resistance of cells 3 and 4 to control the treatment load. For example, the current through the circuit was 120 A. The same number of amperes of electricity was passed through cells 1 and 2, so that the equivalent amount of OCl^- generated in each of these cells was the same. However, the relative amount of oxidation was decreased in cells 3 and 4 because the total current is shared between these cells, depending on the respective resistance in the cells. In this example, the current through cells 3 and 4 was 81 and 39 A, respectively.

Pulp flow through the system was 130 lb/h, corresponding to a retention time of 5 h in the electrolytic cells. A total of 1 ton of ore was processed for each experiment. Liquid–solid separation was performed both in the conventional countercurrent decantation system described earlier and by filtration. Mercury precipitation on zinc and distillation of mercury from the resulting amalgam were accomplished according to the procedure previously described.

A summary of extraction data obtained at 120 and 90 A of current using countercurrent decantation separation is shown in Table 1. At 120 A, a tail of 0.06 lb Hg per ton was achieved, and total recovery of metallic mercury product amounted to 96%. The mercury content of the tails

Table 1. Effect of Amperage on Mercury Extraction

Experiment No.	A	V	Head assay, lb Hg/ton	Power consump-tion, kWh/ton	Tail assay, lb Hg/ton	Soluble Hg lost in tails, lb Hg/ton	Hg extracted from ore, %	NaCl lost in tails, lb/ton ore	Total metallic Hg recovery, %[a]
1	90	12.1	2.3	17	0.1	0.014	94.8	47	94
2	120	12.6	2.3	23	0.06	0.016	97.0	47	96

[a]Based on head assay.

Table 2. Effect of Liquid–Solid Separation Technique on Mercury Extraction

Experiment no.	A	V	Head assay, lb Hg/ton	Power consumption, kWh/ton	Tail assay, lb Hg/ton	Soluble Hg lost in tails lb Hg/ton	Hg extracted from ore, %	NaCl lost in tails, lb/ton ore	Total metallic Hg recovery, %[a]
1	90	12.1	2.3	17	0.1	0.014	94.8	47	94
2	120	12.6	2.3	23	0.06	0.016	97.0	47	96
3	120	12.6	2.3	23	0.07	–[b]	96.9	22	96

[a]Based on head assay.
[b]Included in tail assay.

nearly doubled on decreasing the current to 90 A, and mercury recovery declined to 94%. This ore contained very little slime and settled favorably; consequently, another experiment was conducted using a 3-ft-diam by 1-ft-wide rotary drum filter for liquid–solid separation. The pulp was thickened from the 44% solids used in electrolysis to 60–66% in one thickener before filtration. Other experimental conditions were identical to those used in the previous experiment using 120 A. A thick, porous cake was obtained under the conditions used, and washing of the cake with fresh makeup water was effective.

Table 2 is a summary of extraction data using filtration. Extraction data obtained with CCD liquid–solid separation are repeated as Experiments 1 and 2 for comparison. The filtration experiment is presented as Experiment 3.

The data show that recovery of mercury is essentially independent of the liquid–solid separation technique used. However, filtration of the tails had the effect of decreasing salt loss to one-half that incurred using countercurrent decantation.

Indications are that zinc minerals, such as sphalerite, will yield effectively to dissolution by an electrooxidation sequence. Research is also being conducted on nickel–cobalt sulfide ores and certain copper and antimony ores containing sulfide minerals. Potential also exists for recovering metal values from ores containing organic materials that inhibit dissolution of metals by more conventional acid leaching systems and may otherwise require a roasting treatment. Electrooxidation provides a hydrometallurgical alternate to the smelting of sulfide ores and the consequent emission of SO_2 gases to the environment. Product sulfur, in the form of sulfate, reacts with gangue constituents and is discharged as an innocuous component of the tails. Similarly, elemental sulfur, if formed at lower pH values, remains with gangue constituents.

REFERENCES

[1]F. A. Forward and I. H. Warren, *Metallurgical Rev.* **5** (1960) No. 18.
[2]Hydrometallurgy Applied to Copper Ores, Bulletin No. G3-B146, Deco Trefoil, Fall Issue, 1969.
[3]F. Habashi, *The Recovery of Elemental Sulfur From Sulfide Ores*, Montana College of Mineral Science and Technology, Bulletin 51, March 1966.
[4]L. H. Duschak, *Trans. AIME* **91** (1930) 283.
[5]R. B. Bhappu, D. H. Reynolds, and W. S. Stahmann, Studies on Hypochlorite Leaching of Molybdenite, Unit Processes in Hydrometallurgy, AIME Metallurgical Society Conferences, Vol. 24, p. 95.
[6]U.S. Pat. 3,476,552 (Nov. 4, 1969), G. A. Parks and R. E. Baker.
[7]B. J. Scheiner, R. E. Lindstrom, and T. A. Henrie, *Electrolytic Oxidation of Carbonaceous Ores for Improving Gold Recovery*, U.S. Bureau of Mines Tech. Progress Rept. 8, 1969.
[8]B. J. Scheiner, R. E. Lindstrom, D. E. Shanks, and T. A. Henrie, *Electrolytic Oxidation of Cinnabar Ores for Mercury Recovery*, U.S. Bureau of Mines Tech. Progress Rept. 26, 1970.

INDEX